Sustainability Modeling in Engineering

A Multi-Criteria Perspective

Other Related Titles from World Scientific

Advanced Materials and Energy Sustainability: Proceedings of the 2016 International Conference on Advanced Materials and Energy Sustainability (AMES2016)
edited by Joy Iong-Zong Chen and Qiang Li
ISBN: 978-981-3220-38-6

Materials for Sustainable Energy: A Collection of Peer-Reviewed Research and Review Articles from Nature Publishing Group
edited by Vincent Dusastre
ISBN: 978-981-4317-64-1
ISBN: 978-981-4317-65-8 (pbk)

Sustainability Modeling in Engineering

A Multi-Criteria Perspective

Editors

Prasenjit Chatterjee
MCKV Institute of Engineering, India

Morteza Yazdani
Universidad Loyola Andalucía, Spain

Shankar Chakraborty
Jadavpur University, India

World Scientific

NEW JERSEY • LONDON • SINGAPORE • BEIJING • SHANGHAI • HONG KONG • TAIPEI • CHENNAI • TOKYO

Published by

World Scientific Publishing Co. Pte. Ltd.
5 Toh Tuck Link, Singapore 596224
USA office: 27 Warren Street, Suite 401-402, Hackensack, NJ 07601
UK office: 57 Shelton Street, Covent Garden, London WC2H 9HE

Library of Congress Cataloging-in-Publication Data
Names: Chatterjee, Prasenjit, 1982– editor. | Yazdani, Morteza, editor. |
Chakraborty, Shankar, editor.
Title: Sustainability modeling in engineering : a multi-criteria perspective /
editors, Prasenjit Chatterjee, MCKV Institute of Engineering, India,
Morteza Yazdani, Universidad Loyola Andalucía, Spain,
Shankar Chakraborty, Jadavpur University, India.
Description: Singapore ; Hackensack, NJ ; London : World Scientific, [2020] |
Includes bibliographical references and index.
Identifiers: LCCN 2019027178 | ISBN 9789813276321 (hardcover)
Subjects: LCSH: Sustainable engineering--Data processing--Case studies. |
Multiple criteria decision making--Case studies. | Industrial efficiency--Case studies.
Classification: LCC TA170 .S8683 2020 | DDC 628.0285--dc23
LC record available at https://lccn.loc.gov/2019027178

British Library Cataloguing-in-Publication Data
A catalogue record for this book is available from the British Library.

For any available supplementary material, please visit
https://www.worldscientific.com/worldscibooks/10.1142/11157#t=suppl

Typeset by Stallion Press
Email: enquiries@stallionpress.com

Dr. Prasenjit Chatterjee would like to dedicate this book to his grand-parents, late father Dipak Kumar Chatterjee, mother Mrs. Kalyani Chatterjee, beloved wife Amrita and his little angel Aheli.

Dr. Morteza Yazdani would like to dedicate the book to his father Mr. Mohammad Yazdani, his mother Mrs. Robabeh Saadati, his grandbrother Mani Yazdani, and his nice life partner Dr. Violeta Doval Hernandez.

Prof. (Dr.) Shankar Chakraborty would like to dedicate this book to his late father Phanindra Nath Chakraborty, his late mother Gouri Chakraborty, his beloved wife Rupa, his son Santonab and daughter Sirin. He owes a special dedication to his Gurudev.

Preface

Over the last 20 years, the topic of conservation of environment has gained indispensable importance due to very fast environmental deterioration, global warming, energy consumption and ever increasing population, which have become global threats to the ecosystems. Precipitation due to globalization and the necessity of international relations caused interaction and exchange of sciences, knowledge and technologies in every corner of this world. In the current era of industrial development, sustainability planning has become very complex due to the involvement of multiple benchmarks like technical, social, economic and environmental dimensions. This in turn places major constraints for decision-makers to optimize sustainable alternatives independently and discretely. Engineering decision-making problems frequently involve the consideration and analysis of a group of criteria or attributes that affect the final evaluation. Therefore, the benefit of a multicriteria decision-making (MCDM) approach to solving real-world problems has attracted considerable attention in both academia and practice. The current volume offers a platform for discussion among investigators, industry professionals, stakeholders and economic strategists and then argues the possibility of determining new ways of solving sustainability issues with some comprehensive frameworks. MCDM addresses the challenges in conceptualization of decision-making tools in the framework of sustainable engineering, criteria quantification, selection, simulation modeling, analysis, system approach to energy, manufacturing, waste management, material adaptation, logistics and transportation and agriculture, to name a few. Thus, the present book aimed to collect high-quality chapters that apply different MCDM methods for sustainability modeling in diverse engineering applications that address valuable inputs.

Among operation research (OR) subdisciplines, decision analysis and management have gained crucial importance in economics and are widely recognized as a sound prescriptive theory. Multiple criteria decision making (MCDM) widens a utility foundation for each complex and interconnected decision problem with the kind of approach, strategies, concepts and architectures. This section of OR is most directly characterized by a set of multiple criteria methods and is a famous subdiscipline of OR, which analyzes and selects alternatives from several criteria in real-world applications. It is a systematic quantitative approach, which aims to support decision makers in order to make rational and efficient decisions by considering important objectives and criteria. In today's competitive industrial world, companies seek experts to reduce inevitable risks and add value to the existing system components.

The main advantage of this volume is to deliver a wide range of topics to cater to the requirements of researchers in different engineering domains using modern trends in decision-making disciplines. Discussions about the implementation of sustainable strategies may lead to affordable product price, faster time-to-market, higher quality and advantages relative to enterprise competition. Investigators in area of engineering and design seek to demonstrate advanced tools and models rather than relying on classical and trial-and-error methods in order to enhance quality of engineering process and productivity. In this book, different and constructive demonstration of decision-making modeling has been proposed in order to support a strong framework to finally get the decision makers closer to their need and goals.

Organization of the Book

The complete book is organized into 16 chapters. A brief description of each of the chapters is presented as follows:

Chapter 1 develops a study on logistics center location selection. Due to the increasing significance and need of logistic cities sites in the Andalucía province of Spain, a study was made in order to select best place for logistics and distribution among eight main cities in Andalucia. The authors used BWM and MABAC methods for their objectives.

Chapter 2 designs an MCDM framework in the context of petroleum refinery projects (PRP) in combination with green construction assessment methodologies. The chapter tries to analyze the sustainable performance of these projects using crisp, gray and fuzzy values.

Chapter 3 delivers a perspective of prospective MADM (PMADM) methods to investigate a case study of medical devices while considering critical aspects of sustainability.

Chapter 4 presents a sustainable agricultural practice (SAP) in a fuzzy decision-making environment. In this chapter, 10 indicators of SAP are compared using intuitionistic fuzzy sets (IFS) and decision-making trial and evaluation laboratory (DEMATEL) method to classify the indicators in the cause and effect group while considering their interrelationship.

In Chapter 5, a decision system for solid waste management in Libya is addressed. Four systems for waste management are evaluated using fuzzy analytic hierarchy process (FAHP) and fuzzy technique for order of preference by similarity to ideal solution (FTOPSIS) with regard to four main criteria and 22 subcriteria.

Chapter 6 reports an efficient waste management system using balanced score card and AHP methods for a municipal corporation to facilitate enterprise viewpoint toward an organization's sustainable performance.

Chapter 7 introduces a rough AHP method into a hierarchy manufacturing form to establish a typical productivity model.

Chapter 8 discusses an optimization model in abrasive assisted electrical discharge machining (EDM) of Ti-based biomaterial.

In Chapter 9, an improved algorithm with rank order clustering (ROC) and row and column masking (RCM) methods are proposed for their application in cell formation and production planning in a group technology system.

Chapter 10 suggests a new strategy to improve green machining performance. The chapter conducts the advances on microelectrical discharge machining process with various dielectrics used during micromachining. The chapter presents the evolution of micro-EDM machining and their effects on the variety of performance measures such as machining rate, rate of tool wear and geometric accuracy and sequential development.

Chapter 11 undertakes the advantage of a group decision-making for the selection of a sustainable power transformer design concept using intuitionistic TOPSIS, dynamic intuitionistic fuzzy Einstein geometric averaging operator and intuitionistic fuzzy entropy methods.

Chapter 12 proposes an integrated MCDM model combined with QFD, FMEA and fuzzy AHP to identify the best alternative to minimize the risks of seaweed supply chain within and outside Kei Islands of Indonesia.

Chapter 13 develops a mathematical model to determine the market clearing price and organizational profit by considering bilateral contracts, carbon tax and rational uncertainty in a deregulated power market.

Chapter 14 develops a model for analyzing the causes of smog formation in Pakistan. In this chapter, in order to avoid smog formation due to agriculture fires and hazardous emissions, a fuzzy Vise Kriterijumska Optimizacija I Kompromisno Resenje (FVIKOR) is developed to provide different alternative solutions. The findings show the solutions that can facilitate the government in making of strategies to counter smog and regulation of laws.

In Chapter 15, a decision-making framework using AHP and multiobjective optimization of simple ratio analysis (MOOSRA) methods is propounded in order to select the optimal biodiesel fuel blend for sustainable motorization.

Chapter 16 presents an application of AHP for the selection of renewable energy sources for developing countries with moderate climatic conditions. It is recommended that hydro energy is the most preferred option followed by solar and then wind energy for the presented case study.

About the Editors

Dr. Prasenjit Chatterjee has a Ph.D. in Engineering from Jadavpur University, India and is currently an Associate Professor of Mechanical Engineering at MCKV Institute of Engineering, India. He has authored over 75 research papers in various international journals and peer-reviewed conferences, including serving as Guest Editor on *Algorithms, International Journal of Decision Support System Technology* and *Information*. He has received numerous awards including Best Track Paper Award, Outstanding Reviewer Award, Best Paper Award, Bright Researcher Award, Outstanding Researcher Award and University Gold Medal. He is the author and/or editor of several books on decision-making approaches and sustainability.

Dr. Morteza Yazdani has a Ph.D. in Business and Management Sciences and currently works at Universidad Loyola Andalucia. He has worked at the University of Toulouse and European University of Madrid as post-doctoral and part time lecturer, respectively. He participates as editorial board member of *International Journal of Decision Support System Technology* and reviewer of various journals on expert systems with applications, cleaner production, soft computing, applied soft computing, among others. His main research areas are decision-making modeling and fuzzy decision system in application of supply chain and energy systems. He has published in journals like *Expert Systems with Applications*, *Soft Computing*, *Journal of Cleaner Production*, *Management Decision*, *Clean Technologies and Environmental Policy*.

Prof. (Dr.) Shankar Chakraborty is a Professor at the Department of Production Engineering of Jadavpur University. He graduated in 1986 from University of Calcutta and obtained his post-graduate degree from Jadavpur University in 1989. He obtained his Ph.D. in Engineering from Jadavpur University in 1994. His research interests are in the areas of applications of different multi-criteria decision-making methods in manufacturing environment, control chart pattern recognition, and development of MIS and ERP systems for diverse engineering applications. He has guided several masters and doctoral level theses, and published over 150 research papers in international journals and peer reviewed conferences. He is also a regular reviewer of several journals of international repute.

About the Contributors

Dr. Morteza Yazdani has a Ph.D. in Business and Management Sciences and currently works at Universidad Loyola Andalucia. He has worked at the University of Toulouse and European University of Madrid as post-doctoral and part time lecturer, respectively. He participates as editorial board member of *International Journal of Decision Support System Technology* and reviewer of various journals on expert systems with applications, cleaner production, soft computing, applied soft computing, among others. His main research areas are decision-making modeling and fuzzy decision system in application of supply chain and energy systems. He has published in journals like *Expert Systems with Applications, Soft Computing, Journal of Cleaner Production, Management Decision, Clean Technologies and Environmental Policy.*

Yolanda Muñoz-Ocaña holds a Ph.D. in Legal and Business Sciences. She is currently Professor at the Loyola Andalucía University of the Department of Business Management. She teaches in Operations Management and Models of Excellence, both at graduate and postgraduate levels. Professor Muñoz has research interest in Development Cooperation and Teaching Innovation. Her consulting projects in companies are linked to the rationalization of production processes and warehouse management. In addition, she is an Evaluator of the EFQM Excellence Model.

Vicente Fernández-Rodríguez is a member of the faculty in Loyola Andalucía University, teaching Operations Management and Strategic Management for the Department of Business Administration. He got his Ph.D. in Economic and Business Sciences. He has developed his research and consulting activity in the areas of Operations, Commerce and e-commerce. He has also participated in some projects evaluating Rural Development strategies. In the past few years, he has participated in the Cordoba Social Lab initiative. He is a member of Social Matters, a research group in social economy and social innovation, at the Loyola Andalucía University, where he has contributed to different projects on social impact measurement.

Mercedes Torres-Jiménez received her M.Sc. degree in Business Administration from the University of Cordoba, Spain, in 1993, and Ph.D. degree in Economics and Business Administration from the University of Cordoba in 2000. She was Associate Professor in ETEA, a private Business Administration faculty affiliated at the University of Cordoba, Spain from 1993 to 2013, and since then at the Universidad Loyola Andalucía. Her current research interests include applications of computational intelligence methods to solve real problems in different areas of Business and Economics (especially in Agriculture) as well as International Development Cooperation. She is a member of AYRNA (learning and artificial neural networks) research group since 2000.

Hamidreza Hasheminasab holds a Ph.D. in Construction Engineering and Management (CEM) from Civil Engineering Department at University of Tehran. He earned his B.Sc. and M.Sc. degrees in Civil Engineering from Power and Water University of Technology and Sharif University of Technology, respectively. In recent years, he has been working on decision-making optimization and MCDM techniques in the context of sustainable development. His main focus has been on development of sustainability assessment tools for oil and gas industry projects with different decision environments.

Mohammad Reza Kharrazi got his Ph.D. in Civil and Environmental Engineering from the George Washington University (GWU), US, in 1999. Since then, he has been involved in academic and consultancy work in the US, the UK, the UAE, and Iran. In academia, he has been teaching various Civil Engineering courses in GWU as well as Iranian universities. He has also done research in Sustainability-related fields in various universities and consultancies. Dr. Kharrazi earned his B.Sc. from Isfahan University of Technology and his M.Sc. from Shiraz University in Iran. He headed the Structural Engineering Department at Stantec's Dubai office from 2008 through 2014. He is currently the CEO of SEPID group, a consultancy focused on sustainable design and construction. He has authored various books and papers related to sustainable construction and assessment.

Prof. Yaghob Gholipour graduated from Sharif University of Technology, Iran with a B.Sc. degree in Structural Engineering and obtained his M.Sc. and Ph.D. degrees in Civil Engineering Optimization from the Polytechnic of Mons, Belgium. Now, he is a full professor at the School of Civil Engineering, University of Tehran where he began his academic career in 1990. He is also the head of Engineering Optimization Research Group, a research center in the College of Engineering at the University of Tehran for more than 15 years. During this period, Prof. Gholipour performed several research projects and authored numerous journal articles in the field of engineering design and optimization, energy and sustainability. He has supervised more than 50 masters and doctoral level theses in the field of civil engineering and construction management during the last three decades.

Dalia Streimikiene is a full professor and leading researcher of Lithuanian Energy Institute. She holds a Ph.D. in Economics from Vilnius Gediminas Technical University, Lithuania. Prof. Dr. Streimikiene has published over 200 peer-reviewed papers on multi-criteria decision-making, efficiency analysis, and energy and agricultural economics in such outlets such as applied energy, energy policy, renewable and sustainable energy reviews. She is a member of International Association of Energy Economics, member of editorials boards of many Web of Science journals, including invited Editor for *Energies*.

Sarfaraz Hashemkhani Zolfani is an Assistant Professor at the School of Engineering, Catholic University of the North, Coquimbo, Chile. He was a Research Associate and Adviser at Iranian National Medical Device Directorate.

Dr. Reza Masaeli is an Assistant Professor at the School of Dentistry, University of Medical Sciences, Tehran, Iran. He is also the President of Iranian National Medical Device Directorate.

Lazim Abdullah is a Professor of Computational Mathematics at the School of Informatics and Applied Mathematics, Universiti Malaysia Terengganu. He holds a B.Sc. (Hons) in Mathematics from the University of Malaya, Kuala Lumpur in June 1984 and an M.Ed. in Mathematics Education from University Sains Malaysia, Penang in 1999. He received his Ph.D. in Information Technology Management from the Universiti Malaysia Terengganu, in 2004. His research focuses on the mathematical theory of fuzzy sets, decision-making methods and its applications to social ecology, environmental sciences, health sciences, financial management, technology management and manufacturing engineering. His research findings have been published in more than 300 publications in the fields of social statistics and decision-making. Currently, he is a member of editorial boards of several international journals related to computing and applied mathematics. He is also a regular reviewer for international impact factor journals, member of scientific committees of several symposia and conferences at national and international levels. Dr Abdullah is an associate member of IEEE Computational Intelligence Society, and a member of the International Society on MCDM.

Ibrahim Ahmed Badi completed his B.Sc. in Mechanical Engineering at Garyounis University, Benghazi-Libya. He later completed his M.Sc. in Production Engineering from the Higher Institute of Industry, Libya. He gained a Ph.D. degree in Logistics and Transportation Systems, Germany. He is currently working as Assistant Professor at the Department of Mechanical Engineering, Misurata University, Libya. His research focus is in the areas of Operations Research, Multi-Criteria Decision-Making, Lean Manufacturing and Supply Chain Management.

Dr. Željko Stević is Assistant Professor at Faculty of Transport and Traffic Engineering Doboj, University of East Sarajevo. He received his Ph.D. in Transport and Traffic Engineering from University of Novi Sad, Faculty of Technical Sciences in 2018. His interests are in logistics, supply chain management, transport, traffic engineering, multi-criteria decision-making problems, rough set theory, sustainability and fuzzy set theory. He has authored over 80 papers in his areas of research interest. In his doctoral dissertation he developed a few MCDM models including one universal model for supplier selection in different fields. Dr. Stević has authored/co-authored papers published in refereed international journals including *Applied Soft Computing*, *Sustainability*, *Symmetry*, *Engineering Economics*, *Transport*,

Neural Computing and Applications, *Information*, *Economic Computation and Economic Cybernetics Studies and Research*, *Technical Gazette*, *Studies in Informatics and Control* and others.

Siniša Sremac was born in Ruma in 1983 where he completed elementary and high school with an excellent grade point average. In 2007, he completed an integrated basic and master academic studies in Traffic and Transport at the Faculty of Technical Sciences, University of Novi Sad. As one of the best students, during the studies he was awarded a scholarship by the Fund for Young Talents and the Ministry of Education, Science and Technical Development of the Republic of Serbia. In 2013, he defended his Ph.D. dissertation "Goods flow management model for transport-storage processes" at the Faculty of Technical Sciences in Novi Sad. Upon completion of the integrated studies, he was employed as a Teaching Assistant at the Department for Traffic Engineering and Transport at the Faculty of Technical Sciences. In 2013, he was elected as an Assistant Professor in Organizations and Transport Technologies and in 2018 he was elected to the title of Associate Professor. He is highly rated by students. In 2014, he was awarded the Momcilo Momo Novkovic Award by the Faculty of Technical Sciences for outstanding results in working with students, achieving excellent scientific results and contributing to the promotion of the Faculty. His research interests include logistics, supply chain management, transport economics, inventory management using artificial intelligence, modelling and applications of expert systems. He has authored 97 scientific papers, out of which 15 are published in leading international journals. He co-authored the textbook *Transportation and Logistics Properties and Flows of Goods*, published by the Faculty of Technical Sciences.

Kanika Prasad is an Assistant Professor at the National Institute of Technology, Jamshedpur, India. She graduated from Jadavpur University in Production Engineering after completing her undergraduate study in Mechanical Engineering from SMIT, Sikkim in India, earning distinction in both degrees. She earned a prestigious DST INSPIRE fellowship to pursue her doctoral research from Jadavpur University. She specializes in application of quality function deployment technique and multi-criteria decision-making to develop expert systems for selection and design problems in manufacturing. Of late sustainability in manufacturing interests her and she is working in that area. She has guided several undergraduate and postgraduate dissertations. She is also a regular reviewer of several journals of international repute.

Rishi Dwivedi is currently working as an Assistant Professor in the Department of Finance of XISS, Ranchi. Before joining XISS, Ranchi, he worked as an Assistant Professor in the Management Department at Central University of Jharkhand. He obtained his Ph.D. degree from Jadavpur University, Kolkata. Dr. Rishi has been awarded the esteemed "UGC-BSR Research Fellowship in Science for Meritorious Students". He has one year of corporate experience as an Assistant Manager at ICICI Bank Limited, Kolkata, West Bengal. He has published several research papers in international journals of high repute as well as presented numerous papers in international conferences. He is a regular reviewer of *International Journal of Production Research*, *Journal of Cleaner Production*, *OPSEARCH*, etc. His research interests include development of integrated activity-based costing and quality function deployment models for various industries in order to attain competitive edge. Of late he is also interested in exploring the ways in which MCDM model can be applied to achieve sustainable advantage for assorted industries.

Ayushi Sharma is currently pursuing her MBA in Finance from Xavier Institute of Social Service, Ranchi, India. She ranks among the top 15 percent of students in her class. She is a first division graduate from St. Xavier's College, Ranchi, India. Her interest in finance led her to pursue further studies. Her achievements include 3rd Rank in Computer Olympiad at National Level, as well as completing exams in the field of finance, like MISM 5A examination and foundation program of Company Secretary. Her research interest includes application of multi-criteria decision-making tools in the field of finance.

Bolun Liang is an M.A.Sc. student in the Department of Mechanical, Automotive & Materials Engineering at the University of Windsor, ON, Canada. Mr. Liang received his B.Sc. in Industrial Engineering from the University of Windsor in 2018. Mr. Liang's key research focus is the implementation of network model on production modeling.

Dr. Golam Kabir is an Assistant Professor in Industrial Systems Engineering Program at the Faculty of Engineering and Applied Science, University of Regina, SK, Canada. He was previously an Assistant Professor at the University of Windsor, ON and Post-Doctoral Research Fellow at the University of British Columbia (UBC), BC and University of Waterloo, ON. He received his Ph.D. in Civil Engineering from

UBC in 2016. He also received his M.Sc. and B.Sc. in Industrial and Production Engineering from Bangladesh University of Engineering and Technology (BUET) in 2011 and 2009, respectively. His research interests include system risk, reliability, resilience assessment, interdependent network resilience analytics, multi-criteria decision analysis under risk and uncertainty, sustainable system analytics, data-driven decision making and multi-objective optimization.

Dr. Guoqing Zhang is a Professor in the Department of Mechanical, Automotive & Materials Engineering and Director of Supply Chain and Logistics Optimization Research Centre at the University of Windsor, ON, Canada. He received his Ph.D. in Management Sciences from the City University of Hong Kong. He also received his M.Sc. and B.Sc. in Engineering from Southeast University, Nanjing, China. His research interests include optimization, algorithm design and development, supply chain management and logistics, facility/warehouse layout, mathematic modeling and analyses for manufacturing systems, intelligent decision support systems, computing and parallel optimization.

Dr. Preetkanwal Singh Bains is a faculty member and researcher in the Department of Mechanical Engineering, Beant College of Engineering and Technology, Gurdaspur, India. His current research interests include materials engineering as well as advanced machining processes. He has contributed myriad innovative and novel manufacturing techniques for which three patents are applied in various fields of mechanical engineering. He is a Life Member of the Indian Society for Technical Education (ISTE) and American Society of Mechanical Engineers (ASME) including copious other societies. He has contributed many significant publications and acted as reviewer for prominent journals, as well as chaired sessions and guest lectured.

Dr. Sarabjeet Singh Sidhu is an Assistant Professor in the Department of Mechanical Engineering at Beant College of Engineering and Technology, Gurdaspur, Punjab (India). Sarabjeet graduated first in his Masters of Technology class from IKG Punjab Technical University, Jalandhar. His broad area of research is related to surface modification and residual stresses analysis in metal matrix composites, biomaterials and non-conventional machining process. He has authored more than 70 technical papers in reputed national and international journals and conferences, as well as acted as reviewer for journals and chaired various technical sessions. Most recently, he was awarded a "Contribution Award" by the *Journal of Mechanical Science and Technology*.

Dr. Ali Jahan is active in the development of multi-criteria decision-making (MCDM) techniques for the engineering design process, especially for materials selection. His main research interests are in the application of MCDM and quality tools for improving materials, design and processing. Dr Jahan has been involved in different projects from quality engineering to designing systems based on the requirements of quality management standards and associated decision support systems. He has various publications describing the development of MCDM in design and materials engineering. Dr. Jahan regularly reviews papers in different international journals and is the recipient of awards related to his research and academic activities. He is Associate Professor in the Faculty of Engineering, Semnan Brach, Islamic Azad University, Iran.

Sanaz Souri obtained her B.Sc. degree in Industrial Engineering (Systems Planning and Analysis) in 2009. She then started working in industry and at the moment is director of production planning department in Parizansanat Co. She has been involved in different projects related to quality systems. Mrs Souri has recently completed an M.Sc. in Industrial Engineering.

Dr. Golam Kibria is Assistant Professor in the Department of Mechanical Engineering in Aliah University, Kolkata, India. He completed his M.Tech in Production Engineering from Jadavpur University, Kolkata in 2008 and Ph.D. from Jadavpur University, Kolkata in 2014. He has worked as Senior Research Fellow (SRF) in Council of Scientific & Industrial Research (CSIR) sponsored project from 2008 to 2011. His research interests include non-conventional machining processes, micromachining and advanced manufacturing and forming technology. He is a life member of The Institution of Engineers (IEI), India. He is the author of several book chapters as well as 26 research papers in various reputed journals and 37 research papers in reputed national and international conference proceedings. He is editorial board member as well as reviewer of a number of reputed international journals namely *Optics and Laser Technology*, *International Journal of Advanced Manufacturing Technology*, *Manufacturing Review*, *International Journal of Physical Sciences*, etc. In 2008, he received I.S.T.E. National Award for Best M.Tech. Thesis in Mechanical Engineering and the Institution Prize (Gold Medal) of 2008–2009 from The Institution of Engineers (India) for the best paper.

Ms. Tashi Dolma Sherpa did her B.Tech (Mechanical Engineering) from Sikkim Manipal Institute of Technology. She did her M.Tech (Production Engineering) from Sikkim Manipal Institute of Technology. She worked as Assistant Professor of Mechanical Engineering Department for one year and is working as Junior Engineer for the government of Sikkim.

Mr. Ishwer Shivakoti did his B.Tech (Mechanical Engineering) from Sikkim Manipal Institute of Technology. He did his M.Tech (Production Engineering) from Sikkim Manipal Institute of Technology. He is associated with SMIT since July 2012. He worked as Assistant Professor of Mechanical Engineering Department from July 2012 to date. He has published many research papers in various journal of repute. He is member of various professional or technical societies.

Dr. B. B. Pradhan did his B.E (Mechanical) from MIT, Muzaffarpur under B. R. Ambedkar Bihar University in 1996. He did his M.Tech (Mechanical) with specialization in Manufacturing Process Engineering in 2004 from IIT Kharagpur and Ph.D. (Engineering) from Jadavpur University in 2009. He has been associated with SMIT since July 1997. He worked as Lecturer at Mechanical Engineering from July 1997 to August 2004. He was also the Head of the Department from July 1997 to February 2002. He worked as Reader/ Associate Professor from August 2004 to September 2009. Since 2009, he has been a full Professor. In February 2018 he was appointed Associate Director (Academics). He has guided 4 doctoral students and published many research papers in various journals of repute. He has chaired many sessions at International and National level conferences. He is member of various professional or technical societies in India.

Daniel O. Aikhuele lectures at the College of Engineering, Bells University of Technology, Ota, Nigeria. He holds a Ph.D. in Manufacturing Engineering from the Universiti Malaysia Pahang, Pekan, Malaysia. An MS.C. degree in Advanced Manufacturing Systems from the Teesside University, Middlesbrough, UK and a B.Eng degree in Materials and Production Engineering from the Ambrose Alli University, Ekpoma, Nigeria. His research areas include reliability management, product development and evaluation, energy and fuzzy multi-criteria analysis. He is a seasoned researcher and educator, with more than 30 journal articles published in top-class journal outlets.

Desmond Eseoghene Ighravwe lectures at the College of Engineering, Bells University of Technology, Ota, Nigeria. He holds a Ph.D. in Mechanical Engineering from Ladoke Akintola University of Technology, Nigeria. He is a graduate of the University of Ibadan, Nigeria, where he did his B.Sc. and M.Sc. in Industrial Engineering and Engineering Management, respectively. His research areas include operations research, energy and waste management, and multi-criteria analysis. In 2018, Elsevier ranked him among the topmost published 500 persons in Nigeria between 2013 and 2018. He has co-authored more than 40 journal articles in leading international journals such as *Total Quality Management and Business Excellence, Journal of Renewable Energy, Journal of Building Pathology and Rehabilitation* and *Decision Science Letters.*

Wellem Anselmus Teniwut is Senior Lecturer and Researcher at Tual State Fisheries Polytechnic, Indonesia. His current research interests include Strategy and Policy in Fisheries, Decision Analysis and Decision Support System, Supply Chain Management and the Fisheries Agro industry. Since 2016 he has authored more than 20 scientific articles and holds two intellectual property rights related to his field of research. Wellem is also reviewer for *Ocean & Coastal Management* and *Journal of Natural Resources and Development*, as well as served as guest editor of international conferences.

Mohammad Tolou Askari was born in Mashhad-Iran. He obtained his Bachelor's Degree in Power Electrical Engineering at the University of Applied Science and Technology of Mashhad in 2005. He proceeded to obtain his Master's in Power Electrical Engineering in 2008. His research thesis was on "dynamic thermal model approach to estimate loss of life of mineral oil immersed transformers". He obtained his Ph.D. in Power Electrical Engineering at the University Putra Malaysia in 2011 under the supervision of Prof. Ir. Dr. Mohd Zainal Abidin Ab Kadir, Assoc. Prof. Dr. Hashim Hizam and Dr. Jasronita Jasni. The title of his research was "dynamic investment modeling in the restructured power market with the presence of wind resources". Currently, he is senior lecturer in Islamic Azad University (Semnan Branch). He is also the director of Growth Centre of Islamic Azad University of Semnan. He has supervised 40 M.Sc. students with another 8 doctoral and 6 master's level candidates on the way. His research interests include generation expansion planning, scheduling, uncertainties, loss of life of the transformers.

Mohd. Zainal Abdin Ab-Kadir received his B.Eng. degree in Electrical and Electronic Engineering from Universiti Putra Malaysia and Ph.D. degree in High Voltage Engineering from the University of Manchester, UK, respectively. Currently, he is a Strategic Hire Professor at the Institute of Power Engineering (IPE), Universiti Tenaga Nasional (UNITEN) and a Professor at the Faculty of Engineering, Universiti Putra Malaysia. Administratively, he was the Deputy Dean (Research & Innovation) and the Head of Department between 2014 to 2017 and 2011 to 2014, respectively. He is the Founding Director at the Centre for Electromagnetic and Lightning Protection Research (CELP), Universiti Putra Malaysia. He is also an IEEE Power & Energy Society (PES) Distinguished Lecturer in the field of lightning and high voltage engineering. To date he has authored and co-authored over 300 journal and conference papers. He has supervised 17 Ph.D. and 35 M.Sc. students. Professor Zainal is a Professional Engineer (PEng) and a Chartered Engineer (CEng), as well as a Senior Member of the Institute of Electrical dan Electronics Engineers (IEEE). Currently, he is the Chairman of the National Mirror Committee of IEC TC 81 (Lightning Protection). Apart from that, he is the Local Convener of MNC-CIGRE C4 on System Technical Performance, Past Chair of IEEE Power & Energy Society Malaysia and a Member for various Working Groups. He is also an Advisory Board Member of the National Lightning Safety Institute (NLSI) US, Research Advisor for the African Centre for Lightning and Electromagnetic (ACLE) and Advisor to many other government agencies such as Sustainable Energy Development Authority (SEDA) and the Energy Commission of Malaysia. His research interests include high voltage engineering, lightning protection, electromagnetic compatibility and power system transients.

Professor Ehsan Bolandifar is an Assistant Professor of Operations Management at the Chinese University of Hong Kong. His research mainly focuses on supply chain procurement and healthcare operations. In particular, he is interested in how supply chain procurement interacts with finance and marketing. His research has been featured in journals like *Manufacturing and Service Operations Management*, *Production and Operations Management*, *Omegaand Academic Emergency Medicine.* Professor Bolandifarhas received his Ph.D. in Business Administration from Washington University in St Louis and his M.Sc. from Sharif University of Technology in Tehran.

Yousaf Ali is an Assistant Professor at the Department of Management Sciences, GIK Institute of Engineering Sciences and Technology, Pakistan He received his education from University of Southampton (UK) and University of Macerata (Italy). His main fields of research are Operations Research, Multi-Criteria Decision Modeling, Input-Output Analysis, Policy Modeling, Impact Analysis, Inoperability Input Output Modeling, Disaster Risk Management, Operations & Production Management, Linear Programming, Environmental Economics.

Mahboob Ul Hassan is an Associate Manager of Production Planning at Shibli.

M. Asees Awan did his M.Sc. in Industrial/Engineering Management from Ghulam Ishaq Khan Institute of Engineering Sciences and Technology, Topi, KPK, Pakistan. He has pursued his Bachelor's degree in Mechanical Engineering from the University of Engineering and Technology Peshawar, KPK, Pakistan.

Khawaja Muhammad Faisal Iqbal studied Mechanical Engineering at Ghulam Ishaq Khan Institute of Engineering Sciences and Technology, Swabi. Currently he is serving as a Management Trainee at Thal Limited.

Muhammad Abdul Haseeb Wahla studied at Ghulam Ishaq Khan Institute of Engineering Sciences and Technology, Swabi.

Surya Kanth is currently pursuing his Ph.D. in the Department of Mechanical Engineering, NIT Silchar. His research interest includes Biodiesel, IC Engine and Soft computing. Mr. Kanth has three years of experience in academia and research. He obtained his M.Tech in 2014 from PDA College of Engineering, in the specialization of Thermal Power Engineering.

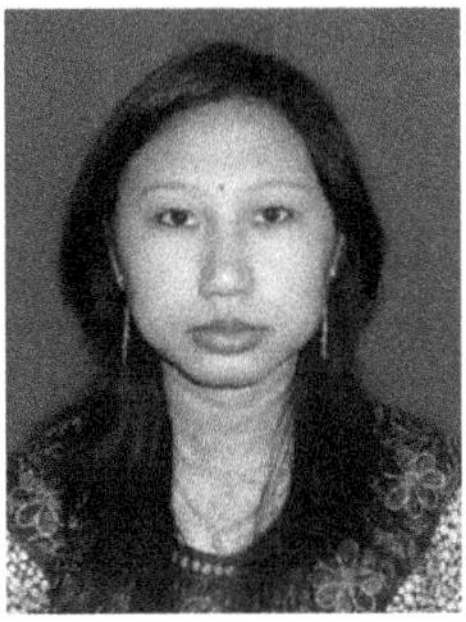

Sumita Debbarma is an Assistant Professor at the Department of Mechanical Engineering, National Institute of Technology Silchar, Assam, India. She obtained her Ph.D. in Mechanical Engineering with specialization in Thermal Engineering from National Institute of Technology Silchar, in 2018. Her areas of interest are Biofuel, Nano fuel, optimization and applied soft computing techniques. She has over 7 years of experience in the field of teaching and research. She has authored more than 10 research papers published in national and international journals, conferences, and book chapters. She is member of various professional bodies like Institution of Engineers of India [IE(I)].

Dr. Jagadish obtained his post-graduation in Product Design and Development from National Institute of Technology Warangal, India and subsequently, Ph.D. in Mechanical Engineering with specialization in Manufacturing Engineering from National Institute of Technology Silchar, India. He has over three years of industrial experience in the field of design and analysis and more than five years of teaching and research experience. He has been an Assistant Professor in the Mechanical Engineering Department of National Institute of Technology Raipur since August 2018. He received an "Institutional Award (Gold Medal)" by the Institution of Engineers India and "Best Innovative Award" by Springer for his outstanding research contribution and published more than 14 research papers, 1 book, and 7 book chapters. He is a regular reviewer of various manufacturing and optimization related journals. He is life member of professional bodies like Associate Member of Institute of Engineers (India), Committee Member of Soft Computing Club SCILAZ at NIT Silchar, Member of American Society of Mechanical Engineers (ASME). His areas of interest are Green Manufacturing, Advanced Manufacturing Process, Rapid Prototyping; Computer Aided

Design/Manufacturing, Composite Machining, Composite Material and Machining Characterization; Applied Soft Computing Techniques and Optimization; Renewable Energy, etc.

Biplab Das is Assistant Professor in the Department of Mechanical Engineering, National Institute of Technology Silchar, India. Dr. Das completed his Ph.D. from NERIST, Itanagar, India, in 2014. Later, he pursued his Post-Doctoral research from University of Idaho, US. He is recipient of the prestigious Bhaskara Advance Solar Energy (BASE) fellowship from Indo-US Science and Technology Forum and Department of Science and Technology, Government of India. He is also awarded with "DBT Associateship" by Department of Biotechnology, Government of India. He has more than 10 years of experience in teaching and research, and authored more than 30 papers in national and international journals and conferences. Presently Dr. Das is actively involved in six projects to develop a solar thermal system for North East India, worth 0.268 billion INR, sponsored by Science and Engineering Research Board, Department of Science and Technology, Ministry of Power, and Ministry of Climate Change, Government of India. He is guiding six doctorate scholars. He has ongoing research activities in collaboration with Jadavpur University, India, IIT Guwahati, India, University of Idaho, US, Ulster University, UK.

Dr. R. Ramaprabha, Associate Professor in the Department of Electrical and Electronics Engineering, SSN College of Engineering, Chennai has 18 years of teaching experience including 10 years of research experience. She obtained her Ph.D. in the area of Solar Photovoltaic Systems from Anna University, Chennai. During her Ph.D. she developed the Solar Energy Research Lab. She has authored 105 research publications in international journals, 130 papers in the proceedings of refereed national and international conferences. She received the Best Teacher Award for 2009–2010 and 2010–2011. She is a reviewer for several referred international journals. She completed an internally funded project worth about ₹13 lakhs. She is a scientist mentor for the project titled "Design and development of flywheel based power conditioning system for a renewable energy fed micro grid" of ₹20 lakhs funded by Department of Science and Technology (DST) under Women Scientists Scheme (WOS-A). She is an active member in IEEE and life member in ISTE.

Dr. S. Malathy, Associate Professor in the Department of Electrical and Electronics Engineering, SSN College of Engineering, Chennai has 14 years of teaching experience including 6 years of research experience. She obtained her Ph.D. in the area of Solar Photovoltaic System from Anna University, Chennai. She has published 14 research publications in international journals, 30 papers in the proceedings of refereed international and national conferences. She completed an internally funded project worth about ₹4 lakhs. She is an active member in IEEE and life member in ISTE.

Acknowledgments

The editors would like to express their gratitude to the many people who cooperated throughout this book; to all those who provided support, talked things over, read, wrote, offered comments, and assisted in the editing, proofreading and design to make this book possible. No undertaking like an edited volume uniting a number of scholarly persons from around the world can be done by one person. This is certainly no exception.

First, the editors owe a lot to the Almighty as well as their family members for their endless support, motivation, guidance and love all through their lives.

Second, the editors wish to acknowledge the valuable contributions of the authors regarding quality, coherence and presentation of content of the chapters.

Third, the editors are very much thankful to the entire World Scientific Publishing team for keeping faith in them and showing the right path to create this book. Their guidance, motivation, positive responses and resources ultimately led to the foundation of this book.

Last but definitely not the least, the editors take this opportunity to thank all the readers for their trust and it is expected that this book will continue to inspire and guide them for their future endeavors.

The Editors

Dr. Prasenjit Chatterjee,
MCKV Institute of Engineering, India

Dr. Morteza Yazdani,
Universidad Loyola Andalucía, Seville, Spain

Dr. Shankar Chakraborty,
Jadavpur University, India

Contents

Preface		vii
About the Editors		xi
About the Contributors		xiii
Acknowledgments		xxxv
Chapter 1.	Logistics Center Location Decision Using a Multi-Attribute Analysis Structure *Morteza Yazdani, Yolanda Muñoz-Ocaña, Vicente Fernández-Rodríguez and Mercedes Torres-Jiménez*	1
Chapter 2.	A Sustainability Assessment Framework for Petroleum Refinery Projects with Multi-Criteria Decision-Making Techniques *Hamidreza Hasheminasab, Mohammadreza Kharrazi, Yaghob Gholipour and Dalia Streimikiene*	27
Chapter 3.	From Past to Present and into the Sustainable Future: PMADM Approach in Shaping Regulatory Policies of Medical Device Industry in the New Sanction Period *Sarfaraz Hashemkhani Zolfani and Reza Masaeli*	73

Chapter 4. Indicators of Sustainable Agriculture: An Evaluation Using Intuitionistic Fuzzy DEMATEL Method 97

Lazim Abdullah

Chapter 5. An Integrated Fuzzy Model for Solid Waste Management in Libya 117

Ibrahim Badi, Željko Stević and Siniša Sremac

Chapter 6. Development of Hybrid Tool using AHP and BSC Models for Analyzing Performance of Waste Disposal Management System of a Municipality 145

Kanika Prasad, Rishi Dwivedi and Ayushi Sharma

Chapter 7. Productivity Modeling of Manufacturing Industry Using the Rough Analytic Hierarchy Process 165

Bolun Liang, Golam Kabir and Guoqing Zhang

Chapter 8. Abrasive Assisted Hybrid EDM of Titanium Alloy for Enhanced Productivity 187

Preetkanwal Singh Bains and Sarabjeet Singh Sidhu

Chapter 9. Cell Formation Based on Similarity in Parts Design Using Multiattribute Decision-Making Algorithm 203

Ali Jahan and Sanaz Souri

Chapter 10. Electrical Discharge Micromachining with Sustainable Dielectrics 229

G. Kibria, T. D. Sherpa, I. Shivakoti and B. B. Pradhan

Chapter 11. Dynamic Decision-Making Method for Design Concept Evaluation Based on Sustainability Criteria 253

Daniel O. Aikhuele and Desmond E. Ighravwe

Chapter 12. Challenges in Reducing Seaweed Supply Chain Risks Arising Within and Outside Remote Islands in Indonesia: An Integrated MCDM Approach 271

Wellem Anselmus Teniwut

Chapter 13. An Approach for Modeling the Deregulated Power Market Based on Cournot Game Theory 293

M. T. Askari, M. Z. A. Ab Kadir and E. Bolandifar

Chapter 14. Fuzzy MCDM Modeling for Controlling Smog Formation: A Case of Pakistan 315

Yousaf Ali, Mahboob Ul Hassan, M. Asees Awan, Khawaja Muhammad Faisal Iqbal and Muhammad Abdul Haseeb Wahla

Chapter 15. Application of MCDM Technique for the Selection of Biodiesel Fuel Blend for Sustainable Motorization 337

Surya Kanth, Sumita Debbarma, Jagadish and Biplab Das

Chapter 16. Selection of Renewable Energy Sources for a Developing Country Using Analytic Hierarchy Process 359

R. Ramaprabha and S. Malathy

Index 381

Chapter 1

Logistics Center Location Decision Using a Multi-Attribute Analysis Structure

Morteza Yazdani[*,‡], Yolanda Muñoz-Ocaña[*,§], Vicente Fernández-Rodríguez[*,¶] and Mercedes Torres-Jiménez[†,||]

[*] *Department of Management, Universidad Loyola Andalucía, Spain*
[†] *Department of Quantitative Methods, Universidad Loyola Andalucía, Spain*
[‡] *myazdani@uloyola.es*
[§] *yolandam@uloyola.es*
[¶] *vicentef@uloyola.es*
[||] *mtorres@uloyola.es*

Logistics and transportation are necessary for modern communities for mobility. Rising population, growing economy and increasing transactions, attractive tourism industry and unpredictable technological change push the regional and provincial governments to put in place an advanced logistics system that fits the society. At present, in big urban regions, locating a logistics and distribution center is drawing considerable attention. The location of strategic logistics centers (LCs) is likely to receive extra focus in the coming years because of global mobility, economic competitiveness and unwanted immigration all over the world. How to deal with that situation is on top of the mind of an executive manager who is charged with decision making. Policymakers at the highest levels of the decision-making pyramid always resort to analytical solutions, guidelines, and a systematic approach.

Multi-attribute decision-making (MADM) refers to a class of decision-making techniques based on which several alternatives are evaluated with respect to a number of criteria. The criteria are defined as either quantitative or qualitative. The objective of this study is to establish a LC and find the best location for Andalucía government in Spain. The question is how the optimal location of LCs can be found in order to facilitate logistics and transport demands in response to increasing growth in industry and tourism. In relation to this problem, a research group evaluated a series of alternatives to get possible solutions.

1. Introduction

One of the issues on which modern communities are assessed and judged is the level of urban development and the availability of infrastructure. This issue is correlated to the strategic planning of governments and provinces in each country. Among the principal infrastructure, the role of distribution and logistics cannot be denied. According to Farahani and Hekmatfar (2009), the "location problem" consists in determining a proper placement of an infrastructural component (ground, site, facility) in a considered area, taking into account the decision-maker's (DM's) preferences and existing constraints. This is a long-term strategic decision having a great impact on the competitiveness of companies. The location problem also plays a crucial role in logistics, where it refers to finding the most desirable location of logistics facilities.

Logistics centers (LCs), which were first seen in the United States during the industrial revolution and were founded in Europe in the 1960s, constitute an important part of organizational strategy (Koldemir *et al.*, 2009). There is no unique term to name LC such as distribution center, freight village, dry port, inland port, load center, logistics node, gateway, central warehouse, freight/transport terminal, transport node, logistics platform, logistics depot, or Distri-park (Higgins and Ferguson, 2011; Rimienė and Grundey, 2007). There is no either a universally accepted definition of an LC. For example, Rimienė and Grundey (2007), as the result of their study that handles the definition and evolution of the LC concept, made the following definition for a LC: "Logistics center (Freight village/Logistics node/Distribution center) is a special intermodal hub (nodal point) in the transportation system, including different logistics facilities, where separate operators are providing number of services, connected to transportation, logistics and distribution in established geographical coverage." Żak and Węgliński (2014) defined the LC as an entity established in a secured area, where all logistics activities including transportation and forwarding, material handling, warehousing, inventory management, cross-docking, intermodal trans-shipment, physical distribution of goods are carried out on a commercial basis. According to Uyanik *et al.* (2018), the most comprehensive definition of LC is enhanced and improved by EUROPLATFORMS, which is an association of approximately 80 transport and LCs all over Europe. This definition is also used by EU, UNECE and OECD and various national authorities. EUROPLATFORMS defines LC as follows: "A Logistics Centre is a center in a defined area within which all activities relating to transport, logistics and the distribution of goods — both for national and international transit, are carried out by various operators on a commercial basis. The operators can either be owners or tenants of buildings and facilities (warehouses, distribution centers, storage areas, offices, truck services, etc.), which have been built here" (www.europlatform.eu).

The chapter is organized in this order; after this introduction, in Section 2, literature review and research impact are described. Some existing researches in area of LC location are compared. Thereafter, the required methodologies for optimal location are delivered as Section 3. In Section 4, the results and implementation of the case are explained. Finally, in Section 5 the conclusion and future research direction are declared.

2. Literature Review and Research Impact

2.1. *Location problem review*

Location problem is a widely discussed topic in transportation and logistics (Owen and Daskin, 1998; Chen *et al.*, 2007; Daganzo, 1996; Özcan *et al.*, 2011) and refers to the selection of specific locations of such facilities as warehouses, distribution centers, transportation hubs, passenger and cargo terminals and many others (Thai and Grewal, 2005; Drezner and Hamacher, 2002; Żak and Fierek, 2007;

Church and Murray, 2009; Farahani and Hekmatfar, 2009). In the majority of cases, the location problem of logistics facilities has a strategic character and the generated solutions and decisions have a long-term impact.

In general, the logistics-oriented location problems can be classified based on different allocation criteria into the following categories (Farahani *et al.*, 2010; Klose and Drexl, 2005):

- The applied approach to the formulation of the decision problem (ranking and mathematical programming — optimization problems).
- The proposed number of evaluation criteria (single and multiple — objective location problems).
- The proposed approach to handling risk and uncertainty (deterministic and non-deterministic problems).
- The topographic character of the network and the resulting approach for selecting the location of facilities (continuous and discrete location problems).
- The number of objects and locations under consideration (single and multiple facility location problems).

Over the years, a great diversity of methodologies has been applied to solve the LC location problem. According to literature review, many mathematical formulations and approaches to solving this problem have been applied, including mathematical programming (Klose and Drexl, 2005), evaluation and ranking of considered sites (Özcan *et al.*, 2011; Korpelaa and Tuominen, 1996; Awasthi *et al.*, 2011). Uyanik *et al.* (2018) found that multi-criteria decision-making (MCDM) techniques were the most frequently used to solve the LC selection location problem. At the same time, this group includes a great variety of methods such as analytic hierarchy process (AHP), analytic network process (ANP), technique for order preference by similarity to ideal solution (TOPSIS), elimination and choice translating reality (ELECTRE), *Vlse Kriterijumska Optimizacija I Kompromisno Resenjetechnique* (VIKOR) — multi-criteria optimization and compromise solution, the preference ranking organization method for enrichment of evaluations (PROMETHEEs), analytic network process/benefits, opportunities, costs and risks (ANP/BOCR), axiomatic fuzzy set (AFS), artificial neural network (ANN), DELPHI method, fuzzy graph theory and goal programming. With respect to the mathematical models and proper solution techniques, the authors found that they were not widespread like MCDM techniques. Anyway, they are used in an integrated manner with fuzzy logic, linear programming, genetic algorithm, greedy heuristic algorithm and fuzzy decision-making model in some publications.

If the decision criteria are taken into consideration, a great variation can be also found. For instance, Turskis and Zavadskas (2010) assumed five fundamental objectives of the government's transport policy like economy, environment, safety, accessibility and integration. The criteria finally they decided to include in the analyze process were investment cost, operation time, expansion possibility and closeness to the demand market. Żak and Wegliński (2014) presented the first level of the application of a MCDM/A-based two-stage procedure for the selection of the best location of a LC. At the first, upper level the macro-analysis of the wider areas — macro-regions — were carried out. They considered nine criteria representing different aspects of evaluation and interests of different stakeholders related to technical, economic, social, market-oriented and environmental aspects (condition of transportation infrastructure; economic development; investment cost; level of transportation and logistics competitiveness; investment attractiveness; transportation and logistics attractiveness; social attractiveness; environmental friendliness and finally safety and security). Kayikci (2010) analyzed the applicability of a combination of AHP and ANN methods in the process of selecting the location of an intermodal LC. The fuzzy extension of the AHP technique was used to derive the priorities (relative weights) of each selected criterion and ANN were used to alleviate the MCDM and to obtain the best model configuration for the location problem. After interviews with decision makers and the intensive literature review, five main evaluation criteria were determined: environmental effect, international market location, intermodal operation and management, national stability and economical scale. These criteria were detailed into different sub-criteria.

Thai and Grewal (2005) studied where to locate a regional distribution center of a company (Schenker International), which in 2001 took the decision of regionalizing its distribution networks in the Nordic region, by combining all its distribution centers into one. They considered eight criteria for the evaluation of a location: (1) proximity to customer "bases"; (2) availability and quality of labor workforce; (3) availability of utilities; (4) local tax environment, especially on inventories; (5) inland transport infrastructure; (6) expansion capability; (7) customs administration and regulations; and (8) local standards of living. In the second stage of selection, the qualitative approach (weighed marking method) was adopted. They also considered alternative seaports and airports to be used as gateways for seaborne and airborne transportation, inbound and outbound cargo flows of the company.

Dey *et al.* (2015) analyzed the selection of a warehouse location by fuzzy MCDM methodologies based on subjective and objective criteria. They studied two cases, where the evaluation criteria were infrastructure (transportation facility; availability of land)); market (proximity to customers, to suppliers, lead times); cost (labor, transportation, taxes); labor characteristics (skilled, availability of labor force); macro-environment (policies of government, industrial regulations, political stability).

Korpelaa *et al.* (2007) analyzed the selection of a warehouse operator by combining AHP and DEA methodologies. The criteria considered were classified into two groups: reliability (delivery time, quality, quantity) and flexibility (urgent deliveries, frequency, and special request add capacity).

Summing up and according to the classification made by Uyanik *et al.* (2018), the criteria normally used to make the location decision of a LC are related to environment (natural resources, environment impact), costs (land, labor workforce, investment, transportation, infrastructure, energy consumption), location (proximity to clients and suppliers, accessibility, proximity to airports, railroad), social factors (standard living, social stability, service level, social benefits, safety and security, availability of labor, political stability) and economic aspects (state subsidies and taxes, volume of trade, development level, investment level, presence of trade barriers).

This literature review will be used for the selection of the evaluation criteria in the analyzed case in our study, which is described later.

2.2. *Research impact*

As stated before, various models have been applied for locating LCs. Rather than the difficulty of modeling, the complexity of a model fitted to the application was high during the survey. We have compared previous investigations and figured out the contribution of our study through the following items:

- The main significance of this study is that we have utilized the real data from Andalucía government to aid them in their investment and future strategies. Through the interview and searching for main characteristics of a LC with respect to the specific region, several factors have been identified. The data was extracted from a platform called Sistema de Informacion Multi-territorial Andalucia (SIMA) and was transferred to the decision process.
- Over time, different MADM techniques have been developed to assign weights to different criteria. The method chosen for this case study is the best worst method (BWM), introduced by Rezaei (2015). The typical pairwise comparison methods like AHP cause two problems. The first one is that it is a lengthy task due to the number of comparisons that have to be made for a full pairwise comparison matrix. The second problem is the inconsistency between the comparisons, which can be caused by several reasons, such as a lack of concentration or information (Rezaei, 2018). The BWM, which is a recently developed technique, contains a structured basis and requires fewer comparisons when they are compared with other full pairwise comparison-based tools. It results in a more reliable evaluation process and enhances the consistency between the comparisons. One advantage of BWM is that reduces the problems of pairwise comparison. To be precise, the advantages

that BWM bring to the MADM is using an understandable way of gathering the data, which results in highly reliable results that are comprehensive to the evaluator and can be easily revised to increase consistency. We have utilized the recent decision-making tools as BWM and Multi-Attributive Border Approximation Area Comparison (MABAC) in the determination of best factor and the alternative. The MABAC methods continue to be frequently applied today within the area of business and economics (Debnath *et al.*, 2017; Andreopoulou *et al.*, 2017; Chatterjee *et al.*, 2018; Mohanty and Gupta, 2015). These applications cover fields as diverse as material selection (Xue *et al.*, 2016), selection of transport and logistics resources (Pamučar and Ćirović, 2015) and evaluation of medical tourism site (Roy *et al.*, 2018). It is a user-friendly instrument and facilitates decision-making process and speeds up the acquisition of results. We report the comparison of the results with other MADM methods to validate the accuracy of our results.

3. Materials and Techniques

3.1. *Best worst method*

Step 1: The decision-maker (DM) determines a set of decision criteria:

$$\{c_1, c_2, \ldots, c_n\}$$

Step 2: The DM chooses the best and the worst criteria. In this step, the DM chooses the best and the worst criteria among the set of identified criteria in the last step. The best criterion represents the most desirable or the most significant one while the worst criterion is the least important one among others.

Step 3: The DM conducts pairwise comparisons between the best criterion and the other criteria. In this step, the goal is to identify the preference of the most important criterion to the other criteria. DM uses a scale from 1 to 9 (1: equally important, and 9: extremely more important). The comparison outcome is described as best-to-other vector:

$$A_B = (a_{B1}, a_{B2}, \ldots, a_{Bn}),$$

where a_{Bj} represents the preference of the best criterion B over the criterion j and $a_{BB} = 1$.

Step 4: The DM conducts a pairwise comparison between the other criteria and the worst criterion. Similar to the last step, the comparison results are expressed by other-to-worst vector:

$$A_W = (a_{1W}, a_{2W}, \ldots, a_{nB})^{\mathrm{T}},$$

where a_{jw} represents the preference of the best criterion j over the criterion W and $a_{WW} = 1$.

Step 5: Calculating the optimal weights

$$(\boldsymbol{W_1^*}, \boldsymbol{W_2^*}, \ldots, \boldsymbol{W_n^*}).$$

For each pair of $\frac{W_B}{W_j}$ and $\frac{W_j}{W_W}$, the optimal weight should meet the requirement that $\frac{W_B}{W_j} = a_{Bj}$ and $\frac{W_j}{W_W} = a_{jW}$. To satisfy the conditions, the maximum absolute differences $\left|\frac{W_B}{W_j} - a_{Bj}\right|$ and $\left|\frac{W_j}{W_W} - a_{jW}\right|$ for all j is minimized. Also, taking into consideration the non-negativity characteristic and sum condition of the weights, the following problem can be formulated:

$$\text{Min} \max_j \left\{ \left| \frac{W_B}{W_j} - a_{Bj} \right|, \left| \frac{W_j}{W_W} - a_{jW} \right| \right\}$$

s.t:

$$\sum_j W_j = 1, \quad W_j \geq 0 \quad \text{for all } j. \tag{1}$$

The model can be transformed as

$$\min \xi$$

s.t:

$$\left| \frac{W_B}{W_j} - a_{Bj} \right| \leq \xi, \quad \text{for all } j,$$

$$\left|\frac{W_j}{W_W} - a_{jW}\right| \leq \xi, \quad \text{for all } j,$$

$$\sum_j W_j = 1, \quad W_j \geq 0 \quad \text{for all } j. \tag{2}$$

By solving the optimization problem (2), the optimal weights $(W_1^*, W_2^*, \ldots, W_n^*)$ and ξ^* can be obtained. As such, the following two models are used to calculate the lower and upper bounds of the weight of criterion j. These models are solved after solving model (2) and finding ξ^* (Rezaei, 2015).

$$\begin{aligned}
&\text{Min } W_j,\\
&\text{s.t:}\\
&\left|\frac{W_B}{W_j} - a_{Bj}\right| \leq \xi^*, \quad \text{for all } j,\\
&\left|\frac{W_j}{W_W} - a_{jW}\right| \leq \xi^*, \quad \text{for all } j,\\
&\sum_j W_j = 1, \quad W_j \geq 0, \quad \text{for all } j.
\end{aligned} \tag{3}$$

$$\begin{aligned}
&\text{Max } W_j\\
&\text{s.t:}\\
&\left|\frac{W_B}{W_j} - a_{Bj}\right| \leq \xi^*, \quad \text{for all } j,\\
&\left|\frac{W_j}{W_W} - a_{jW}\right| \leq \xi^*, \quad \text{for all } j,\\
&\sum_j W_j = 1, \quad W_j \geq 0, \quad \text{for all } j.
\end{aligned} \tag{4}$$

Solving these two models for all the criteria, we can find the upper and lower bounds of the weights of the criteria. So, we have an optimal weight interval for each criterion. One way is to calculate the center of the interval as a representative weight of criterion j as

Table 1: BWM consistency index.

a_{BW}	1	2	3	4	5	6	7	8	9
Consistency index	0.00	0.44	1.00	1.63	2.30	3.00	3.73	4.47	5.23

follows:

$$W_j^* = (\min W_j + \max W_j)/2. \tag{5}$$

After finding the final results, we should calculate the consistency level of the comparisons. The consistency is defined as follows (Rezaei, 2015).

The consistency ratio of BWM can be expressed by using ξ^* and the corresponding consistency index (Table 1) as follows:

$$\text{Consistency ratio} = \frac{\xi^*}{\text{Consistency index}}. \tag{6}$$

It can be seen that the smaller the ξ^*, the smaller the "consistency ratio" and the more consistent the vectors are.

3.2. *MABAC*

The MADM family of tools were advanced and updated over the years. The MABAC method is one of the recent designed methods for decision-making (Pamučar and Ćirović, 2015). The MABAC method has been both widely applied to solve numerous MADM problems (Stevic *et al.* 2017; Gigović *et al.* 2017; Pamučar *et al.*, 2018; Xue *et al.*, 2016; Bozanic *et al.*, 2016, 2018; Roy *et al.*, 2016; Yu *et al.*, 2016; Peng and Dai, 2017; Peng and Yang, 2016; Chatterjee *et al.*, 2018; Bojanic *et al.*, 2018). The conceptual origin of the MABAC method is reflected in the definition of the distance of the criterion function from each of the observed alternative border approximation areas. The next section shows the process of implementing the MABAC method, consisting of six steps.

Step 1: Formation of the initial decision matrix (X). The first step is to evaluate m alternatives according to n criteria. We show the alternatives in the form of vectors $A_i = (x_{i1}, x_{i2}, \ldots, x_{in})$, where

x_{ij} is the value of the ith alternative according to the jth criterion $(i = 1, 2, \ldots, m;\ j = 1, 2, \ldots, n)$.

$$X = \begin{array}{c} \\ A_1 \\ A_2 \\ \ldots \\ A_m \end{array} \begin{array}{c} \begin{array}{cccc} C_1 & C_2 & \ldots & C_n \end{array} \\ \begin{bmatrix} x_{11} & x_{12} & \ldots & x_{1n} \\ x_{21} & x_{22} & & x_{2n} \\ \ldots & \ldots & \ldots & \ldots \\ x_{m1} & x_{m2} & \ldots & x_{mn} \end{bmatrix} \end{array} \tag{7}$$

where m indicates the number of the alternatives and n indicates the total number of criteria.

Step 2: Normalization of the elements from the initial matrix (X):

$$N = \begin{array}{c} \\ A_1 \\ A_2 \\ \ldots \\ A_m \end{array} \begin{array}{c} \begin{array}{cccc} C_1 & C_2 & \ldots & C_n \end{array} \\ \begin{bmatrix} n_{11} & n_{12} & \ldots & n_{1n} \\ n_{21} & n_{22} & & n_{2n} \\ \ldots & \ldots & \ldots & \ldots \\ n_{m1} & n_{m2} & \ldots & n_{mn} \end{bmatrix} \end{array} \tag{8}$$

The elements of the normalized matrix (N) are determined using the equation:

(a) For benefit-type criteria (a higher value of the criterion is preferable):

$$n_{ij} = \frac{x_{ij} - x_i^-}{x_i^+ - x_i^-}. \tag{9}$$

(b) For cost-type criteria (a lower value of the criterion is preferable):

$$n_{ij} = \frac{x_{ij} - x_i^+}{x_i^- - x_i^+}, \tag{10}$$

where x_{ij}, x_i^+ and x_i^- are the elements from the initial decision matrix (X), for which x_i^+ and x_i^- are defined as

$$x_i^+ = \max(x_1, x_2, \ldots, x_m)$$

which is the maximum value of the observed criterion according to the alternatives.

$$x_i^- = \min(x_1, x_2, \ldots, x_m)$$

which is the minimum value of the observed criterion according to the alternatives.

Step 3: Calculation of the elements from the weighted matrix (V). The elements from the weighted matrix (V) are calculated on the basis of the expression

$$v_{ij} = w_i * (n_{ij} + 1), \tag{11}$$

where n_{ij} are the elements of the normalized matrix (N), w_i is the weight coefficients of the criteria. Using equation (11) we obtain the weighted matrix,

$$V = \begin{bmatrix} v_{11} & v_{12} & \ldots & v_{1n} \\ v_{21} & v_{22} & & v_{2n} \\ \ldots & \ldots & \ldots & \ldots \\ v_{m1} & v_{m2} & \ldots & v_{mn} \end{bmatrix}$$

$$= \begin{bmatrix} w_1 \cdot (n_{11} + 1) & w_2 \cdot (n_{12} + 1) & \ldots & w_n \cdot (n_{1n} + 1) \\ w_1 \cdot (n_{21} + 1) & w_2 \cdot (n_{22} + 1) & \ldots & w_n \cdot (n_{2n} + 1) \\ \ldots & \ldots & \ldots & \ldots \\ w_1 \cdot (n_{m1} + 1) & w_2 \cdot (n_{m2} + 1) & \ldots & w_n \cdot (n_{mn} + 1) \end{bmatrix},$$

where n is the total number of criteria and m the total number of alternatives.

Step 4: Determining the border approximation area matrix (G). The border approximation area (BAA) for each criterion is determined according to the equation

$$g_i = \left(\prod_{j=1}^{m} v_{ij} \right)^{1/m}, \tag{12}$$

where v_{ij} are the elements of the weighted matrix (V) and m is the total number of alternatives.

After calculating the value g_i for each criterion, a border approximation area matrix $G(13)$ is formed with the format $n\,x\,1$ (n is the total number of criteria according to which the selection is made from the alternatives offered).

$$\begin{matrix} & C_1 & C_2 & \dots & C_n \\ G = & [g_1 & g_2 & \dots & g_n]. \end{matrix} \tag{13}$$

Step 5: Calculation of the distance of the alternative from the border approximation area for the matrix elements (Q):

$$Q = \begin{bmatrix} q_{11} & q_{12} & \dots & q_{1n} \\ q_{21} & q_{22} & & q_{2n} \\ \dots & \dots & \dots & \dots \\ q_{m1} & q_{m2} & \dots & q_{mn} \end{bmatrix}. \tag{14}$$

The distance of the alternatives from the border approximation area (q_{ij}) is determined as the difference between the elements in the weighted matrix (V) and the value of the border approximation area (G).

$$Q = V - G = \begin{bmatrix} v_{11} & v_{12} & \dots & v_{1n} \\ v_{21} & v_{22} & & v_{2n} \\ \dots & \dots & \dots & \dots \\ v_{m1} & v_{m2} & \dots & v_{mn} \end{bmatrix} - \begin{bmatrix} g_1 & g_2 & \dots & g_n \\ g_1 & g_2 & & g_n \\ \dots & \dots & \dots & \dots \\ g_1 & g_2 & \dots & g_n \end{bmatrix}, \tag{15}$$

$$Q = \begin{bmatrix} v_{11} - g_1 & v_{12} - g_2 & \dots & v_{1n} - g_n \\ v_{21} - g_1 & v_{22} - g_2 & \dots & v_{2n} - g_n \\ \dots & \dots & \dots & \dots \\ v_{m1} - g_1 & v_{m2} - g_2 & \dots & v_{mn} - g_n \end{bmatrix} = \begin{bmatrix} q_{11} & q_{12} & \dots & q_{1n} \\ q_{21} & q_{22} & & q_{2n} \\ \dots & \dots & \dots & \dots \\ q_{m1} & q_{m2} & \dots & q_{mn} \end{bmatrix}, \tag{16}$$

where g_i is the border approximation area for criterion C_i, v_{ij} is the weighted matrix of the elements (V), n is the number of criteria and m is the number of alternatives.

Alternative A_i could belong to the border approximation area (G), upper approximation area (G^+) or lower approximation area (G^-), that is, $A_i \in \{G \vee G^+ \vee G^-\}$. The upper approximation area (G^+) is the area that contains the ideal alternative (A^+), while the lower approximation area (G^-) is the area that contains the anti-ideal alternative (A^-).

The belonging of alternative A_i to the approximation area (G, G^+ or G^-) is determined on the basis of the equation

$$A_i \in \begin{cases} G^+ & \text{if } q_{ij} > 0 \\ G & \text{if } q_{ij} = 0 \\ G^- & \text{if } q_{ij} < 0 \end{cases} \tag{17}$$

In order for alternative A_i to be selected as the best in the set, it is necessary for it to have as many criteria as possible belonging to the upper approximate area (G^+). If, for example, alternative A_i has five criteria (out of a total of six criteria) belonging to the upper approximate area, and one criterion belonging to the lower approximate area (G^-), it means that according to five criteria, the alternative is near or equal to the ideal alternative, while for one criterion, it is near or equal to the anti-ideal alternative. If the value $q_{ij} > 0$, that is $q_{ij} \in G^+$, then alternative A_i is near or equal to the ideal alternative. If the value $q_{ij} < 0$, that is $q_{ij} \in G^-$, it shows that alternative A_i is near or equal to the anti-ideal alternative.

Step 6: Ranking the alternatives: A calculation of the values of the criterion functions for the alternatives is obtained as the sum of the distance of the alternatives from the border approximation areas (q_i). By calculating the sum of the elements of matrix Q by rows, we obtain the final values of the criterion functions of the alternatives:

$$S_i = \sum_{j=1}^{n} q_{ij},\ j = 1, 2, \ldots, n,\ i = 1, 2, \ldots, m, \tag{18}$$

where n is the number of criteria and m is the number of alternatives.

4. Implementing the Proposed Model and Results

In this section, the case study and implementation of the proposed model will be explained. The model will be solved by a decision-making approach using BWM and MABAC tools in order to gain the weights of the criteria and alternatives ranking, respectively.

4.1. *The procedure and algorithm toward logistics center location*

To obtain the solution of a multi-attribute decision-making (MADM) problem, a procedure introduced here: (a) identification of the objectives of the model including determination of decision factors, alternative recognition with a meeting by group of experts, (b) designing questionnaires in order to detect the importance of each criterion, (c) gathering information and the required data from the Andalucia province information system, (d) measuring the weights using the BWM tool, (e) forming the primary decision matrix containing alternatives and criteria, (f) demonstrating the ranking of location centers using the MABAC tool, (g) comparing the alternative rankings with traditional models and other MADM tools, (h) making final decision and reporting to the responsible office.

4.2. *Application of decision-making model in logistics center development in Andalucia*

The objective of this study is to find the location of establishing a LC in Andalucia, a big province of Spain . The study is dedicated to finding the best location (city) for according to the factors and characteristic we considered. The main cities in this province are A_1 (Almeria), A_2 (Cadiz), A_3 (Cordoba), A_4 (Granada), A_5 (Huelva), A_6 (Jaen), A_7 (Malaga) and A_8 (Sevilla) which we call them alternative LCs. To compare and analyze each location, several factors must be involved. For this, we chose the following parameters: C_1 (population active in economy) by 1,000 persons, C_2 (approximately % of tourists), C_3 (per capita income), C4 (unemployment), C5 (value added), C6 (number of active businesses), C7 (ports and nautical activities), C8 (communication network) and C9 (traffic level). In MADM methodology, decision makers must distinguish the optimization direction of each factor. Therefore, C_9 and C_4 are classified as cost factor and the rest of them are placed as the benefit factor.

As mentioned earlier, in the MADM problem, it is important how to obtain the weight of each decision factor (criterion). There are plenty of tools designed to find the weights. Among them AHP, SWARA, ANP, Entropy and DEMATEL are used widely by investigators based on the conditions and needs. However, the method we are going to utilize is BWM developed by Rezaei (2015). The method is quite well defined and has a different logic from other weighting tools.

As described in the methodology section, at first, the best and worst criteria are determined; for our case study, per capita income (C_3) and traffic level (C_9) are the best and worst items, respectively. The weights are shown in Table 2. As the table states, the C_3 (per capita income) weight has the value of 0.26 while the value of weight for C_9 is 0.036. These weights are a vector that will be utilized for the MABAC method. Figure 1 exhibits the schematic view for criteria weights. The reliability of data (consistency ratio) is 0.151, which denotes an acceptable level.

The next task in locating best place to establish the LC is application of a MADM tool called MABAC. The first step in this method is to form a decision matrix (as equation (7) indicates) containing decision alternatives and criteria. This matrix is obtained by adding the opinion of decision makers' judgment usually and sometime quantitative data as well. In this case study, we have extracted data from Andalucía government, which are shown in Table 3. The next step is to normalize data using equations (9) and (10) for benefit and non-benefit attributes in that order. The normalized table is computed and shown in Table 4.

Table 2: The weights determined by BWM.

Weights	C_1	C_2	C_3	C_4	C_5	C_6	C_7	C_8	C_9
	0.137	0.103	0.260	0.059	0.137	0.082	0.082	0.103	0.036

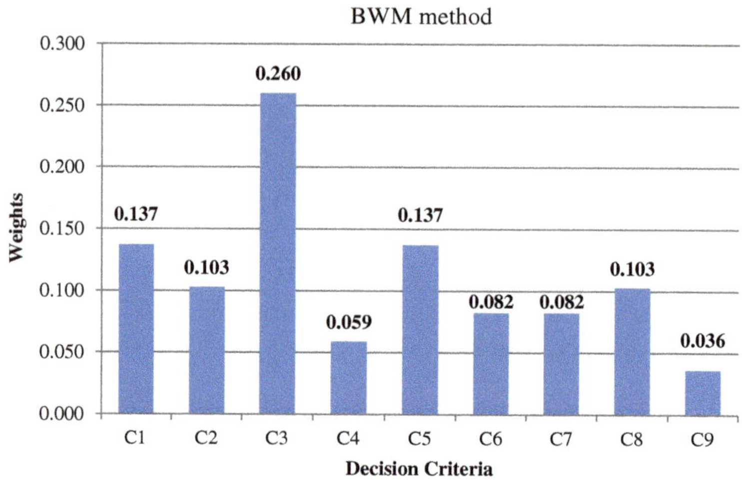

Figure 1: The decision criteria weights.

Table 3: The main decision matrix including criteria and alternatives.

Alternative location	C_1	C_2	C_3	C_4	C_5	C_6	C_7	C_8	C_9
A_1 (Almeria)	249.2	8.13	9.0367	62.8333	10.667732	40.127	9.7	8	8.6
A_2 (Cadiz)	484.7	15.61	6.93852	157.8795	19.055652	58.073	8.7	7.8	7.7
A_3 (Cordoba)	340.6	6.25	7.89651	77.0074	11.852014	46.603	0.0	8	8.1
A_4 (Granada)	334.5	14.94	6.97274	86.4077	13.498596	54.528	8.5	8.2	7.6
A_5 (Huelva)	185.5	7.46	8.07214	50.3387	8.396120	23.936	7.3	7.6	7.4
A_6 (Jaen)	265.6	2.73	7.9102	51.7421	9.507230	33.390	0.0	7.6	8.1
A_7 (Malaga)	596.4	31.91	8.15126	155.6047	25.439979	110.116	9.2	8.1	7.7
A_8 (Seville)	780.4	12.97	6.85049	207.6469	31.470665	106.886	8.5	8.4	7.9

Table 4: The normalized decision matrix.

Alternatives	C_1	C_2	C_3	C_4	C_5	C_6	C_7	C_8	C_9
Almeria	0.107	0.185	1	0.921	0.098	0.188	1	0.5	0
Cadiz	0.503	0.441	0.04	0.316	0.462	0.396	0.897	0.25	0.75
Cordoba	0.261	0.121	0.478	0.83	0.15	0.263	0	0.5	0.417
Granada	0.25	0.418	0.056	0.771	0.221	0.355	0.876	0.75	0.833
Huelva	0	0.162	0.559	1	0	0	0.753	0	1
Jaen	0.135	0	0.485	0.991	0.048	0.11	0	0	0.417
Malaga	0.691	1	0.595	0.331	0.739	1	0.948	0.625	0.75
Seville	1	0.351	0	0	1	0.963	0.876	1	0.583

Table 5: The weighted normalized decision matrix.

Alternatives	C_1	C_2	C_3	C_4	C_5	C_6	C_7	C_8	C_9
Almeria	0.1517	0.1221	0.52	0.1133	0.1505	0.0974	0.164	0.1545	0.036
Cadiz	0.2059	0.1485	0.2705	0.0777	0.2003	0.1145	0.1555	0.1288	0.063
Cordoba	0.1727	0.1154	0.3844	0.108	0.1575	0.1036	0.082	0.1545	0.051
Granada	0.1713	0.1461	0.2745	0.1045	0.1673	0.1111	0.1539	0.1803	0.066
Huelva	0.137	0.1197	0.4053	0.118	0.137	0.082	0.1437	0.103	0.072
Jaen	0.1554	0.103	0.386	0.1175	0.1436	0.091	0.082	0.103	0.051
Malaga	0.2316	0.206	0.4147	0.0785	0.2382	0.164	0.1598	0.1674	0.063
Seville	0.274	0.1391	0.26	0.059	0.274	0.1609	0.1539	0.2060	0.057

Table 6: Border approximation area matrix.

	C_1	C_2	C_3	C_4	C_5	C_6	C_7	C_8	C_9
G	0.1828	0.1346	0.3547	0.0945	0.1783	0.1123	0.1322	0.1457	0.0563

After this step, the weighted normalized matrix must be computed. This is carried out using equation (11) and is indicated as Table 5. To achieve the border approximation area matrix, equation (12) is used. This gives a reference; in order to measure the distance, equation (15) is applied. The calculation of the distance of the alternative from the border approximation area for the matrix is as shown in Table 6.

The final step of this decision problem is to deliver the final evaluation of the alternatives. This action is performed using equation (18). The results show that Malaga is best place and after that Seville can

Table 7: Ranking of LCs.

Alternative	Si	Rank
Almeria	0.1181	3
Cadiz	−0.0268	5
Cordoba	−0.0622	6
Granada	−0.0164	4
Huelva	−0.0737	7
Jaen	−0.1588	8
Malaga	**0.3318**	**1**
Seville	0.1926	2

be a suitable option. Among the alternatives, Jaen obtained the minimum score, which made it the worst option.

The order of alternatives is presented as

Malaga > Seville > Almeria > Granada > Cadiz > Cordoba > Huelva > Jaen.

5. Discussion

The transportation and distribution of goods, for all types of companies and for final consumers, has become a key element in the economy of countries and regions. This factor influences not only costs for different agents, but also in location or even relocation decisions. With this, public administrations have a very important responsibility in the economic growth and ultimately in the development and degree of well-being of the citizens. On the other hand, there are so many interests, both individual and collective, of different industries, different productive sectors, and different population groups. The information to be handled exceeds the capacity of the administration itself in many cases. Selecting the most pertinent information for the purpose that is pursued, as well as processing it adequately to ensure that it contributes to better decision- making is a vitally important task.

With this chapter, we have not tried to address the problem in its entirety, nor solve it in all its complexity. For example, other factors could have been included for decision-making, while some important details should have been specified related to the problem of locating the LC (such as the type of merchandise to be moved, the geographical area that would encompass the market,). Rather, it has been used as an excuse to show the power and versatility of the applied analysis procedure. Even so, the solution obtained makes sense in some aspects. According to the socio-economic report 2017–2018 of the province of Málaga, carried out by the confederation of businessmen of Málaga (CEM, 2017), Málaga is one of the most dynamic provinces of Andalusia. It concluded 2017 with a 3.3% increase in GDP (two-tenths above the regional figure) and this good growth is expected to be maintained in 2018 with a forecast close to 3.1%, which could put the province back in the leadership position in the community. For Málaga, which leads the province in creation of companies, the tourism sector is very important, and it continues to grow in a positive manner, with increase in both demand and supply. Foreign demand has grown around 6.2% and the airport has registered a record 18.6 million international passengers. Imports and exports continue growing as in the previous year. One of the main potentials of Málaga is in communications and the transport networks, since it has road infrastructure, railway, port, and airport. Therefore, all these factors related to tourism (C_2), number of active business (C_6) and ports and nautical activities position Málaga in the first place in the ranking. Sevilla is also in a very good position, due to its high economic activity (C_1 and C_6), and it has the highest activity rate in Andalucía (the only one that exceeds 60% according to the active population survey, INE 2016). Seville also has the best score in communication networks (C_8). The position occupied by Almería (3rd) is due to their highest income per-capita (C_3), and ports and nautical activities (C_7). It is also one of the provinces with the lowest number of unemployment in absolute terms. On the other hand, Jaen and Huelva are the worst positioned. Jaen has in general low values in all the criteria, reaching the minimum in ports and nautical activities (C_7), and communication networks (C_8). The position of Huelva is explained by its reduced values in economic activity (C_1 and C_6) as well as in communication networks (C_8).

Therefore, this investigation demonstrates that once again MABAC is a very valuable method to analyze decision problems such as the current one, with lack of accuracy in some respects (lack of weight of the factors) and incomplete information. Researchers can trust the method and feel comfortable to implement it effectively. In this instance, it once again demonstrates its applicability and usefulness in order to achieve the objective pursued in a very efficient manner: ordering the alternatives as an aid to decision-making, despite the low level of concreteness of the proposed problem. It is expected that as more concrete problems are addressed with this methodology, the choice of criteria will be more accurate, making the result obtained more reliable. The contribution of the BWM allows establishing in a consistent way the weight of the factors to be used, and it presents operative advantages with respect to some of the techniques used until now. The effect on the final solution of using different techniques for identifying weights is an interesting line of research to pursue.

6. Conclusion

This study designs the application of MADM techniques in terms of supply chain and logistics. The vital role of urban and city LCs is increasing over the years. While regional policymakers are concerned with the optimal location of LCs, academics must take the opportunity of developing advanced models in order to help decision makers in this area. In this chapter, we have built the evaluation and selection process of locating LCs in Andalucia, Spain. There are eight main cities that are potentially able to establish a center for logistics and distribution purposes. A two-step process was designed to reach that goal. Nine decision-making factors were assumed and then we utilized the BWM method to gain the importance of the criteria. The results indicated per capita income was the main factor influencing our evaluation matrix.

Regarding the procedure used here, MABAC, which has been validated in previous studies, it shows its versatility and efficiency, generating the ranking of the alternative LCs. In the resolution of the important problem of determining weights, the contribution of the WBM proves to be a very valuable aid with a high consistency level. The relationship between the weights obtained reaches a maximum eight times the lowest, thus offering a good degree of differentiation between the factors.

Based on our analysis, the best place (city) is Málaga with the highest-ranking rate. The final order obtained, although susceptible to sensitivity analysis in general, in this case offers a very clear superiority of the first option (Málaga) in comparison with the second (Seville), with a final score of just over 50% of the first option. It is thus shown how the technique used has a capacity and utility to be applied by public administrations in the support of decision-making on a broad scale such as the one analyzed in this case. This study seeks effective future investigations where we need more concrete, robust and advanced models to deal with complex variables and parameters.

References

Awasthi, A., Chauhan, S. and Goyal, S. (2011). A multi-criteria decision-making approach for location planning for urban distribution centers under uncertainty. *Math and Comput Model*, 53, 98–109.

Andreopoulou, Z., Koliouska, C. and Zopounidis, C. (2017). *Multicriteria and Clustering: Classification Techniques in Agrifood and Environment.* Springer, Cham. https://doi.org/10.1007/978-3-319-55565-2.

Bojanic, D., Kovač, M., Bojanic, M. and Ristic, V. (2018). Multi-criteria decision-making in a defensive operation of the guided anti-tank missile battery: an example of the hybrid model fuzzy AHP — MABAC. *Decision-Making: Appl Manage Eng*, 1(1), 51–66.

Božanić, D., Pamučar, D. and Karovic, S. (2016). Application the MABAC method in support of decision-making on the use of force in a defensive operation. *Technics.* 71, 97–104.

Božanić, D., Tešić, D. and Milićević, J., (2018). A hybrid fuzzy AHP-MABAC model: Application in the serbian army — The selection of the location for deep wading as a technique of crossing the river by tanks. *Decision-Making: Appl Manage Eng*, 1, 143–164.

CEM (2017). *Informe socio-económico de la provincia de Málaga.* Confederación de empresarios de Málaga, 2017–2018.

Chatterjee, K. and Tamosaitiene, J. (2018). A hybrid MCDM technique for risk management in construction. *Proj Symm*, 10(2), 46.

Chen, T. W. and Yuan, W. C. (2007). Lee Multi-criteria fuzzy optimization for locating warehouses and distribution centers in a supply chain network. *J Chin Instit Chem Eng*, 38, 393–407.

Church, R. L. and Murray, A. T. (2009). *Business Site Selection, Location Analysis and GIS*. John Wiley and Sons, New York.

Daganzo (1996). *Logistics System Analysis*. Springer Verlag, Berlin.

Debnath A., Roy, J., Kar, S., Kazimieras, E. and Antucheviciene, J. (2017). Hybrid MCDM Approach for strategic project portfolio selection of agro by-product. *Sustainablility* 9, 1302.

Dey, B., Bairagi, B., Sarkar, B. and Sanyal, S. K. (2016), Warehouse location selection by fuzzy multi-criteria decision making methodologies based on subjective and objective criteria. *Int J Manage Sci Eng Manage*, 11(4), 262–278.

Drezner, Z. and Hamacher, H. (2002). *Facility Location: Applications and Theory*. Springer, Berlin.

EUROPLATFORMS, Corporate Presentation, October 2015 http://www.europlatforms.eu/wp-content/uploads/2016/01/Corporate-Presentation-2015-Europlatforms-Final_20151229.pdf.

Farahani, R. Z. and Hekmatfar, M. (eds.) (2009). *Facility Location: Concepts, Models, Algorithms and Case Studies*. Physica-Verlag, Heidelberg.

Farahani, R., SteadiSeifi, M. S. and Asgari, N. (2010). Multiple criteria facility location problems: A survey. *Appl Math Model*, 34, 1689–1709.

Gigović, L., Pamučar, D., Bajić, Z. and Drobnjak, S. (2017). Application of GIS-interval rough AHP methodology for flood hazard mapping in urban areas. *Water*, 9, 360.

Higgins, C. D. and Ferguson, M. R. (2011). *An Exploration of the Freight Village Concept and its Applicability to Ontario*. McMaster Institute of Transportation and Logistics.

Kayikci, Y. (2010). A conceptual model for intermodal freight logistics centre location decisions. *Procedia — Soci Behav Sci*, 2(3), 6297–6311.

Klose, A. and Drexl, A. (2005). Facility location models for distribution system design. *Eur J Oper Res*, 162(1), 4–29.

Koldemir, B., Çancı, M. and Gönüler, E. (2009). *Buǥyuǥk Ölçekli Kent Planlamasında Lojistik Köyler*. İzmir Ulaşım Sempozyumu, İzmir, pp. 457–464.

Korpelaa, J. and Tuominen, M. (1996). A decision aid in warehouse site selection. *Int J Product Econom*, 45, 169–180.

Korpelaa, J., Lehmusvaarab, A. and Nisonenc, J. (2007). Warehouse operator selection by combining AHP and DEA methodologies. *Int J Product Econom*, 108, 135–142.

Mohanty, B. K. and Gupta, M. (2015). Product selection in e-commerce under fuzzy environment: A MADM game theoretic model. *Int Game Theory Rev*, 17(01), 154–208.

Özcan, N., Çelebi, S. and Esnaf (2011). Comparative analysis of multi-criteria decision making methodologies and implementation of a warehouse location selection problem. *Expert Syst Appl*, 38, 9773–9779.

Owen, S. H. and Daskin, M S. (1998). Strategic facility location: A review. *Eur J Oper Res*, 111(3), 423–447.

Pamučar, D. and Ćirović, G. (2015). The selection of transport and handling resources in logistics centers using multi-attributive border approximation area comparison (MABAC). *Expert Syst Appl*, 42(6), 3016–3028.

Pamučar, D., Stević, Z. and Zavadskas, E. K. (2018). Integration of interval rough AHP and interval rough MABAC methods for evaluating university web pages. *Appl Soft Comput*, 67, 141–163.

Peng, X. and Yang, Y. (2016). Pythagorean fuzzy choquet integral based MABAC Method for multiple attribute group decision making. *Int J Intell Syst*, 31(10), 989–1020.

Peng, X., Dai, J. and Yuan, H. (2017). Interval-valued fuzzy soft decision making methods based on MABAC, similarity measure and EDAS. *Fundam Inform*, 152, 373–396.

Rezaei, J., Wang, J. and Tavasszy, L. (2015). Linking supplier development to supplier segmentation using Best Worst Method. *Expert Syst Appl*, 42, 9152–9164.

Rezaei, J., Roekel, W. S., Tavasszy, L. (2018). Measuring the relative importance of the logistics performance index indicators using Best Worst Method. *Transp Policy*, 68, 158–169.

Rimienė, K. and Grundey, D. (2007). Logistics center concept through evolution and definition. *Eng Econom*, 4(54), 87–95.

Roy, J., Ranjan, A., Debnath, A. and Kar, S. (2016). An extended MABAC for multi-attribute decision making using trapezoidal interval type-2 fuzzy numbers, *Artif Intell*, arXiv:1607.01254.

Roy, J., Chatterjee, K., Bandyopadhyay, A. and Kar, S. (2018). Evaluation and selection of medical tourism sites: A rough analytic hierarchy process based multi-attributive border approximation area comparison approach. *Expert Syst*, 35(1), e12232.

Stević, Ž., Pamučar, D., Vasiljević, M., Stojić, G. and Korica, S. (2017). Novel integrated multi-criteria model for supplier selection: Case study construction company. *Symmetry*, 9, 279.

Thai, V. V. and Grewal, D. (2005). Selecting the location of distribution centre in logistics operations: A conceptual framework and case study. *Asia Pacific J Market Logist*, 17, 3–24.

Turskis, Z. and Zavadskas, E. K. (2010) A new fuzzy additive ratio assessment method (ARAS-F). Case study: The analysis of fuzzy multiple criteria in order to select the logistic centers location. *Transport*, 25(4), 423–432.

Uyanik, C., Tuzkaya, G. and Oğuztimur, S. (2018). A literature survey on logistics centers' location selection problem. *Sigma J Eng Nat Sci*, 36(1), 141–160.

Xue, Y. X., You, J. X., Lai, X. D. and Liu, H. C. (2016). An interval-valued intuitionistic fuzzy MABAC approach for material selection with incomplete weight information. *Appl Soft Comput*, 38, 703–713.

Yu, S. M., Wang, J. and Wang, J. Q. (2016). An interval type-2 fuzzy likelihood-based MABAC approach and its application in selecting hotels on a tourism website. *Int J Fuzzy Syst*, 1–15.

Żak, J. and Fierek, Sz. (2007). Design and evaluation of alternative solutions for an integrated urban transportation system. In *Selected CD Proceedings of the World Conference on Transport Research*, Berkeley, 24–28 June.

Żak, J. and Węgliński, S. (2014). The selection of the logistics center location based on MCDM/A methodology. *Transport Res Procedia*, 3, 555–564.

Chapter 2

A Sustainability Assessment Framework for Petroleum Refinery Projects with Multi-Criteria Decision-Making Techniques

Hamidreza Hasheminasab[*,¶], Mohammadreza Kharrazi[†,||], Yaghob Gholipour[‡,**] and Dalia Streimikiene[§,††]

[*] *School of Civil Engineering, College of Engineering, University of Tehran, Tehran, Iran*
[†] *Department of Sustainability, SEPID Group Ltd., Tehran, Iran*
[‡] *Engineering Optimization Research Gr., College of Engineering, University of Tehran, Tehran, Iran*
[§] *Lithuanian Energy Institute, Breslaujos 3, Kaunas, LT-44403, Lithuania*
[¶] *hasheminasab@ut.ac.ir*
[||] *mohammad.kharrazi@sepidgroup.com*
[**] *ygholipour@ut.ac.ir*
[††] *dalia@mail.lei.lt*

In petroleum refinery projects (PRPs) various engineering disciplines are involved; also, many types of stakeholders are dealt with such as vendors and governments. Therefore, PRPs are complex and multi-disciplinary; and in order to assess the sustainability of these projects in each and all of the triple bottom lines (economic, social and environmental), an integrated framework and a comprehensive approach is needed. To assess the sustainability of PRP *vis-à-vis* each pillar, many criteria will be defined.

In this chapter, life-cycle model of PRP will be introduced and used as a basis for developing sustainability assessment tools based on multiple-criteria decision-making (MCDM) techniques in combination with green construction assessment methodologies.

A set of sustainability indicators and factors are defined at first as a representation of sustainability pillars. The gist of the assessment tool is a normalized summation of sustainability factors in different life-cycle phases. However, the framework needs to be dynamic to adapt to the continuous changes in the PRP. It is important to note that the dynamic methodology would be more ambiguous than its static counterpart. In this chapter, the dynamic methodology is defined at different levels of decision-making, with crisp, gray and fuzzy numbers as required.

1. Introduction

The role of petroleum refinery projects (PRPs) in sustainable development has always been controversial. On one hand, they consume crude oil, which is a nonrenewable natural resource, to produce various products, most of which pollute the environment to various degrees. Moreover, resource-based economies have to deal with many challenges in terms of transparency, integrity and rule of law. On the other hand, however, oil as a natural resource can be utilized to accelerate sustainable growth of developing countries. No matter on which side of the argument one lands, it is important to note that by improving the sustainability metrics of a PRP, one will be taking positive steps toward a more sustainable economy, society and environment, as the three pillars of sustainability.

Development can have different meanings in various fields. However, in general, development can be defined as the process of gradual growth, enhancement, improvement and advancement. Prior to mid-1970s, in the industrial context, at the core of the conventional development was the process of transforming raw materials to final products, with "peace and security", "economic development", "social development" and "national governance" being the four concepts forming the bigger picture (Dernbach, 1998, 2002). In this picture, the environmental concerns were largely ignored. As the environment deteriorated in various parts of the world, environmental concerns gradually came to the forefront, and nations came to the conclusion that, in order to achieve sustainable development, environmental considerations could no longer be ignored. This changing mindset led to an international movement, the start of which was the

United Nations conference on the integrity of the environment and development in 1992.

The concept of "sustainable development" was proposed by Brundtland Commission (alternatively known as World Commission on Environment and Development (WCED)) in 1987 (Brundtland, 1987). In Brundtland's definition of sustainable development, it is stated that future generations should be able to fulfill their needs from the earth's remaining resources, at least as much as our potential to meet our needs. In other words, sustainable development is the combination of conventional development and environmental protection (Hall, 2006):

$$\begin{aligned}\text{Sustainable Development} = {} & \text{Conventional Development} \\ & + \text{Environmental Protection.}\end{aligned}$$

The relationship between sustainable development and PRP is viewed from two contrasting perspectives. From the favorable vintage point, PRP are regarded as crucial contributors to advance sustainable development goals, such as eradication of poverty and education for all. However, the negative view perceives PRP as contributing to development at the expense of depletion of Earth's crude oil reserves and polluting the air, water and soil on the other hand, while they can cause social degradation as well.

To help reconcile between these opposing views, we may borrow two terms that are commonly used in environmental economics: weak sustainability and strong sustainability. While in weak sustainability, the value of the capital stock (natural plus physical capital) should not decline, individual components of the aggregate could decline in value as long as other components are increased in value to leave the aggregate value unchanged. In strong sustainability, on the other hand, the value of the remaining stock of natural capital (not total capital) should not decrease. This implies that natural and physical capitals (such as the built environment) are not interchangeable for the most part. While PRP can be justifiable in a weak sustainability context, it is more difficult to see them in line with strong sustainability views. Regardless of how PRPs are viewed versus sustainable development, such projects are undeniably crucial for nations to satisfy their needs. Hence, sustainable development of PRP primarily can mean focusing on the improvement of positive impacts and decreasing the negative ones simultaneously at the local, national and international levels (Mahmoud and Shuhaimi, 2013).

The perception of sustainability can vary from one individual to another, based on each individual's needs. For instance, environmentalists, consumers and workers desire a sustainable environment, consumption and jobs respectively (Norgaard, 1988). Accordingly, any sustainability framework should address and consider these varying viewpoints. Primary studies on sustainable development demonstrate a wide range of interpretations for sustainability concepts in different fields. Lack of a powerful theoretical framework can cause problems in sustainability assessment process (Dixon and Fallon, 1989; Holdren *et al.*, 1995; Holmberg and Sandbrook, 1992; Lélé, 1991; Redclift, 1991; Shiva, 1992; Toman, 1992). Thus, to resolve this issue, most sustainability assessment standards try to take an integrated approach in the development of their methodology.

2. Literature Review

Infrastructure projects can greatly influence the development of economies and societies (Shen *et al.*, 2010; Yao *et al.*, 2011). As such, sustainability assessment of infrastructure projects has been the subject of many previous studies (Papajohn *et al.*, 2016; Sierra *et al.*, 2015). Sustainability assessment tools usually introduce sustainability indicators that are used to represent the criteria to be met to satisfy the requirements for each of the triple bottom lines: economic, environmental and social.

Various studies have been carried out, introducing and using indicators for sustainability assessment. For instance, an evaluation methodology based on sustainability indicators was developed to evaluate the sustainability of infrastructure projects (Ugwu *et al.*, 2006). Transportation projects have been the subject of another study, which utilized sustainability indicators (Mihyeon Jeon and Amekudzi, 2005). The relationship between energy and environment is of great importance to governments and nations (Cucchiella and D'Adamo, 2013); sustainable energy supply has been investigated in many studies. As an example, with the aim of 80% reduction in GHGs emissions by 2070, different scenarios were evaluated for the UK (Stamford and Azapagic, 2014). A life-cycle assessment for environmental footprint in Tanzania electricity generation was evaluated and the worst points in the life-cycle were predicted (Felix and Gheewala, 2012). Also, sustainability of energy was evaluated using sustainability indicators (Angelis-Dimakis *et al.*, 2012).

Despite the amount of investigation on renewable and green energies in the world, oil and gas industry and related products have had a growing market as well. Based on a report from the US environmental protection agency, the amount of oil and gas and other products consumed in 2010 was 176 quadrillion BTU and this will increase to 197 quadrillion BTU in 2020 (EIA, 2014). Also, oil and gas industry products fulfilled 32.9% of global energy demand in 2015 (BP, 2016).

Given the importance of oil and gas projects, their sustainability assessment has been undertaken in a variety of previous studies with different viewpoints and approaches. From off-shore oil and gas projects (Silvestre and Gimenes, 2017), to impacts of external elements in the supply chain in order to reach sustainability goals (Ahmad *et al.*, 2016), to study of barriers and enablers for incorporation of sustainability into an oil and gas company (George *et al.*, 2016), and analysis of different scenarios to reduce SO_2 emission in PRPs (Ba-Shammakh, 2010), have all addressed sustainability of oil and gas projects from various perspectives.

The complexities of sustainability assessment are created to a large extent because of involvement of different fields of science that underlie the three pillars of sustainability (Ciegis *et al.*, 2009). Some previous studies in the field of sustainability assessment have put a greater emphasis on one of the sustainability aspects (de Castro Hilsdorf *et al.*, 2017; Ferreira *et al.*, 2015; Jovanovic*et al.*, 2010; Missimer*et al.*, 2017a, 2017b; Silajdžić *et al.*, 2015). However, the sustainability gains a greater importance if a comprehensive assessment combines all three pillars in its process (Bansal and DesJardine, 2014; Labuschagne *et al.*, 2005).

A comprehensive and integrated evaluation based on sustainability indicators, which would incorporate all triple bottom line considerations in a balanced and seamless way, would require the use of MCDM methodologies similar to what has been done in several previous studies (Heravi *et al.*, 2015, 2017; Poveda and Lipsett, 2011, 2014).

3. Petroleum Refinery Projects

PRPs are among the largest and most complicated construction projects. Different stakeholders are involved from manufacturers to vendors, suppliers, designers, constructors and licensors, which increases the complexity of these projects. The diversity of stakeholders results in diversity of needs and expectations, and consequently, different interpretations and attitudes toward sustainable development of these projects. Also, it adds to the importance of taking an integrated approach in the sustainable development of PRP.

PRPs involve different disciplines such as mechanical engineering, chemical engineering, economics, management, environmental engineering and information technology, all of which could constitute the use of latest technologies, as long as those technologies can be made an integral part of PRI projects. However, technology is a double-edged sword. On one hand, it assists the society with resolving disputes, making policies, distributing money and power and so on, and on the other hand, it is mainly responsible for natural resource destruction so that the ecosystem and wildlife are threatened with extinction (Carson, 1962). Besides, rapid changing technology is one of the chief concerns confronting PRP (Streeten, 2001). This high technology dependency, along with continuing changes, adds to the mentioned complexity of PRP in developing an integrated framework for their sustainability assessment. That is why flexibility in this framework has a great importance. The very first step toward a flexible sustainability assessment framework for PRP is getting a general understanding of petroleum refineries and different units that have varying environmental footprints, as well as social and economic impacts in order to develop a comprehensive methodology for the assessment process.

A petroleum refinery can be as simple as a physical separator unit in atmospheric pressure to complex units of cracking, reforming, unifying and enhancing the products (Hinton *et al.*, 1999). Besides, sustainability instinctively has to measure every positive and negative effect on all three pillars of sustainability before and after the construction project. Hence, in addition to being integrated across various disciplines, units and stakeholders, sustainability assessment has to be defined along the entire lifecycle of the raw material as well as the final product. In this chapter, a petroleum refinery model has been developed based on the above two considerations.

3.1. *Existing sustainability tools for PRP*

PRPs are environmental pollutants by nature and because of super heavy equipment, as well as flammable or explosive materials, are hazardous projects whether in construction or operation phases. Hence, traditionally, various tools have been created in the past to address environmental and safety concerns. Some of the most important of these tools are listed below:

- **Environmental Impact Assessment (EIA) Report:** Although the name of this type of report implies its environmental aspect, usually other pillars of sustainability are included in such reports as well. An EIA can be used for conceptual design and feasibility studies. The assessments are qualitative as well as quantitative to some extent, and the integrity of the assessments is generally minimal.
- **Hazard and Operability (HAZOP) and Hazard Identification (HAZID):** These two types of analysis are mainly developed for process hazards and failures and are considered for design purposes.
- **Safety Impact Level (SIL):** SIL analysis evaluates the probability of functionality for a system in different conditions and time

intervals. In hazardous projects like PRP, development and intensification of an incident may increase the scope of risk and damage. Thus, comprehensiveness and mutual effects of incidents have to be taken into account, which is the purpose of this type of analysis.

- **Consequence Analysis (CA):** This type of analysis is about site zoning with regard to the project layout. Different hazardous events are considered including fire, explosion and toxic emissions to limit the scope of risk. Some internal analyses are conducted such as protected area analysis, distances of area of fire and affected area analysis.
- **Quantitative Risk Analysis (QRA):** This analysis is carried out based on the previous lessons learned, organizational process assets and enterprise environmental factors. These items, in combination with experts' judgment, will result in the identification of risk scenarios. The outputs are probabilistic and provide a great advantage for the project manager in his/her proactive management.
- **Emergency Response Plan (ERP):** In emergencies, people often make mistakes and choose wrong choices, which will slip into complete chaos and as a result, the risk of consequent incidents will be increased. To minimize such additional risks, ERP details responsibilities and instructions for people to reduce the consequences caused by different hazardous scenarios.
- **Reliability, Availability, Maintainability, and Safety (RAMS):** This constitutes a probabilistic analysis of different systems and equipment, which will result in probabilistic functions of systems performance in safety.
- **Sustainability Reports (SRs):** Giant companies working in the field of oil and gas projects and operations, issue annual reports and present their scorecards for their commitments to sustainability. The template of these reports is mainly based on a guideline developed by an NGO for this purpose (GRI, 2013).

As it can be seen from the above list, these reports and analysis types are mainly developed for environment and safety of personnel and equipment. However, there is a lack of comprehensiveness of these tools, when it comes to assessment of all three pillars of sustainability in an integrated way. In what follows, we will develop a sustainability assessment framework, which is designed to fix the above deficiencies as much as possible.

3.2. *Petroleum refinery units and lifecycle modeling*

At the first level, PRP modeling is divided into three phases as follows:

(1) **Cradle-to-Gate (refinery entrance gate):** This part is about raw materials, from crude oil, its extraction and its transportation to the refinery.
(2) **Gate (refinery entrance gate)-to-Gate (refinery product gate):** The petroleum refinery, its units and components.
(3) **Gate (refinery product gate)-to-Grave:** The refinery's final products, their consumption and resulting emissions.

The first phase as well as the last one are outside the boundary of the PRP. The second phase can be detailed based on the refinery units as follows:

Petroleum refinery consists of different process flows, which will be defined based on their properties such as pressure, temperature and their phases. Figure 1 presents a distillation column with its furnace. The more detail entered into the refinery model, the more complexity and time would be involved when conducting the assessment process. Hence, three levels of accuracy are defined to be used in different situations as follows:

(1) **Production and consumption:** Some of the previous studies in sustainability assessment regard an entire project as a single alternative and choose the best scenario based on the outcome of sustainability indicators. However, PRP are large and complicated projects, which may not be suitable for consideration as a single alternative in the assessment. Consequently, in this study, a project is regarded as two parts at its most general level. Units that are making products and utilities and units that are consuming the utilities and products.

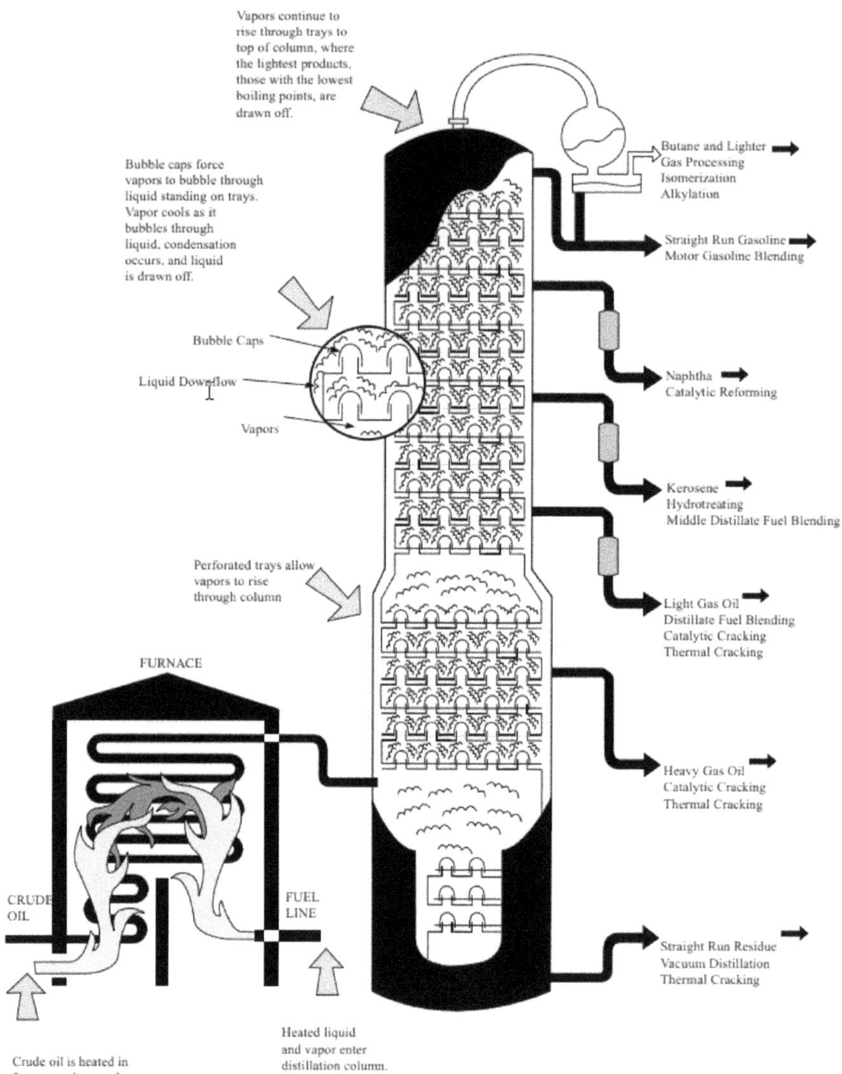

Figure 1: A distillation column scheme.

(2) **Refinery units:** Refinery units are categorized as shown in Figure 2. The main unit in every refinery is the distillation unit in atmospheric and/or vacuum pressure, which makes the final products. If the crude oil has salt, it may increase the furnace

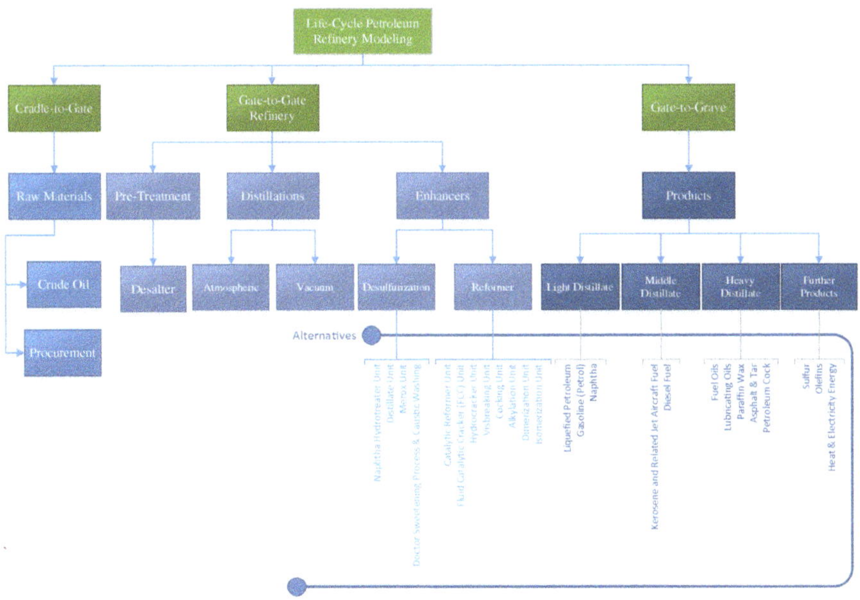

Figure 2: Lifecycle model of a petroleum refinery with the second level of accuracy.

fuel use, so the pretreatment unit includes the de-salter unit. Moreover, production of distillation units can be enhanced or if it contains sulfur, it may be desulfurized to avoid polluting the environment. Thus, de-sulfurization can be added as an enhancer unit.

(3) **Refinery components:** A refinery may contain thousands of components. This is the most detailed level. For instance, a furnace consumes fuel and produces emissions. Refinery components can be categorized into various engineering disciplines as follows. This type of grouping is particularly useful, because the documentation for each discipline is done separately, and this categorization can be easily used; in this way documents such as effluent summary and electricity demand summary can be used much more conveniently.

a. Structure and civil works
b. Mechanics and equipment

c. Rotary equipment
d. Piping
e. Electricity
f. Instrument
g. Health, safety, environment [HSE]
h. Others.

Different parameters contribute to choosing the desired level of accuracy. Level of available information in the assessment time period, level of accuracy required by managers and level of ambiguity and unknowns involved in the measurement process can be some of these parameters.

4. Methodology

Managers need to make quick and accurate decisions regarding the project at different levels. Strategic decisions have to consider various criteria from different fields and are oftentimes hard to make. Besides, in the decision-making process, managers sometimes need to form a coalition in a decision group or a decision chain in different levels of management to guide the process toward a particular outcome. Realization of sustainable development in PRP would require dealing with multiple criteria and relies on a decision chain, which starts at a strategic level, maybe in the government or in the private sector and ends at small decisions in day-to-day activities of the project. Hence, in developing a sustainability assessment tool, it is important to take these different levels of decision-making with their criteria, properties, requirements and ambiguities into account.

In sustainability studies, definition of sustainability indicators from all three pillars will help deal with the inherent complexities of the problem in a more systematic way. Indicators are representatives of different criteria defined for sustainability pillars. Thus, in sustainability problems, scientists always deal with a set of indicators as criteria that would be used to make decisions about gas emission reduction scenarios (Hadidi *et al.*, 2016) and alternative selection (Heravi *et al.*, 2017). For this situation, which involves multiple criteria and various alternatives to choose from, MCDM methodologies are very suitable. That is why there is a wide variety of sustainability studies carried out using different MCDMs as their methodology of choice.

4.1. *Decision chain in sustainability assessment of PRP*

Sustainability assessment is a chain of decisions starting from developing a sustainability framework to sustainability assessment tools with different applications in the industry. As described in Figure 3, this chain can be categorized in two levels.

(1) **Sustainability framework:** In previous studies, frameworks were developed only based on literature reviews. However, PRP

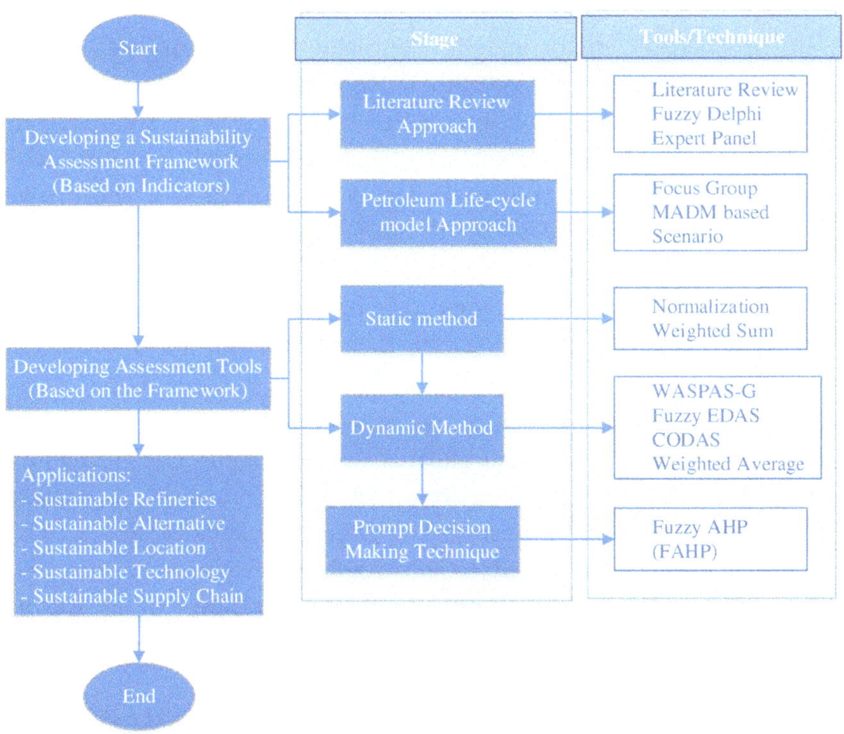

Figure 3: The hierarchy of phases and methodologies.

in particular are multidisciplinary projects and different stakeholders' needs and expectations have to be considered. Hence, it is inevitable to combine the literature review with industrial viewpoints in order to define the sustainability indicators and factors and develop the framework.

(2) **Sustainability assessment tools:** Based on the developed sustainability framework from the previous step, various assessment tools to cope with the specific environment of the decision-making process have to be developed. For this reason, a static methodology is developed in the simplest form to assess the alternatives based on the developed framework of indicators and factors. However, PRP can be extremely different in their project-specific conditions, such as policies, budget, project location, environmental local resiliency, local access to the infrastructure and workforce. Thus, the developed framework needs to be customized for specific conditions of the project. Toward this end, a dynamic assessment tool is developed to take different levels of sustainability assessment and their requirements into account. In other words, dynamic assessment is a method of weighing the static tool to factor in the specifics of the project environment. Finally, as the dynamic method combined with the top level of accuracy required in PRP's lifecycle model, becomes time-consuming, another methodology is developed based on the Pareto principle, in which only the important parts would be assessed.

As it is explained, sustainability is modeled in an environment of indicators as the assessment criteria. Thus, at first, a set of available criteria can be defined based on the literature review. Sustainability indicators are developed in a three-layer framework, from the relatively abstract level of pillars to qualitative level of indicators, and quantitative level of factors in order to ensure a comprehensive coverage of the scope as follows (see Table 1):

(1) **Sustainability Pillars:** Pillars, namely the environment, the society and the economy, are taken care of at this layer.

Table 1: Selected indicators' references.

Pillars	Indicators	References						
		((UNSC), 2015)	((UNCSD), 2005)	(Ugwu et al., 2006)	(Labuschagne et al., 2005)	(Shen et al., 2010)	(Heravi et al., 2015)	(Sierra et al., 2015)
Social	Poverty & Equality	X	X		X		X	X
	Health	X	X		X	X		X
	Safety & Security	X	X	X	X	X		X
	Education & Training	X	X		X			X
	Welfare	X			X	X	X	X
Environment	Atmosphere	X	X	X	X	X	X	
	Water	X	X	X	X	X	X	
	Land and Soil Pollution	X	X	X	X	X	X	
	Natural Resource	X		X	X	X	X	
	Biodiversity	X	X	X		X		
Economic	Energy Consumption	X						
	Financial	X			X	X	X	
	Economic Performance	X	X	X	X	X	X	
	Occupation	X					X	
	Earning	X					X	

Developing the sustainability framework in a hierarchical structure would definitely improve the comprehensiveness and would minimize the likelihood of having blind spots in the assessment process.

(2) **Sustainability Indicators:** Qualitative indicators are defined to cover the main concerns in the intended pillar of sustainability. Therefore, main concerns are somewhat independent from the selected industrial field and project. Thus, they can be defined through a comprehensive literature review as will follow.

(3) **Sustainability Factors:** Quantitative factors are defined for a specific indicator in a specific pillar of sustainability to best

describe the concerns behind the indicator in a quantitative manner. Factors get a dual evaluation as mentioned before with literature and industrial viewpoints.

4.1.1. *MCDM techniques*

As presented in Figure 3, a sequence of different MCDM methodologies are used in the sustainability assessment of PRP, which will guide the process from its conception to the end. Every methodology is developed to have the most suitability with the specific problem's requirement, with the following details:

- **Fuzzy Delphi:** This method is used to help select the most important factors from a list of candidate factors that would be used in subsequent steps to quantify the extent to which various sustainability indicators have been realized. Since these quantitative factors are defined in relationship to the lifecycle model of PRP, given the complexities in such a model, experts need to share their opinions, toward reaching consensus. As experts may fall in bias, the problem in this part needs to be solved in a reciprocal and repetitive process, in which experts collaborate and get notified about other experts' opinions. Delphi method's features lend themselves easily to the requirements of this process. Moreover, since the prioritization of sustainability factors is being done in general and not for a specific project, there is always a level of uncertainty in experts' judgment about the importance of each factor. Thus, Fuzzy Delphi method is selected for this part as it is more consistent with these characteristics (Chang *et al.*, 2000).
- **Hashemkhani's MADM-based Scenario:** in order to measure the connectivity of the developed factors in the literature review phase with the lifecycle model of a petroleum refinery, the problem is a bit different from other traditional MCDM methods because not all of the defined factors necessarily connect to all parts of the project model. The best methodology in which a set of alternatives are evaluated against a set of criteria by a scenario-based approach is Hashemkhani's methodology in which

the decision matrix linking criteria with alternatives is not fixed (Hashemkhani*et al.*, 2016).

- **WASPAS-G:** Now we turn our attention to assessment of a specific refinery project. Evaluation of sustainability pillars for a specific project is expected to be done by top managers and the most influential stakeholders, as it constitutes the most strategic level of sustainability assessment. For this exercise, an enhanced weighted average methodology is used in combination with gray numbers (Zavadskas *et al.*, 2015); using gray numbers would enable managers to give a range of values instead of fixed figures for each question that they are asked. This is more suitable for managers given their high level mindset.
- **Fuzzy EDAS:** After finishing the assessment of pillars, fuzzy EDAS method is used for the specific refinery project to rank the importance of indicators, which are generally qualitative (Ghorabaee *et al.*, 2016). This is a comparative process, for which EDAS is used. EDAS is a new methodology that works based on averages and is more efficient with faster convergence than its predecessors (Keshavarz *et al.*, 2015). The fuzzy component has been added here because this part is expected to be done by the middle management of the project and the problem is in a qualitative environment; so some level of ambiguity exists in this layer. Hence, considering the more specific mindset of the middle management (compared to the top management), triangular fuzzy numbers, not gray numbers, are used for this part.
- **CODAS:** At the next level, the process of weighting the importance of the quantitative factors is expected to be done at an expert level, who are involved with details and day-to-day activities of refinery projects. CODAS is a great methodology for problems with a quantitative nature because it works with a negative ideal solution and by using a creative dual distance measurement methodology, it accelerates the convergence (Keshavarz *et al.*, 2016). However, quantitative factors for the specific project at hand, for an expert have the lowest level of ambiguity, and that is why crisp numbers are used in this case.

- **FAHP:** Analytical hierarchy process (AHP) is greatly suitable for a quick assessment, which involves measuring the relative importance of the sustainability criteria in the hierarchical structure of the developed framework and finding the most influential criteria. Moreover, considering the strategic high level of the problem and its inherent ambiguities as well as the more specific level of decision makers in this problem [middle managers to experts], fuzzy numbers are combined with the selected methodology (Aghdaie *et al.*, 2013; Chang, 1992).

4.2. *Develop sustainability assessment framework*

As stated before, sustainability assessment framework has been developed by taking the viewpoints from the literature as well as the industry.

4.2.1. *Literature viewpoints*

Pillars and indicators for this series of methodologies are solely defined based on the literature review. However, due to their specificity to each project, sustainability factors have to be defined based on the available literature as well as the model of the specific PRP simultaneously. Some of the main references used in the definition of sustainability factors is listed below:

(4) UN sustainability assessment goals and indicators. Indicators in this reference are general and need to be translated and interpreted for PRP (UNSC, 2015).
(5) Some of the main effluents of PRP, which are of interest in sustainability assessments, are taken from US Environment Protection Agency's publications (EPA, 2015).
(6) Sustainability reports of oil and gas companies are issued annually (BP, 2016). These reports are developed based on GRI's template, which is used internationally (GRI, 2013).
(7) Previous studies and papers are also used in defining sustainability factors (Hall, 2006; Shen *et al.*, 2010).

The process of defining sustainability factors is as follows:

Step 1 — Literature Review: developing a set of relative factors.

Step 2 — Fuzzy Delphi: The developed set of factors are evaluated using the selected methodology and a list of the most relevant factors for PRP is developed in three rounds. For more information on this methodology, previous studies can be consulted (Chang *et al.*, 2000). As an example, Factor No. 30 achieved consensus and its relative importance was measured as given in Tables 2–4:

Step 3 — Expert Panel: A group of experts from oil and gas industry were selected to translate the chosen factors to quantitative factors, which are also highly relevant to PRP.

Table 2: Expert judgment about GHG emissions.

No.	Pillar	Indicator	Factor	Experts opinion
30	Environment	Air	GHG emission ...	#3: 4–5 #7: 5–5

Table 3: Calculating average based on experts' opinion.

Expert (i)	$\delta_1^{(i.30)}$	$\delta_2^{(i.30)}$	$\delta_3^{(i.30)}$	$\delta_4^{(i.30)}$	$\delta_5^{(i.30)}$
1	0	0	0	1	1
2	0	0	0	1	1
3	0	0	0	1	1
4	0	0	0	0	1
5	0	0	0	0	1
6	0	0	0	0	1
7	0	0	0	0	1
8	0	0	0	0	1
9	0	0	0	0	1
10	0	0	0	0	1
	$y_1^{(30)}$	$y_2^{(30)}$	$y_3^{(30)}$	$y_4^{(30)}$	$y_5^{(30)}$
	0	0	0	3	10
	$Y^{(30)}(x_1)$	$Y^{(30)}(x_2)$	$Y^{(30)}(x_3)$	$Y^{(30)}(x_4)$	$Y^{(30)}(x_5)$
	0	0	0	0	0.3
			$C\bar{O}G^{(30)}$		
			4.77		

Table 4: Convergence control.

Expert (i)	$COG^{(i.30)}$	$ABS(COG^{(i.30)} - \overline{COG^{(30)}})$
1	4.5	0.27
2	4.5	0.27
3	4.5	0.27
4	5	0.23
5	5	0.23
6	5	0.23
7	5	0.23
8	5	0.23
9	5	0.23
10	5	0.23

4.2.2. *Industrial viewpoints*

The scenarios in Hashemkhani's MADM-based scenario methodology, which is used for this section, constitute lifecycle phases of PRP; that is (Gholipour *et al.*, 2018):

(8) Cradle-to-[refinery entrance] Gate
(9) Gate [refinery entrance] -to- [refinery production] Gate
(10) Gate [refinery production] -to-Grave

In this methodology, various parts of the refinery model are considered as assessment alternatives under the above three scenarios, which are evaluated against sustainability factors, the latter considered as assessment criteria herein. A table is developed with the details presented in Table 5.

4.3. *Sustainability assessment*

After developing sustainability framework in the previous section, in this section two sustainability assessment tools are developed:

4.3.1. *Static sustainability assessment*

In this method, the framework consists of sustainability criteria which are used to evaluate life-cycle alternatives. This method is static since various factors, indicators, and pillars remain unchanged

Table 5: Decision table for MADM-based scenario.

	Cradle-to-Gate		Gate-to-Gate					Gate-to-Grave			
	Raw material		Pre-treatment	Distillations		Enhancers		Products			
Factors	Crude oil	Procurement	De-salter	Atmospheric	Vacuum	De-sulfurization	Re-former	Light Distillate	Middle Distillate	Heavy Distillate	Further Products
F_i											

from one project to another. The developed framework is simply a set of criteria with fixed weights. Consequently, the measured quantitative factors are normalized in order to calculate the average value as follows (Hasheminasab *et al.*, 2018b):

$$X_p^{i-j}(T) = \frac{F - \left(\frac{X_{\max}+X_{\min}}{2}\right)}{\left(\frac{X_{\max}-X_{\min}}{2}\right)}, \tag{1}$$

$$X_p^{i-j}(T) = -\left(\frac{F - \left(\frac{X_{\max}+X_{\min}}{2}\right)}{\left(\frac{X_{\max}-X_{\min}}{2}\right)}\right), \tag{2}$$

where $X_p^{i-j}(T)$ is the normalized value of the pth pillar, ith indicator and jth factor in time interval of T; $X_{\max}$ is the upper limit of the measured factor, which can be based on standards or rules: $X_{\min}$ is the lower limit of the measured factor, which can be based on best practices, and F is the value of the measured factor.

If the optimal extreme value is maximum, the first formula is applied; otherwise the second formula is used to normalize the factors. Since the factors can be discrete and integer or continuous and decimal, the evaluation of pillars' sustainability can follow either of two following formula:

$$S_p = \begin{cases} \sum_{i=1}^{I} \sum_{T=1}^{4} \sum_{j=1}^{J} X_p^{i-j}(T) \\ \int_1^I \int_1^4 \int_1^J X_p^{i-j}(T) dJ \cdot dT \cdot dI \end{cases} ; \quad \forall p \in \{1.2.3\}. \tag{3}$$

Next, the normalized sustainability of pillars is calculated as follows:

$$\overline{S_p} = \frac{S_p}{4 \times N_p}, \tag{4}$$

where N_p is the number of factors that have a contribution in the pth pillar.

Finally, the sustainability profile is calculated as follows:

$$\text{MOS} = \sum_{p=1}^{3} \overline{S_p}. \tag{5}$$

4.3.2. *Dynamic sustainability assessment*

In the dynamic assessment of sustainability, the developed framework would be weighted and customized for each individual project. For this purpose, the three decision levels are recognized with different environments, decision makers and requirements. All three levels are defined and modeled according to the following procedure, and the results are finally combined to return an integrated result.

Step 1 — Weighing the Pillars: As it can be seen from Figure 4, at first the priorities will be determined based on project stakeholders, rules, regulations and contract. Afterward, the most important and influential stakeholders or their representatives need to participate in the decision-making (Table 6) to measure the weights of the pillars.

Step 2 — Weighing the Indicators: In this step, an expert panel is formed and using the fuzzy-EDAS methodology, the weights of the indicators are found in the form of a decision table (Table 7). For each of the three pillars of sustainability, which were evaluated in the

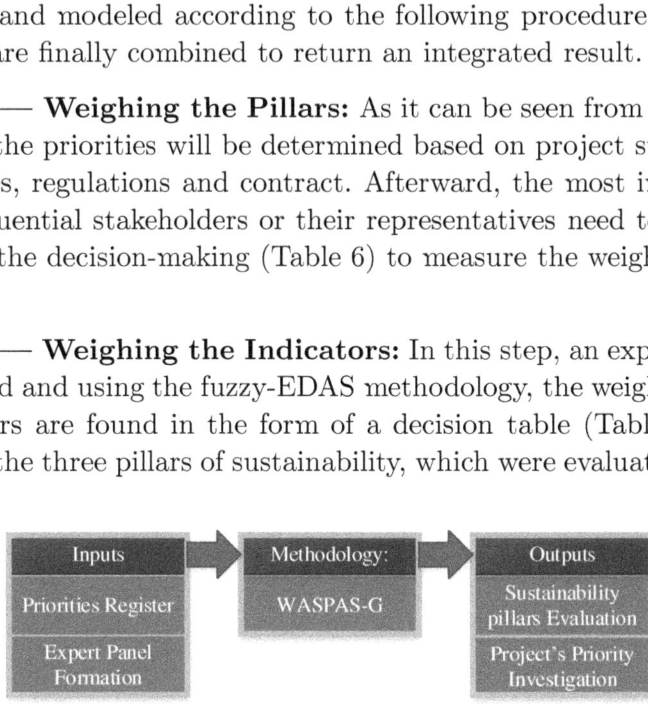

Figure 4: Evaluating sustainability pillars.

Table 6: Decision table for WASPAS-G.

	Political (C_1)	Strategical (C_2)	Military (C_3)	Financial (C_4)	Resiliency (C_5)
Environment (P_1)	$[\overline{X}_{11} \cdot \underline{X}_{11}]$		$\cdots$		$[\overline{X}_{15} \cdot \underline{X}_{15}]$
Economic (P_2)	$\vdots$		$\ddots$		$\vdots$
Social (P_3)	$[\overline{X}_{31} . \underline{X}_{31}]$		$\cdots$		$[\overline{X}_{35} . \underline{X}_{35}]$

Table 7: Decision table for fuzzy EDAS.

Indicators Based on the pillars	Pre-construction (C_1)	Construction (C_2)	Post-construction (C_3)	Consumption (C_4)
(I_1)	$[\tilde{x}_{11}]$	$\cdots$	$[\tilde{x}_{14}]$	
$\cdots$	$\vdots$	$\ddots$	$\vdots$	
(I_n)	$[\tilde{x}_{n1}]$	$\cdots$	$[\tilde{x}_{n4}]$	

Figure 5: Evaluating sustainability indicators.

Figure 6: Evaluating sustainability factors.

previous step, there is a separate decision table, which needs to be independently solved in this part (see Figure 5).

Step 3 — Weighing the Factors: Figure 6 illustrates the decision process for weighing the sustainability factors. Decision Table 8 is defined for this stage. It should be noted that this step has to be repeated for all 15 different indicators separately.

Step 4 — Integrated tool: Finally based on the outputs from previous steps, and by using a weighted average, sustainability profile would be calculated as presented in Figure 7.

Table 8: Decision table for CODAS.

Factors based on the indicators	Pre-construction (C_1)	Construction (C_2)	Post-construction (C_3)	Consumption (C_4)
(F_1) ... (F_n)	$\begin{bmatrix} x_{11} & \cdots & x_{14} \\ \vdots & \ddots & \vdots \\ x_{n1} & \cdots & x_{n4} \end{bmatrix}$			

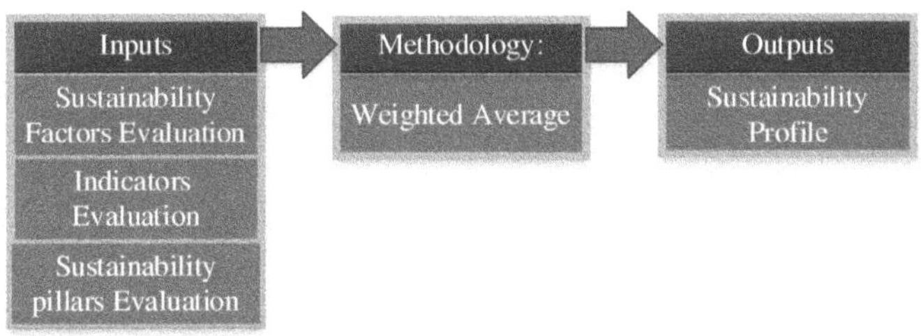

Figure 7: Integrated sustainability assessment.

$$\text{Sus. Profile} = \sum_{P=1}^{3} w_p \times \frac{\left(\sum_{i=1}^{I} w_i^p \times \left(\sum_{f=1}^{F} w_f^{pi} \times \left(\sum_{t=1}^{3} r_t^{pif}\right)\right)\right)}{3 \times N_p}, \tag{6}$$

where w_p is the pth pillar's weight measured in the first step, $(\sum_{p=1}^{3} w_p = 1)$, w_i^p is the weight of ith indicator and pth pillar measured in the second step $(\sum_{i=1}^{I} w_i^p = 1. \forall p)$, w_f^{pi} is the weight of fth factor, ith indicator and pth pillar measured in the third step $(\sum_{f=1}^{F} w_f^{pi} = 1.\forall p.i)$ r_t^{pif} is the normalized weight of fth factor, ith indicator and pth pillar just like the static method for normalization of sustainability factors, and N_p is the number of factors in a given pillar of sustainability.

4.4. *Express decision-making method*

As explained, the dynamic method in combination with the high accuracy of PRP lifecycle model would be a useful customized tool

Table 9: Prompt method decision table.

		Pillars of sustainability					
		Social		Environment		Economy	
		Indicators	···	Indicators	···	Indicators	···
LC Phases	Pre-Construction						
	Design & Construction						
	Operation & consumption						

for sustainability assessment of these projects. However, the only drawback in this assessment process is the time it consumes. In this section, the proposed methodology seeks to determine the top 20% of the items with 80% of the importance and influence, based on the Pareto principle. Hence, the FAHP method is used in the following hierarchy of criteria (Hasheminasab *et al.*, 2018a):

(1) Sustainability profile;
(2) Sustainability pillars;
(3) Sustainability indicators.

The decision table for this method is shown in Table 9, which would be solved three times for three independent problems.

The outputs reveal the important indicators, which, if the management decides, can be used with an acceptable level of accuracy in the assessment of sustainability.

5. Case Study and Conclusion

For different phases and stages, four independent case studies are carried out as represented in Figure 8 in order to ensure the accuracy and reliability of the proposed methodologies. These four case studies show how the methodologies can be applied to refineries that are under design and construction or refineries under operating conditions. Additional information about stages of methodologies and calculations are presented in more detail in previously published papers (Gholipour *et al.*, 2018; Hasheminasab *et al.*, 2018a, 2018b). In

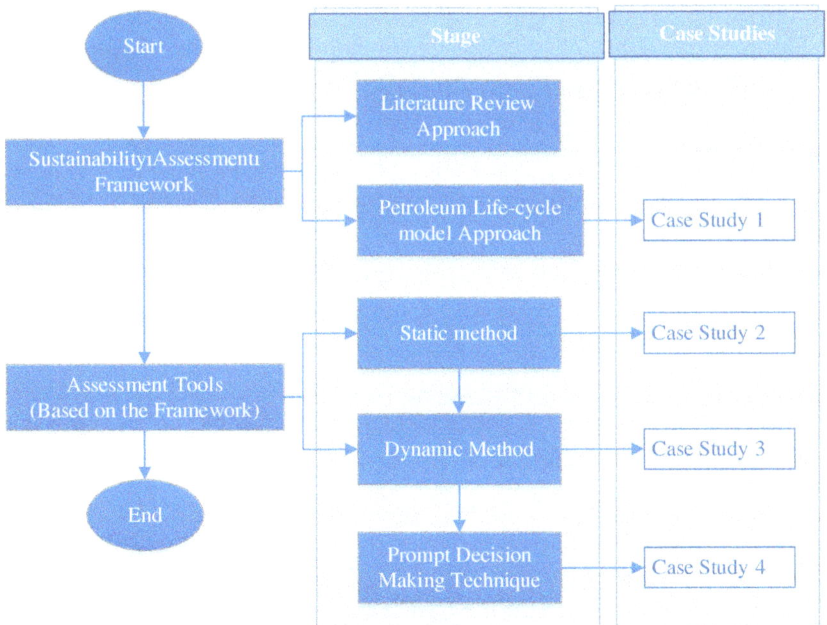

Figure 8: Hierarchy of case studies.

the following section, only the static method of assessment is briefly explained.

5.1. *Static sustainability assessment of PRP*

The petroleum refinery that is selected for the case study is in its early design and construction phases and the available data at this stage is minimal. Thus, the data shortage imposes the following constraints, which will limit the scope of the case study:

(1) Since no exact supplier or producer is selected at this stage, study of the Cradle-to-Gate scenario is not done here.
(2) Since the final decision on the targeted market hasn't been made yet, the Gate-to-Grave scenario will be eliminated.

Thus, the proposed stage for the case study is the Gate-to-Gate stage (scenario).

Sustainability indicators constraints:

By forming a panel of highly experienced experts in the selected case refinery, the most important indicators are selected based on the current environment and available data of the case study as follows:

(1) In the construction and operation phases, the largest impacts are caused on the local air and water; hence, these two indicators are selected from the list of indicators in the pillar of environment.
(2) At this stage, education and training, as well as safety and security, have the decisive role among the indicators listed for the social pillar of sustainability.
(3) The role of economy is prominent in all PRP. In the chosen stage, the economic performance, energy consumption and occupation are the most important economic indicators that are selected to be assessed in this study.

Sustainability standards like LEED and BREEAM specify limits and thresholds for their criteria based on best practices and various types of analysis. If a project can show that for a specific credit (criterion), it has passed the specified threshold and it will gain that credit. Developing such a standard for assessing PRP would require much more work involving various stakeholders, with large databases of best practices and different types of analysis, which is beyond the scope of the methodologies that have been discussed here. In the absence of such a process, experts' opinions have been used for this case study to determine the required thresholds.

Minimum and maximum extremes are developed and measured for the Gate-to-Gate phase are shown in Table 10.

Factors are measured based on experts' opinion within the threshold range and the project environment. For instance, with regard to the mentioned limitations, a similar PRP can make up to 500 job opportunities on average, in all parts of the lifecycle model. However, in this project, only 325 positions would be existing in the vacuum distillation unit for the mentioned circumstances.

With regard to different formulations for optimal nature of the factors, normalized factors (scores) are calculated by the measured

Table 10: Threshold definition and factors' scoring.

No. of factors	Indicator	Threshold Min	Threshold Max	Unit	Factors measurement: Vacuum distillation	Atmospheric distillation	De-salter	De-sulfurization
14	Safety and	0	50	People	20	30	15	10
15	Security	0	100	Percent	55	55	55	65
—	—	—	—	—	—	—	—	—
96	Occupation	0	500	Number	325	340	275	275

factors and their threshold. Results are presented in Table 11. For a complete list of sustainability factors, the reader can refer to the appendix.

Factor scores are then summed up, to find out sustainability pillars' scores, similar to Table 12:

The results of calculations are presented as a summary in Figure 9.

Undeniably, a comprehensive study with coverage of the entire lifecycle of projects could result in a more accurate sustainability framework. Despite the limitations imposed on this study, the presented results can be used by managers for decision-making and for sensitivity analysis as well. The results presented in Figure 9 can be concluded as follows:

(1) The negative scores in the environmental column show the negative impacts of the listed units in the gate-to-gate life-cycle phase. Looking at these figures, the manager would realize the large negative environmental impact of the distillation units in the case refinery, which would help him focus on finding solutions for the same.
(2) Distillation units can be regarded as the main units of petroleum refineries. The results show the positive impact that they have on the economic pillar. Although the desalting and desulfurization units have negative economic impacts, their large positive social impacts in improving the health of adjacent communities and users makes their presence justifiable.

Table 11: Calculate normalized scores for static assessment.

No.	Factor	Indicator	Optimality nature	Score			
				Vacuum distillation	Atmospheric distillation	De-salter	De-sulfurization
14	Number of people having accident in the project/plant per annum	Safety and Security	Min	0.20	–0.20	0.40	0.60
15	Level of health and educational facilities in order to prevent accidents' occurrence		Max	0.10	0.10	0.10	0.30
—	—	—	—	—	—	—	—
96	Number of illiterate and unskilled workforce who acquire skills in the refinery project/plant	Occupation	Max	0.30	0.36	0.10	0.10

Table 12: Sustainability pillars' assessment.

	Score		
	Social	Environmental	E conomical
Vacuum Distillation	0.190	−0.417	0.136
Atmospheric Distillation	0.150	−0.312	0.202
De-Salter	0.233	−0.168	−0.051
De-Sulfurization	0.398	−0.182	−0.110

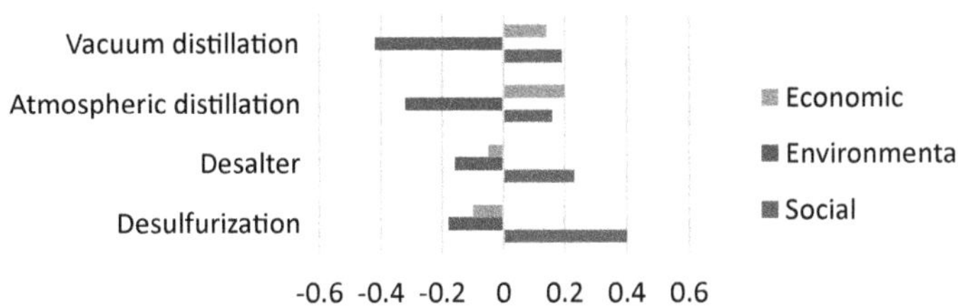

Figure 9: Static scores of PRP.

(3) With regard to the social pillar of sustainability, the results show positive impacts of the refinery in general, which can be attributed to increasing the livelihood of the local residents and improving the lifestyles of end users, while eliminating the potential health hazards by installation of desalting and desulfurization units.

References

BP (2016). *Statistical Review of World Energy.* Retrieved from United Kingdom: http://www.bp.com/en/global/corporate/energy-economics/statistical-review-of-world-energy.html

US EIA (2014). *International Energy Outlook 2014: World Petroleum and Other Liquid Fuels.* Retrieved from Washington, DC: http://www.eia.gov/forecasts/ieo/pdf/0484%282014%29.pdf

EPA (2015). Emissions Estimation Protocol for Petroleum Refineries.

GRI (2013). Sustainability Reporting Guidelines (G4), Global Reporting Initiative. In. UNCSD (2005). *Indicators of Sustainable Development, CSD Theme Indicator Framework.* Retrieved from New York:

UNSC (2015). Technical report by the Bureau of the United Nations Statistical Commission on the process of the development of an indicator framework for the goals and targets of the Post-2015 Development Agenda. *New York: Author. Retrieved July, 23*, 2015.

Aghdaie, M. H., Zolfani, S. H. and Zavadskas, E. K. (2013). Market segment evaluation and selection based on application of fuzzy AHP and COPRAS-G methods. *Journal of Business Economics and Management*, 14(1), 213–233.

Ahmad, W. N. K. W., Rezaei, J., de Brito, M. P. and Tavasszy, L. A. (2016). The influence of external factors on supply chain sustainability goals of the oil and gas industry. *Resources Policy*, 49, 302–314.

Angelis-Dimakis, A., Arampatzis, G. and Assimacopoulos, D. (2012). Monitoring the sustainability of the Greek energy system. *Energy Sustain Develop*, 16(1), 51–56.

Azapagic, A. and Perdan, S. (2005). An integrated sustainability decision-support framework Part II: Problem analysis. *Int J Sustain Develop World Ecol*, 12(2), 112–131.

Ba-Shammakh, M. S. (2010). Generalized mathematical model for SO_2 reduction in an oil refinery based on arabian light crude oil. *Energy Fuels*, 24(6), 3526–3533.

Bansal, P. and DesJardine, M. R. (2014). Business sustainability: It is about time. *Strat Organiz*, 12(1), 70–78.

Carson, R. (1962). *Silent Spring*. New York: Houghton Mifflin Company.

Chang, D.-Y. (1992). Extent analysis and synthetic decision. *Optimiz Techniq Appl*, 1(1), 352–355.

Chang, P.-T., Huang, L.-C. and Lin, H.-J. (2000). The fuzzy Delphi method via fuzzy statistics and membership function fitting and an application to the human resources. *Fuzzy Sets Syst*, 112(3), 511–520.

Ciegis, R., Ramanauskiene, J. and Martinkus, B. (2009). The concept of sustainable development and its use for sustainability scenarios. *Eng Econom*, 62(2).

Cucchiella, F. and D'Adamo, I. (2013). Issue on supply chain of renewable energy. *Energy Conver Manage*, 76, 774–780.

de Castro Hilsdorf, W., de Mattos, C. A. and de Campos Maciel, L. O. (2017). Principles of sustainability and practices in the heavy-duty vehicle industry: A study of multiple cases. *J Clean Product*, 141, 1231–1239.

Dernbach, J. C. (1998). Sustainable development as a framework for national governance. *Case W Res L Rev*, 49, 1.

Dernbach, J. C. (2002). Making sustainable development happen: From Johannesburg to Albany. *Alb L Envtl Outlook*, 8, 173.

Dixon, J. A. and Fallon, L. A. (1989). The concept of sustainability: origins, extensions, and usefulness for policy. *Soc Nat Resour*, 2(1), 73–84.

Felix, M. and Gheewala, S. H. (2012). Environmental assessment of electricity production in Tanzania. *Energy Sustain Develop*, 16(4), 439–447.

Ferreira, G., López-Sabirón, A. M., Aranda, J., Mainar-Toledo, M. and Aranda-Usón, A. (2015). Environmental analysis for identifying challenges to recover used reinforced refractories in industrial furnaces. *J Clean Product*, 88, 242–253.

George, R. A., Siti-Nabiha, A., Jalaludin, D. and Abdalla, Y. A. (2016). Barriers to and enablers of sustainability integration in the performance management systems of an oil and gas company. *J Clean Product*, 136, 197–212.

Gholipour, Y., Hasheminasab, H., Kharrazi, M. and Streimikis, J. (2018). Sustainability criteria assessment for life-cycle phases of petroleum refinery projects by MADM technique. *Econom Manage.*

Ghorabaee, M. K., Zavadskas, E. K., Amiri, M. and Turskis, Z. (2016). Extended EDAS method for fuzzy multi-criteria decision-making: An application to supplier selection. *Int J Comput Commun Control*, 11(3), 358–371.

Hadidi, L. A., AlDosary, A. S., Al-Matar, A. K. and Mudallah, O. A. (2016). An optimization model to improve gas emission mitigation in oil refineries. *J Clean Product*, 118, 29–36.

Hall, R. P. (2006). *Understanding and Applying the Concept of Sustainable Development to Transportation Planning and Decision-making in the US.* Massachusetts Institute of Technology,

Hasheminasab, H., Gholipour, Y., Kharrazi, M. and Streimikiene, D. (2018a). Life cycle approach in sustainability assessment for petroleum refinery projects with fuzzy-AHP. doi:10.1177/0958305X18772425

Hasheminasab, H., Gholipour, Y., Kharrazi, M. and Streimikiene, D. (2018b). A novel Metric of Sustainability for petroleum refinery projects. *J Clean Product*, 171, 1215–1224.

Hashemkhani Zolfani, S., Maknoon, R. and Zavadskas, E. K. (2016). Multiple attribute decision making (MADM) based scenarios. *Int J Strat Prop Manage*, 20(1), 101–111.

Heravi, G., Fathi, M. and Faeghi, S. (2015). Evaluation of sustainability indicators of industrial buildings focused on petrochemical projects. *J Clean Product*, 109, 92–107.

Heravi, G., Fathi, M. and Faeghi, S. (2017). Multi-criteria group decision-making method for optimal selection of sustainable industrial building options focused on petrochemical projects. *J Clean Product*, 142, 2999–3013.

Hinton, D., Masterson, N., Lippert, A. and King, R. (1999). *Petroleum: An Energy Profile.* US Energy Information Administration (EIA).

Holdren, J. P., Daily, G. C. and Ehrlich, P. R. (1995). The meaning of sustainability: biogeophysical aspects. *Defining and Measuring Sustainability: The Biological Foundations* (eds. Munasinghe M. and Shearer W.), pp. 3–17.

Holmberg, J. and Sandbrook, R. (1992). Sustainable development: What is to be done?

Jovanovic, J., Jovanovic, M., Jovanovic, A. and Marinovic, V. (2010). Introduction of cleaner production in the tank farm of the Pancevo Oil Refinery, Serbia. *J Clean Product*, 18(8), 791–798.

Keshavarz Ghorabaee, M., Zavadskas, E. K., Turskis, Z. and Antucheviciene, J. (2016). A new combinative distance-based assessment (codas) method for multi-criteria decision-making. *Econom Comput Econom Cybernet Stud Res*, 50(3).

Keshavarz Ghorabaee, M., Zavadskas, E. K., Olfat, L. and Turskis, Z. (2015). Multi-criteria inventory classification using a new method of evaluation based on distance from average solution (EDAS). *Informatica*, 26(3), 435–451.

Labuschagne, C., Brent, A. C. and Van Erck, R. P. (2005). Assessing the sustainability performances of industries. *J Clean Product*, 13(4), 373–385.

Lélé, S. M. (1991). Sustainable development: A critical review. *World Develop*, 19(6), 607–621.

Mihyeon Jeon, C. and Amekudzi, A. (2005). Addressing sustainability in transportation systems: definitions, indicators, and metrics. *J Infrastruct syst*, 11(1), 31–50.

Missimer, M., Robèrt, K.-H. and Broman, G. (2017a). A strategic approach to social sustainability–Part 1: exploring the social system. *J Clean Product*, 140, 32–41.

Missimer, M., Robèrt, K.-H. and Broman, G. (2017b). A strategic approach to social sustainability–Part 2: a principle-based definition. *J Clean Product*, 140, 42–52.

Norgaard, R. B. (1988). Sustainable development: a co-evolutionary view. *Futures*, 20(6), 606–620.

Papajohn, D., Brinker, C. and El Asmar, M. (2016). MARS: Metaframework for assessing ratings of sustainability for buildings and infrastructure. *J Manage Eng*, 33(1), 04016026.

Poveda, C. A. and Lipsett, M. G. (2011). A Rating system for sustainability of industrial projects with application in oil sands and heavy oil projects: Areas of excellence, sub-divisions, and management interactions. *J Sustain Develop*, 4(4), 3.

Poveda, C. A. and Lipsett, M. G. (2014). The Wa-Pa-Su project sustainability rating system: a simulated case study of implementation and sustainability assessment. *Environ Manage Sustain Develop*, 3(1), 1.

Redclift, M. (1991). The multiple dimensions of sustainable development. *Geography*, 36–42.

Shen, L., Wu, Y. and Zhang, X. (2010). Key assessment indicators for the sustainability of infrastructure projects. *J Construct Eng Manage*, 137(6), 441–451.

Shiva, V. (1992). Recovering the real meaning of sustainability. *The Environ Question*, 187–193.

Sierra, L. A., Pellicer, E. and Yepes, V. (2015). Social sustainability in the lifecycle of chilean public infrastructure. *J Construct Eng Manage*, 142(5), 05015020.

Silajdžić, I., Kurtagić, S. M. and Vučijak, B. (2015). Green entrepreneurship in transition economies: a case study of Bosnia and Herzegovina. *J Clean Product, 88*, 376–384.

Silvestre, B. S. and Gimenes, F. A. P. (2017). A sustainability paradox? Sustainable operations in the offshore oil and gas industry: The case of Petrobras. *J Clean Product*, 142, 360–370.

Stamford, L. and Azapagic, A. (2014). Life cycle sustainability assessment of UK electricity scenarios to 2070. *Energy Sustain Develop*, 23, 194–211.

Streeten, P. (2001). *Globalisation: Threat or Opportunity?* Copenhagen: Copenhagen Business School Press.
Toman, M. A. (1992). The difficulty in defining sustainability. *Resources*, 106 (Winter).
Ugwu, O., Kumaraswamy, M., Wong, A. and Ng, S. (2006). Sustainability appraisal in infrastructure projects (SUSAIP): Part 1. Development of indicators and computational methods. *Autom Construct*, 15(2), 239–251.
Yao, H., Shen, L., Tan, Y. and Hao, J. (2011). Simulating the impacts of policy scenarios on the sustainability performance of infrastructure projects. *Autom Construct*, 20(8), 1060-1069.
Zavadskas, E. K., Turskis, Z. and Antucheviciene, J. (2015). Selecting a contractor by using a novel method for multiple attribute analysis: Weighted Aggregated sum product assessment with grey values (WASPAS-G). *Stud Inform Control*, 24(2), 141–150.

Appendix A. List of PRP Sustainability Indicators and Factors

No	Indicator	Factor
1	Poverty & Equality	Proportion of project human resource living below national poverty line
2		Fraction of project human resource protected against impoverishment by out-of-pocket health expenditures
3		Staff forums and grievance procedures forming in the project and operation stage
4		Gender diversity in all different levels of leadership, management, and expertise
5		Proportion of working women with computational skills to men with similar skills

(*Continued*)

(*Continued*)

No	Indicator	Factor
6		Ratio of basic salary and remuneration of women to men by employee category, by significant locations
7	Health	Annual number of project/plant manpower getting sick because of poisoned and unhealthy food and water
8		Annual number of project/plant manpower death because of cardiovascular disease, cancer, or chronic respiratory disease as a result of exposure to toxic gasses and other substances in the refinery
9		Total recordable occupational illness frequency (TROIF)
10		Tobacco consumption per capita among project manpower
11		Population in local areas in the vicinity of the project/plant exposed to outdoor air, water, and soil pollutions caused by the refinery
12		Percentage of project/plant manpower and local people using safely managed sanitation services
13		Percentage of project/plant manpower and local people having health insurance coverage
14	Safety & Security	Number of people having accident in the project/plant per annum

(*Continued*)

(*Continued*)

No	Indicator	Factor
15		Level of health and educational facilities in order to prevent accidents' occurrence
16		The extent of damage to local cultural heritage as a result of petroleum refinery project/plant construction/operation
17		Lost days due to accidents leading to injury or death
18		Frequency rates of fatal and non-fatal occupational injuries
19	Education & Training	Number of people who are trained and develop practical skills in the project
20		Participation rate of owner/operator in formal and non-formal education and training of staff
21		Percentage of staff who are computer and information literate in the project
22		Literacy rate among project/plant staff
23		Number of schools and training centers established in project/plant accommodation and local city
24	Welfare	Amount of increase in water resource infrastructure access, which has been made by the project/plant
25		Percentage of project/plant staff using safely managed drinking water services

(*Continued*)

(*Continued*)

No	Indicator	Factor
26		Amount of increase in electricity infrastructure access, which has been made by the project/plant
27		Percentage of project/plant staff living in unsuitable accommodation and informal settlements
28		The amount of happiness and national pride among staff due to their occupation
29		Scale of serviceability of refinery and its production for local community
30	Atmosphere	Emissions of greenhouse gases per ton of refinery product (CO_2, CH_4, N_2O, HFCs, CCl_4, CH_3CCl_3, CCl_3F, CCl_2F_2, $C_2Cl_3F_3$)
31		Net carbon intensity per kilowatt electricity production
32		Level of ambient particulate matter per ton of refinery product
33		Level of hazardous chemicals (HAPs) and wastes per ton of refinery production
34		Levels of selected contaminants in air caused by to refinery operation: Acid Gas (SOX, NOX), VOCs, CO, Ozone, Hydrocarbons, H_2S, NH_3
35		Level of noise pollution due to project/plant construction/ operation

(*Continued*)

(*Continued*)

No	Indicator	Factor
36	Water (Fresh Water, Ocean, Sea, Coast)	Percentage of wastewater safely treated to standard levels: cooling water, process water, sanitary sewage water, storm water
37		The proportion of wastewater contamination in the refinery output to absorbable pollution by water resource (sea, river, etc.)
38		The refinery efficiency in water resource usage
39		Level of integrity in water resource management in different process units
40		Average acidity (PH) of the water resource and water stress value
41		Average levels of contaminants in water per ton of refinery production (Azapagic & Perdan)
42		Number of local households and businesses affected by water contamination
43		Annual withdrawal of ground and surface water as a percentage of refinery demanded water separately
44	Land & Soil Pollution	The amount of solid waste produced by refinery per ton of production
45		Per capita waste generation (in kg) for refinery employees
46		Recycling rate and percentage of industrial waste reuse

(*Continued*)

(*Continued*)

No	Indicator	Factor
47		Level of soil degradation due to construction/operation of the project/plant
48		Level of contaminants in soil per ton of refinery production
49		Residence status of the project area (the number of households within a radius of 50 km)
50		Level of Protection to landscape and historical sites
51	Natural Resource	The extent to which the refinery can satisfy the demand of "non-solid fuel"
52		Renewable energy share in the total energy of refinery final energy consumption
53		The extent of documentation issued in the refinery project/plant on the utilization of renewable energy
54		Indicator of crude oil efficiency (ratio of crude oil to product)
55		Percentage of electricity demand of the refinery which is accommodated by the crude oil in the refinery processes
56		Natural resource utilization per refinery worker, as well as per unit production in the refinery

(*Continued*)

(*Continued*)

No	Indicator	Factor
57		Percentage of required refinery procurement with approved LCA sustainability assessment documentations (Equipment, material, etc.)
58		Certified wood used in construction/operation of refinery project/plant
59	Biodiversity	Level of damage to the flora by refinery construction and operations
60		Level of damage to fauna by refinery construction and operation
61		Percentage of vegetation coverage in the refinery site and local region
62	Energy consumption	Achievement rate of the designed energy usage in the operation phase
63		The efficiency of electricity production in the refinery (output to input)
64		The energy economic efficiency in the refinery (the economic return per unit of energy consumed)
65		Required energy intensity per unit of value added
66	Financial	The amount of initial investment or forecasted budget of the project
67		Payback period (investment & interest) based on analysis of market supply and demand

(*Continued*)

(*Continued*)

No	Indicator	Factor
68		Net present value of refinery life-cycle costs (including operating costs, maintenance, overhaul, etc.)
69		Net present value of refinery life-cycle profits (including product sales, etc.)
70		Refinery internal return ratio (IRR)
71		The foreign direct investment (financing) or mobilizing domestic financial resources in order to achieve more advanced technologies and cleaner fuel production
72		The level of international cooperation in the refinery project in order to facilitate access to clean and sustainable energy that is consumed in the project
73		Number of man-hours in the research and development in order to promote technologies in clean energy utilization in the refinery project/plant construction/ operation
74		Investment program (Annual credit flow) in the refinery project
75		Percentage of financiers on the short list who are interested in participation and are willing to pay for the refinery project

(*Continued*)

(*Continued*)

No	Indicator	Factor
76		Level of investment risks in the refinery project
77		Amount of spending on R&D by owner to achieve sustainable production and consumption in the refinery project
78		The amount of reward which is intended in the contract to cover costs of patent in sustainable production and consumption of refinery project
79	Economic Performance	Level of refinery participation in increasing GDP
80		Level of refinery participation in inclusive wealth
81		Growth rate of GDP per capita per employed person in the refinery
82		Export diversification of the refinery products in terms of products and markets
83		Amount of financial losses due to climate change and natural disasters risks in the refinery project/plant
84		Percentage of growth in GDP per capita by the refinery among other available alternative projects
85		Ratio of created jobs in the refinery project/plant to job opportunities in other available alternative projects

(*Continued*)

(*Continued*)

No	Indicator	Factor
86		Intensity of raw material use per unit of value-added
87		Percentage of decrease in value added in exchange for technology upgrades and stability enhancement in the refinery project/plant
88		Amount of refinery products' subsidies per total refinery expenditure
89		Average index of Economic Welfare of project manpower
90		The financial and time losses resulting from accidents in the refinery project/plant
91		The amount of each refinery product and its means of transportation (kg.km)
92	Occupation	Number of job opportunities created by the refinery and local economic boom within different phases of the project
93		Number of small and medium firms that have developed during the construction and operation of the refinery project
94		Employment to working-age population (15 years and above) ratio by gender and age group

(*Continued*)

(*Continued*)

No	Indicator	Factor
95		Unemployment rate in different phases of the project due to different human resource demand (hiring and firing)
96		Number of illiterate and unskilled workforce who acquire skills in the refinery project/plant
97	Earning	Refinery project/plant participation in the promotion of regional labor income
98		Proportion of labors living below 50% of median income
99		Proportion of labors that spends more than 30% of its income on accommodation
100		Average health insurance expenditure to income ratio
101		Ratios of standard entry level wage by gender compared to local minimum wage at significant locations

Chapter 3

From Past to Present and into the Sustainable Future: PMADM Approach in Shaping Regulatory Policies of Medical Device Industry in the New Sanction Period

Sarfaraz Hashemkhani Zolfani[*,§] and Reza Masaeli[†,‡,¶]

[*] *Catholic University of the North, School of Engineering, Larrondo 1281, Coquimbo, Chile*
[†] *School of Dentistry, Tehran University of Medical Sciences, Iran*
[‡] *Center for Research in Biomaterials and Medical Devices (CRBMD), National Medical Device Directorate (IR. FDA), Ministry of Health and Medical Education, Iran*
[§] *sa.hashemkhani@gmail.com*
[¶] *dr.rezamasaeli@yahoo.com*

Both decision-making or policymaking and regulatory measures have an undeniable role in directing and leading a community, society or a country. New methods and tools are being developed to ensure a brighter future because our future hinges on our current decisions. Prospective multiple-attribute decision-making (PMADM) as a new approach in multiple-attribute decision-making (MADM) structure has been developed as a powerful tool for making critical decisions and policies. PMADM as an approach with a flexible structure or a pilot for developing policies and making decisions can cover more uncertainties and ambiguities in real situations. The government, regulatory bodies, policymakers and organizations are in constant endeavor to apply the best policies and decisions in their activities and responsibilities.

The Iranian National Medical Device Directorate is a regulator of the medical device industry of Iran. This organization considers ensuring a sustainable future (critical aspects of sustainability) by managing the present conditions, which is the main aim of this study. Iran is planning to increase the share of medical device market ten times more as a sustainable market for the country. For this aim, the PMADM approach was adopted as a pilot for the policymaking procedure. Max capacity, ideal directed scenario and supportive backup criteria are presented as the new approaches for applying the PMADM structure. Supportive backup criteria open a new window for illustrating different scenarios and vision in response to a complicated future. In the end, the directorate and policymakers are concerned with how is it possible to see uncertainties of the future in deeper scenarios and challenges. As an example and a pilot, this study can open up new horizons for similar studies in the future.

1. Introduction

Regulation of medical devices as a responsibility plays a critical role and many regulatory policies are in place throughout the world. Similar to Food and Drug Administration in the United States, Iranian National Medical Device Directorate is the regulator for medical device equipment in Iran. Recently, with the new managerial changes and changing attitudes within the Ministry of Health and Medical Education of the country, the regulatory system has been changing to a more effective structure. This study is aimed at illustrating new achievements and outputs about policymaking to ensure a better sustainable future in terms of regulating the manufacture of medical device equipment. After the new sanctions period came into force (2018), a big opportunity opened for some companies and countries to work with Iran. It is a mutual arrangement of importing or exporting medical devices, especially software-related industry, which can be an exceptional market niche because of sanctions (Madsen, 2018). Making new and sustainable policies presents a big challenge and the path toward a sustainable future seems to be difficult for policymakers. Therefore there is a need to apply suitable approaches and methodologies (Dinarvand, 2009).

At present, multiple criteria decision-making (MCDM) has emerged as a powerful tool for use in the decision processes in several areas, including health issues, especially medical devices (Taghipour *et al.*, 2011; Akdag *et al.*, 2014; Peijia *et al.*, 2017). It is necessary to see how MCDM methods can enable decisions very appropriately. The MCDM framework is not limited to simple decision-making issues, but it is capable of providing suitable inputs for policies and even for future policies. Recently, prospective multiple-attribute decision-making (PMADM) as a new multi-disciplinary approach has been introduced. PMADM supports future decisions and policies in complicated situations. This new approach is being used for the policymaking and regulatory measures in critical and substantial scenarios. Therefore, this approach has been selected for issuing regulatory policies of medical devices. So, PMADM can be considered a suitable tool for decision-making at the highest level of policymaking for different issues, especially for the medical device industry, which is main topic of this study. Policymaking aims to usher in a sustainable future that will be able see things and illustrate them in an applicable framework. Eventually, it can be forecast how the selected approach will play this role in the process of policymaking toward a better and more sustainable future of the medical device industry.

Sustainable future is a new perspective from a different paradigm. Ensuring a sustainable society in different aspects is an indication of good governance. Governments are responsible for policymaking and planning for a better and sustainable future in different aspects of governance. In this regard, as mentioned above, PMADM is applied for policymaking to ensure a sustainable future and Iranian National Medical Device Directorate is using this new approach for long-term decision-making for the medical device industry in imports and exports as well as their quality and structure to guarantee a more sustainable future for that industry in the country. This chapter depicts how PMADM can be useful in the process of policymaking for the medical device industry and its regulations. The main aspects of this chapter will be regulatory measures on medical devices, which cover international organizations and regulations, and medical devices in Iran, which encompass the history of regulation of medical devices in Iran and all new procedures and policies that will be put in place by using PMADM, the policymaking tool, and introducing new criteria in the policymaking structure based on PMADM and eventually discussion about all the procedures.

2. Regulation on Medical Devices

Medical devices and drugs have evolved into large high-tech markets and they have been growing too fast in the modern era. These markets mostly rely on R&D activities of companies and related research institutes. Developing a new drug or a medical device has been increasing in the past years and the new drug/device industry is worth more than US$1 billion. The need for investing money posed a new barrier and challenge for the progress of the industry and caused a delay in the provision of new pharmaceuticals and devices for patients all over the world. Furthermore, a spurt in introducing and developing drugs and medical devices means increase in the final product prices for the patients and eventually more allocation in national budgets for public healthcare systems. The main reason for this rise of prices is clinical trials. The drug and medical device development has two sections: preclinical and clinical trials. As obvious, about 97% of total cost is associated with clinical trials. Prices increased tenfold from 1991 to 2003. Finally, cost-effective methods for performing clinical trials remain a big barrier and challenge (DiMasi and Grabowski, 2007; Paul *et al.*, 2010; Matsushita *et al.*, 2 018).

Governments regulate medical devices in the market through various legislations, which set the responsibility for their manufacturers by specifying the related technical requirements. It is common to check technical requirements based on available documents of manufacturers as technical standards. There are different criteria for evaluating medical devices by a regulator and regulatory measures with respect to medical devices vary across the world. Furthermore, the complexity, variety and other new features have been increasing over the previous two decades and new regulatory policies need to be developed. As mentioned above, there are different regulatory structures in different countries and effort to cope with them is becoming more challenging. Mostly FDA and EU regulations are accepted more widely between countries (Steinberg *et al.*, 2015; De Maria *et al.*, 2018).

The Global Harmonization Task Force (GHTF) was established with the aim of standardizing medical device regulations worldwide.

The GHTF started its work in 1993 based on an international partnership between European Union (EU), the United States, Canada and Japan. Since 2011 the International Medical Device Regulators Forum (IMDRF) replaced GHTF as the body responsible for future directions of medical device regulations and harmonizing them. The main partners of IMDRF are regulatory authorities of countries including Australia, Brazil, Canada, China, EU, Japan, Russia, Singapore, South Korea and the United States with an observer, which is the World Health Organization (WHO) (Tsai *et al.*, 2017; International Medical Device Regulators Forum, 2018; De Maria *et al.*, 2018).

Still there is a gap in the regulatory structures to have a practical harmonization; although several African countries have been working on regulations based on the EU system in medical device regulations, a number of other African countries are representatives of the African Biomedical Engineering Consortium (ABEC), which was founded in 2012 (African Biomedical Engineering Consortium, 2018). Another important regulatory organization is the Asian Harmonization Working Party (AHWP), which was founded in 1998. AHWP is a voluntary organization and currently has 24 members from different continents. The Republic of Korea, Chile, Hong Kong, Singapore, South Africa, China, Abu Dhabi and Taiwan can be considered as key players of the party (http://www.ahwp.info). Switzerland isn't a member of EU and European Economic Area. The Swiss drug regulatory body, Swissmedic, is responsible for medical device regulations in Switzerland (http://www.swissmedic.ch). Some of the largest pharmaceutical companies are located in the country. Switzerland is the host of WHO, International Organizations of Medical Sciences Working Group, the International Standard Organization and the Council for International Organizations of Medical Sciences Working Group, all of which have a considerable role in the procedure relating to regulatory systems across the world (Billiones, 2013). Japan, as a regulator, is working on regulations on medical devices based on the Ministry of Health, Labor and Welfare (MHLW). The latest regulations called the Pharmaceutical and Medical Device Law (PMDL) were announced on November 25, 2014 and all medical devices and products were categorized at three levels of risk (Kondo, 2014; Wu *et al.*, 2016).

3. Medical Devices in Iran

This section is presented based on the newest report of the Iranian National Medical Device Directorate, which is responsible for medical device regulations (National Medical Device Directorate, 2018). Based on estimates in the literature, more than 500,000 medical devices are available across the globe. Iran produces more than 280,000 types of medical devices. The annual share of the medical device market in Iran is US$2 billion and US$500 million, in which the share of import is about US$1 billion, and US$800 million, which is accomplished by 2,200 firms and 5,500 international companies registered with the Ministry of Health and Medical Education of Iran. In the field of medical device manufacturing, the policy is to support production, and now nearly 1,000 companies with national and international investments are engaged in the production of medical devices and the combined annual turnover of these companies is US$800 million in 2018. Interestingly, according to the policies of the government of the Islamic Republic of Iran, 400 products in the C and D classes are manufactured. Iran has targeted that in the year 2020, the quality assessment system of medical equipment will be managed in the same way as the European Union (EU) by notified bodies and accredited laboratories certified by ISO 17025. Iran has intended to cooperate with the notified bodies in the EU to facilitate the export of medical equipment qualified by the EU standards with the CE mark to European countries. The manufacture of medical devices in Iran has been growing rapidly and the level of production has assumed a more complicated structure and medical devices with higher class of risks are no longer difficult to manufacture. Iran is able to produce high-tech products with higher quality in comparison to the past.

The Ministry of Health and Medical Education and National Medical Device Directorate in Iran are planning for a brighter future for the medical device industry. As new policies about increasing levels of quality and attaining international standards are emerging, Iran is positioning for a bigger role in this industry globally. In the new vision, Iran is planning to increase exports from US$24 million to US$240 million annually. Toward this aim, three new MoUs for increasing exports have been signed with Brazil, Iraq and Oman. In October 2018, there was a meeting between Iranian National Medical Devices and The Department of Welfare, Public Health and Family of Belgium. The first outcome of that primary agreement was a contract between two medical device companies from Iran and Belgium and the value of this contract was worth €15 million. This cooperation between two countries is planned to be developed at different levels and services in the future.

Subsidy has to be paid to the producers to cooperate in international exhibitions such as Medica, Oman Health, Arab Health and exhibitions in Brazil. The new expectations of the Iranian Ministry of Health and Medical Education concluded a planned cooperation between Iranian embassies and the ministry. The embassies now have a defined role to follow these responsibilities:

1. Identifying regulations about importing and registering medical devices in the target country.
2. Meeting with managers and regulators about their importing priorities.
3. Facilitating and helping to register Iranian medical device products in the target country.
4. Participating in tenders and all other opportunities for exporting medical device products.
5. Identifying and introducing all important related events, exhibitions, congress and conferences to the ministry for informing Iranian companies and medical device producers.

In the previous year, Iran exported medical devices to 55 countries from all continents including two from North America, one from South America, 15 from Europe, 11 from Africa, one from Oceania and others from Asia. There were 23 active companies in exporting medical devices. Throughout this year, 105 products were exported from all the Iranian manufacturers and companies. It should be observed that importing medical devices from the United States and Germany has the highest share of market in the country.

4. Prospective Multiple-Attribute Decision Making

There is a trend in the literature about MCDM and future studies as two separate multidisciplinary fields for considering the future and for the process of decision-making (Trutnevyte *et al.*, 2012; Arms *et al.*, 2012; Zhang, 2012; Durbach and Stewart 2012; Hashemkhani Zolfani *et al.*, 2013; Wang *et al.*, 2014; Tadic *et al.*, 2014; Gonzalez-Prida *et al.*, 2014; Ondrus *et al.*, 2014; Jassbi *et al.*, 2014; Hashemkhani Zolfani *et al.*, 2016a). These two multidisciplinary fields have created a new interdisciplinary field, which is called PMADM. In facing the future, flexible decision-making for solving problems and making the right decision can be considered as a new field of study. This "perspective and paradigm" from the developing field, future studies, are now part of the curriculum in different universities and organizations all around the world such as the United States, Finland, Australia, Taiwan and Iran. In between, MCDM has been progressing to be a separate field from operations research (OR), which led to the introduction of new specialized journals such as *Journal of Multi-Criteria Decision Analysis* and *International Journal of Multi-criteria Decision-making.* PMADM can be considered as a new field of study in the future in a more flexible structure. This new approach resembles a platform that is able to solve more complicated decision problems in the future because it just needs to be more updated about what we would like or need to do. Also, the approach can be a suitable tool for thinking and asking about what the future will be (Hashemkhani *et al.*, 2016b; Hashemkhani Zolfani, 2018).

PMADM, as mentioned, is a new platform having more items in comparison to the classic type of MADM. Hashemkhani Zolfani *et al.* (2016b) presented the first new item in the PMADM framework, which can be called limiters and boosters. Limiters and boosters can have the role of pay-offs of the future scenarios for evaluating the alternatives in their positions. It can be illustrated through where they are located the structure of a classic MADM framework as an example in Table 1.

The table is designed for the limiters. Boosters can be put the same in the table. It should be kept in mind that limiters and boosters can both be applied at the same time in the process of decision-making for the future because it can happen in reality. Future probability also can be added to the process and it can be measured as an introduction to the PMADM. Other items can be added to the classic structure of PMADM to have a better outcome in evaluating the criteria, relative weights and other probable calculations of the MADM framework.

PMADM approach has this capacity to be used with common MADM methods as the easiest way of calculations for the decision-making process. It is possible to see new methods based on PMADM in the future as an upgraded infrastructure for decision-making. In this study, new items will be introduced as new tools for a current new item and an independent new item, which can be presented on the criteria side. These new tools and item can help policymakers to consider and see the deeper side of their decision in a real situation.

5. A Big Picture for Policymaking Based on PMADM about Developing Market of Medical Device

In this section, a preview from the future is designed based on experts' opinion. As mentioned above, Iran is planning to increase the export of medical devices 10 times more than the previous year. It can be a long-term policy but government and related organizations such as Iranian National Medical Device Directorate decided to make it happen as fast as possible. PMADM structure as a powerful policymaking method for the future is applied to help formulate policies to achieve the set goal, which forms a bright vision for the industry, both in private and governmental sectors. An extended PMADM model as a new modified framework is considered for this study and two new tools have added to the approach. These new tools have this ability to consider more uncertainty conditions in reality and make the decision-making process in a rational structure.

Table 1: PMADM model based on future limitations.

		C_1		C_{n+1}		C_n
Weights						
Limiters (L)/ Boosters (B)		$L_{1-1} \ldots L_{1-n}$		$L_{n+1-1} \ldots L_{n+1-n}$		$L_{n-1} \ldots L_{n-n}$
	Based on C_1	Average	Based on C_{n+1}	Average	Based on C_n	Average
A_1 without L						
A_1 based on L_{1-1}						
A_1 based on ...						
A_1 based on L_{1-n}						
A_{n+1} without L						
A_{n+1} based on L_{n+1-1}						
A_{n+1} based on ...						
A_{n+1} based on L_{n+1-n}						
A_n without L						
A_n based on L_{n-1}						
A_n based on ...						
A_n based on L_{n-n}						

5.1. *Max capacity and Ideal directed scenario*

Max capacity is a new tool for evaluating limiters and boosters better. In the previous model for limiters and boosters, probabilities were considered. In this new approach, maximum capacity for considering future effects is examined. This new tool can direct us to see different scenarios based on realities and see how much our decision problem can be changed or upgraded based on new limiters and boosters. When there is a chance to control the future, an ideal scenario to shape and control the future seems necessary. "Ideal directed scenario" is another tool that can control and direct future limiters and boosters with this aim to maximize the boosters and minimize the limiters. It is possible to consider them based on main driving forces in the future. Their impacts are similar to each other. The differences and illustrations for supporting the idea and approach can be seen in Table 2.

As a regulator, controlling and directing the future need to have enough powers to make the best policy in tune with reality. The Iranian National Medical Device Directorate, as a regulator with responsibility from the ministry side and by considering all political and economic aspects, is planning an ideal directed scenario for the medical device industry. The key point is there are so many players and pay-offs as regards regulatory measures on the medical device industry. Being rational is a key point here and main players and regulators should consider all obvious and hidden sides of a challenge and a problem. A team of experts, high-level managers and advisors from different fields such as international law, future studies, economic, etc. cooperated in the process of evaluating and policymaking about directing the export goal of medical devices.

There is a direct influence of limiters and boosters on each other. The difference between B_1 and L_1 is 50% as regards the overall capacity in managing exporting medical devices from the country. L_1, B_2 and B_3 are completely related to each other. In general, all those limiters and boosters can be a part of bigger scenarios and they can happen randomly or as a package in reality. The boosters and limiters should be explained carefully and deeply to complete a big picture

Table 2: List of limiters and boosters.

	Sanction situation	Limiters/ Boosters	Max capacity (%)	Ideal directed scenario (%)
1	Supportive countries, societies and unions	B_1	+25	+20
2	Limited markets based on sanctions	L_1	−25	−5
3	International insurance	L_2	−20	−5
4	New markets (supportive)	B_2	+15	+15
5	Considering as an exceptional market	B_3	+20	+10
6	Future opportunities for using other currencies	B_4	+25	+15
7	New policies for exporting products easier	B_5	+25	+15
8	Conflict in the region (Middle East)	L_3	−20	−5
9	New conflicts in future with USA	L_4	−40	−10
10	Limitation for transferring money (mutual problem)	L_5	−30	−10
11	Tariff policy from both sides	B_6	+15	+10
	Final tolerance		−10	50

of the situation in which policymaking takes place. The definitions of boosters and limiters presented as follows:

B_1 — Supportive countries, societies and unions: EU, 5+1 countries except the United States, Russia, China and some other countries consider acting against the new sanctions of the United States on Iran. The quality of this support can help the country to import and export better depending upon the priority and need. Increasing the quality of products can be useful because developed countries can stand more effectively against the sanctions, but they are looking for the best quality materials and products.

B_2 — New markets (supportive): Recently and as mentioned before, some MoUs were signed between Iran and some other countries like Brazil, Oman and Iraq. This kind of cooperation is developing. There are some other markets and countries that are ripe for

Iran to enter. As mentioned, in B_1 some countries are supporting as much as they can. It can be an opportunity for Iranian producers to export more products.

B_3 — Considering as an exceptional market: Generally, importing medicine and medical devices will be possible even in the sanctions period but for exporting, there are some limitations such as transferring US dollars and problems with international bank systems. This industry is somehow an exception and eventually, negotiating about medical devices can be an appropriate way for imports and exports.

B_4 — Future opportunities for using other currencies: Recently, some countries have started to use their currencies directly in their deals for importing and exporting. Iran also has this chance to work with Turkey, Russia, China and EU with other currencies. This issue would be a great opportunity for Iran to have a better access to markets across the world or at least in special countries.

B_5 — New policies for exporting products easier: Generally, Iranian regulations and law for exporting products need a revision. It has been working already, but it needs better and more fundamental changes and the process should be clearer and be free for all producers to be more involved.

B_6 — Tariff policy from both sides: Tariff has two sides: internal and between countries. Government and related sections should engage in negotiations about facilities for exporting products from and those that are responsible for health structure need to follow procedures of different countries for considering a share for Iranian producers. Negotiating for tariffs will be really helpful because Iran isn't still a permanent member of World Trade Organization (WTO). These activities can cause a radical change in exporting medical devices from Iran because the rate of producing devices is considerable and it would be in the best interests of Iran to have a better output for the country.

L_1 — Limited markets based on sanctions: Although medicine and medical devices are out of sanctions for importing, exporting them has some challenges and limitations. It depends on the situation in the future and how many countries will follow the US sanctions. The United States has not reacted to exporting medical devices based on a future situation. It can be a better situation for exports or worse. The general relationship between the United States and Iran somehow can be a direct topic in between.

L_2 — International insurance: International insurance for importing and exporting products is a challenge itself and isn't directly related. It can be solved in different ways. For example, special measures for exporting can be put in place for the aimed markets and countries, and for the neighboring countries, exports would be easier. Different scenarios can be designed, which probably increase the cost of transportation. The chances of buying products from the origin and port is still possible, but it needs better international cooperation.

L_3 — Conflict in the region (Middle East): Iran faces a big challenge with Saudi Arabia and also with some other countries that are cooperating with Saudi Arabia. Saudi Arabia enjoys a good relationship with the United States and this issue caused another big challenge for Iran. Iran needs to find a more stable position in the region and decrease these conflicts faster with countries in the region. At the same time, Iran enjoys good relationships with Oman, Qatar, Iraq and Syria in the Middle East, but it should be extended to other countries of the region, which have a negative reaction to the policies. Controlling these conflicts can lead Iran to a better position in the markets and undoubtedly medical device industry could be better positioned in the international market.

L_4 — New conflicts in future with the United States: A conflict with the United States has a fast growing impact on the economy and currency rate of a country. Recently, Turkey faced a severe challenge with the United States just in a simple case. Iran should try to make the conflict more stable and other countries can play a great role in that. Iranian politicians also prepare a win-win scenario for Iran-US conflict, although the United States must show a more reliable face in the negotiations. Eventually, Iran needs to decrease the pressure and find the best scenario to have more profits from that. A new conflict will be really dangerous for both import and export and the resolved conflict can open more doors toward Iran and the markets.

L_5 — Limitation for transferring money (mutual problem): As mentioned above, transferring money itself is a critical challenge for both imports and exports. In this case, exporting to other countries poses different challenges including transferring money. There are some legal ways for to avoid US sanctions. Regulators, policymakers, governments and politicians should prepare suitable ways to overcome this challenge using negotiations. Iran should consider all probable ways for solving this challenge to provide a stable position for producers.

5.2. *Supportive backup criteria*

In the uncertainty atmosphere of the future space, we need to think about things differently. Decision-making and process of selecting or prioritizing something would be really complicated and future state needs more parameters to be determined, which also leads decision makers to seek reliable answers. The classic structure of MADM models need to be developed because there are so many elements that can shape the future and not necessarily the future we want or we are supposed to plan. Since 2016, PMADM has been developing alternatives and limiters and boosters that can make a difference in the process of decision-making and policymaking. One of the main elements and items of a MADM structure is criteria that a decision-making problem establishes based on their characteristics and features. A main and key question here is, are they enough for facing the future or not? What is going to be presented here is a new item for empowering criteria to face the future. "Supportive backup criteria" is a new item that can prepare and support a clearer future atmosphere for the process of decision making and policymaking.

In the council of Iranian National Medical Devices, a new item was defined for supporting criteria in the PMADM outline. "Supportive backup" criteria are supporting classic criteria for decision-making process and future can be seen in the new model of solving a MADM decision-making problem. "Supportive-backup" can come instead of current/considered criteria and can be added as a new criterion or added as a subcriterion for current or considered criteria. It means new criteria will be added when another scenario, event, or limiter of booster happens. Certain "supportive-backup" can be determined for a predictable future, but always it is possible to consider some criteria for unpredictable scenarios and situations. They can be determined as criteria for somehow managing uncertainty or open boxes (empty criterion) for prepare for what can happened as a wildcard scenario.

It should be kept in mind that some scenarios can happen based on other scenarios. Some other methodologies such as game theory can be applied to this part of study. For instance, considering Iran in new sanctions period from the US side can have some other scenarios or games based on the scenario because it is important and vital to know what will happen. The final pay-off based on mutual political activities will show how can we shape the decision-making question in reality. This new approach can lead policymakers to have all scenarios together at the same time, by posing different research questions about challenges, and there will be answers (like if–then), but policymakers can answer those when they would like or when faced with reality. It can be applied as a support system to have answers for all questions about the future or those situations and scenarios that can predict something happening and they need to find the right answers at the shortest time based on priority.

"Supportive backup" criteria should add to the criteria side in the decision-making matrix. An example can illustrate their positions and locations. Table 3 depicts position of "supportive-backup" criteria as an example with varieties of scenarios.

*A key point about alternatives will be determining them based on those scenarios predicted for the criteria. An alternative should be able to cover all prerequisites and needs of those criteria that can happen. All criteria won't be predictable, especially criteria based on wildcard scenarios. Alternatives should be the closest to the reality,

Table 3: Position of "supportive-backup" criteria.

	C_1	C_2	C_{n-1}	C_n
Supportive/backup criteria	C_{s1-1}	C_{2^*-s1}	$C_{s1-n-1-sb1}$	...
	C_{s2-1}	C_{s1-2}	C_{n-1^*-s2}	...
	C_{u1-1}	...	...	...
A_1	...	...	...	...
Reserved A_1	...	...	...	...
Reserved A_1	...	...	...	...
Reserved A_1	...	...	...	...
...	...	...	...	...
...	...	...	...	...
...	...	...	...	...
A_n	...	...	...	...

which we are preparing for. In some exceptional cases and scenarios, deleting an alternative will be possible when we face a complicated scenario, which we aren't prepared for. Eventually, alternatives will be an important issue in considering "Supportive/backup" criteria.

***C_{s1-1}**: It means a criterion based on the first scenario that can be a "limiter", "booster" or something else will be added as an independent criterion for supporting criterion 1.

C_{u1-1}: It is about an unpredictable scenario as an output of a wildcard scenario. When we know there is a chance of something occurring but it is only out of imagination. It should be considered that it works for really rare cases but can be intended for events of really low probability.

$C_{2^*-s^1}$: When based on a scenario, another criterion related to current criterion can be changed. This is happening when prerequisites and needs will be changed based on a new situation. It should be more complicated when policymakers want to consider more than one scenario at the same time and they should consider the new criterion to support this changed criterion as something not independent completely. As it can be seen, the second criterion that is defined supports the changed second criterion.

$C_{n-1^*-s^2}$: The same as $C_{2^*-s^1}$ and a similar way of definition.

$C_{s1-n-1-sb^1}$: This is another case as mentioned above. When in a scenario it feels that a sub-criterion is needed, it is possible to add a sub-criterion for that. So, this is scenario 1, number of criteria and sub-criterion 1.

Reserved A_1: It means pay-off and general evaluation of an alternative based on a certain criterion. It depends on whether we want to consider different scenarios, mixed scenarios or some of them. Pay attention to this point that all scenarios considered actively in this study as an example. It depends on the research question and what policymakers decide to focus on. The process and result will be supporting the research questions undoubtedly. The A_1 row is the alternative without considering other scenarios in facing the future. It means ideal future that we are planning for, which would be a rational and logical scenario and it isn't necessarily the best thing that could happen.

It is possible to consider probability for the scenarios, chance of events occurring or using stochastic calculations for developing scenarios and decision-making about the future. They can easily add scenarios and "supportive-backup" criteria to be considered for a final evaluation.

Table 3 is established based on a long-term plan prepared by Iranian National Medical Devices for the medical device industry market in future and in the new sanction period, which will start in November 2018 against Iran from the United States. Because of confidential reasons, it won't be possible to publish all information.

The easiest way to calculate final results will be based on the classic way, which was presented in the first model of PMADM. The current classic MADM methods used are AHP (Saaty, 1980), ANP (Saaty, 1996), BWM (Rezaei, 2015), SWARA (Kersuliene *et al.*, 2010), COPRAS (Zavadskas and Kaklauskas, 1996), TOPSIS (Hwang and Yoon, 1981), VIKOR (Opricovic, 1998), ARAS (Zavadskas *et al.*, 2010), WASPAS (Zavadskas *et al.*, 2012), EDAS (Keshavarz *et al.*, 2015), CODAS (Keshavarz *et al.*, 2016) and other methods.

6. Conclusion

Regulatory and policymaking for a better sustainable future is a really complicated activity, which a trustee and regulator should intend to do and differences between countries, organizations and other related positions arise because of those outputs. A sustainable future needs a descriptive paradigm, which brings an explorative prediction about the future. A future that considers better economic, social and environmental scenario for all should be extracted from a suitable approach and methodology. In this study, policymaking and regulations for the medical device industry was the main goal taking into account the politico-economic situation of Iran. As mentioned in the study, the National Medical Device Directorate of Ministry of Health and Medical Education is the regulator of medical devices in Iran. Policymaking for the devices market and for export of more medical devices by the country is the job of this governmental organization. An expert team consisting of top managers created an opportunity by applying a new developing approach in the multiple attribute decision-making (MADM) field, which is called PMADM. In policymaking research, PMADM has intended to cover new dimensions of policymaking and regulations in a real world case. As a flexible structure, PMADM has enough capacity to consider different new items in its structure. To cover all needs and prerequisites, new items were introduced by the expert team of National Medical Device Directorate of the country. This complicated regulatory decision was made for the new sanction periods from the United States against the economy of Iran.

According to the primary model of PMADM approach, the most critical Boosters and Limiters identified as follows:

B_1: Supportive countries, societies and unions;

- B_2: New markets (supportive);
- B_3: Considering an exceptional market;
- B_4: Future opportunities for using other currencies;
- B_5: New policies for exporting products easier;
- B_6: Tariff policy from both sides;

- L_1: Limited markets based on sanctions;
- L_2: International insurance;
- L_3: Conflict in the region (Middle East);
- L_4: New conflicts in future with the United States;
- L_5: Limitation for transferring money (mutual problem).

Two new tools intended for evaluating boosters and limiters are called "Max capacity" and "Ideal directed scenario". These two new tools for evaluating the situations based on boosters and limiters can prepare a more comprehensive perspective about the future that we are trying to achieve. A general perspective can lead the policymakers for a bigger picture of what they are looking for that and eventually better and more reliable decisions can be made based on them in reality, undoubtedly.

A new another item has introduced, in the study, which is called "supportive-backup" criteria. As mentioned in this chapter, instead of current/considered criteria, add a new criterion or add a sub-criterion for current/considered criteria. It means based on different future scenarios, both predictable and unpredictable, a picture (at least) can be imagined for the future. Criteria can be in highest flexibility level in comparison to the classic style of criteria before. "Supportive-backup" criteria can be determined as a criteria for somehow managing uncertainty or open boxes (empty criterion) to be prepared for what can happen as a wildcard scenario. This item can change the future of PMADM about all issues related to the criteria for evaluating alternatives and it can be intended as a new window to the future needs of policymakers and regulators in any level of decision-making. National Medical Device Directorate considers developing the exports of medical devices of the country by applying good policies and this study illustrated that there is a bright future and perspective about the reality of the industry while considering all limited choices. From other point of view, methodological perspective, PMADM has been developed and advanced by applying new perspectives and visions to the real world. The mentioned items can prepare more possibilities for have better decisions and policies in different complicated future-based studies.

References

African Biomedical Engineering Consortium (2018). http://abec-africa.org [verified on August 2018].

Akdag, H., Kalayci, T., Karagoz, S., Zulfikar, H. and Giz, D. (2014). The evaluation of hospital service quality by fuzzy MCDM. *Appl Soft Comput*, 23, 239–248.

Arms, H., Wiecher, M. and Kleiderman, V. (2012). Dynamic models for managing big decisions. *Strategy Leadership*, 40(5), 39–46.

Billions, R. (2013). Regulatory medical writing in Switzerland. *Med Writing*, 22(2), 119–120.

De Maria, C., Di Pietro, L., Díaz Lantada, A., Madete, J., Ngaju Makobore, P., Mridha, M., Ravizza, A., Torop, J. and Ahluwalia, A. (2018). Safe innovation: On medical device legislation in Europe and Africa. *Health Policy Technol*, 7, 156–165.

DiMasi, J. A. and Grabowski, H. G. (2007). The cost of biopharmaceutical R&D: Is biotech different? *Manag Decis Econ*, 28, 469–479.

Dinarvand, R. (2009). New national drug policy in Iran leading to expanded pharmaceutical market and extended access of public to medicines. *Iran J Public Health*, 38(1), 158–161.

Durbach, I. N. and Stewart, T. J. (2012). Modeling uncertainty in multi-criteria decision analysis. *Eur J Oper Res*, 223(1), 1–14.

International medical device regulators forum (2018). http://www.imdrf.org [verified on August 2018].

Gonzalez-Prida, V., Viveros, P., Barbera, L. and Crespo Marquez, A. (2014). Dynamic analytic hierarchy process: AHP method adapted to a changing environment. *J Manufact Technol Manage*, 24(4), 457–475.

Hashemkhani Zolfani, S., Aghdaie, M. H., Derakhti, A., Zavadskas, E. K. and Morshed Varzandeh, M. H. (2013). Decision-making on business issues with foresight perspective; an application of new hybrid MCDM model in shopping mall locating. *Expert Syst Appl*, 40(17), 7111–7121.

Hashemkhani Zolfani, S., Maknoon, R. and Zavadskas, E. K. (2016a). Multiple attribute decision-making (MADM) based scenarios. *Int J Strat Prop Manage*, 20(1), 101–111.

Hashemkhani Zolfani, S., Maknoon, R. and Zavadskas, E. K. (2016b). An introduction to prospective multiple attribute decision-making (PMADM). *Technol Econ Develop Econ*, 22(2), 309–326.

Hashemkhani Zolfani, S. (2018). *Futures Studies based on Decision-making Methods*, PhD thesis, Amirkabir University of Technology.

Hwang, C. L. and Yoon, K. (1981). *Multiple Attribute Decision-Making: A State-of-the-Art Survey*, Lecture Notes in Economics and Mathematical Systems, vol. 186. Springer-Verlag, Berlin.

Jassbi, J. J, Ribeiro, R. A. and Varela, L. R. (2014). Dynamic MCDM with future knowledge for supplier selection. *J Decis Syst*, 23(3), 232–248.

Keršulienė, V., Zavadskas, E. K. and Turskis, Z. (2010). Selection of rational dispute resolution method by applying new step-wise weight assessment ratio analysis (SWARA). *J Bus Econ Manage*, 11(2), 243–258.

Keshavarz Ghorabaee, M., Zavadskas, E. K., Olfat, R. and Turskis, Z. (2015). Multi-criteria inventory classification using a new method of evaluation based on distance from average solution (EDAS). *Informatica*, 26(3), 435–451.

Keshavarz Ghorabaee, M., Zavadskas, E. K., Turskis, Z. and Antucheviciene, J. (2016). A new combinative distance-based assessment (CODAS) method for multi-criteria decision-making. *Econ Comput Econ Cybernet Studies Res*, 50(3), 25–44.

Kondo, H. (2014). Revision of pharmaceutical affairs law (PAL) and implementation of PMD act. In *2nd Joint Conference of Taiwan and Japan on Medical Products Regulation*, Tokyo, 2014.

Madsen, K. (2018). *Entering the Iranian Market with Software as a Medical Device*. FORCE Technology Report.

Matsushita, S., Tachibana, K. and Kondoh, M. (2018). The clinical innovation network: A policy for promoting development of drugs and medical devices in Japan. *Drug Discovery Today*, https://doi.org/10.1016/j.drudis.2018.05.026.

National Medical Device Directorate (2018). *Report on Current Status of Medical Devices in Iran*. National Medical Device Directorate, Ministry of Health and Medical Education (in Persian).

Ondrus, J., Bui, T. and Pigneur, Y. (2014). A foresight support system using MCDM methods. *Group Dec Negotiat*, 24(2), 333–358.

Opricovic, S. (1998). *Multiple Criteria Optimization of Civil Engineering Systems*. Faculty of Civil Engineering, Belgrade (in Serbian).

Paul, S. M., Mytelka, D. S., Dunwiddie, C. T., Persinger, C. C., Munos, B. H., Lindborg, S. R. and Schacht, A. L. (2010). How to improve R&D productivity: The pharmaceutical industry's grand challenge. *Nat Rev Drug Disc*, 9(3), 203–214.

Peijia, R., Zeshui, X., Huchang, L. and Xiao-Jun, Z. (2017). A thermodynamic method of intuitionistic fuzzy MCDM to assist the hierarchical medical system in China. *Inform Sci*, 420, 490–504.

Rezaei, J. (2015). Best-worst multi-criteria decision-making method. *Omega*, 53, 49–57.

Saaty, T. L. (1980). *The Analytic Hierarchy Process*. McGraw-Hill, New York.

Saaty, T. L. (1996). *Decision-making with Dependence and Feedback: The Analytic Network Process*. RWS Publications, Pittsburgh.

Steinberg, D., Horwitz, G. and Zohar, D. (2015). Building a business model in digital medicine. *Nat Biotechnol*, 33(9), 910–920.

Tadic, S. R., Zecevic, S. M. and Krstic, M. D. (2014). Ranking of logistics system scenarios for central business district. *Promet — Traffic Transport*, 26(2), 159–167.

Taghipour, S., Banjevic, D. and Jardine, A. K. S. (2011). Prioritization of medical equipment for maintenance decisions. *J Oper Res Soc*, 62(9), 1666–1687.

Tsai, W. W., Tu, P. W., Lin, H. H., Wang, Y. H., Wu, S. I. and Fan, Y. T. (2017). A comparison study of Taiwan regulation and GHTF regulatory model on *in vitro* diagnostic medical devices. *Expert Rev Med Dev*, 14(4), 285–296.

Trutnevyte, E., Stauffacher, M. and Scholz, R. W. (2012). Linking stakeholder visions with resource allocation scenarios and multi-criteria assessment. *Eur J Oper Res*, 219(3), 762–772.

Wang, Y., Shi, X., Sun, J. and Qian, W. (2014). A grey interval relational degree-based dynamic multi-attribute decision-making method and its application in investment decision-making. *Mathematical Problemsin Engineering*, Article ID 607016, 1–6.

Wu, Y. H., Li, F. A., Fan, Y. T. and Tu, P. W. (2016). A study of medical device regulation management model in Asia. *Expert Rev Med Dev*, 13(6), 533–543.

Zavadskas, E. K. and Kaklauskas, A. (1996). Determination of an efficient contractor by using the new method of multicriteria assessment. In D. A. Langford, A. Retik (Eds.), *International Symposium for "The Organisation and Management of Construction". Shaping Theory and Practice 2: Managing the Construction Project and Managing Risk.* CIB W 65. London, Weinheim, New York, Tokyo, Melbourne, Madras — London: E and FN SPON, pp. 94–104.

Zavadskas, E. K., Turskis, Z. and Vilutiene, T. (2010). Multiple criteria analysis of foundation instalment alternatives by applying additive ratio assessment (ARAS) method. *Arch Civil Mech Eng*, 10(3), 123–141.

Zavadskas, E. K., Turskis, Z., Antucheviciene, J. and Zakarevicius, A. (2012). Optimization of weighted aggregated sum product assessment. *Electron Electr Eng*, (6), 3–6.

Zhang, Z. (2012). An approach to dynamic multi-attribute decision-making for choosing green supplier. *J Conver Inform Technol*, 7(21), 261–269.

Chapter 4

Indicators of Sustainable Agriculture: An Evaluation Using Intuitionistic Fuzzy DEMATEL Method

Lazim Abdullah

School of Informatics and Applied Mathematics,
Universiti Malaysia Terengganu, Malaysia
lazim_m@umt.edu.my

A sustainable agricultural practice (SAP) is vital to preserving the environment, and at the same time, this practice would increase agricultural productivity. The importance of SAP and its indicators is indisputable. However, the importance of one indicator over other indicators of sustainable agriculture is not conclusive. Moreover, the multiple indicators of environment, economic and social aspects of agriculture sustainability are vague and interrelated. This research aims to investigate the interdependence among indicators that would affect the SAP. Ten indicators of SAP are selected and analyzed using the integration of intuitionistic fuzzy sets (IFS) and decision-making trial and evaluation laboratory (DEMATEL) method. The IFS-DEMATEL method is known as an effective tool to investigate the interrelationships among indicators via a causal diagram and digraph. Unlike fuzzy DEMATEL, the proposed IFS-DEMATEL used a centroid method to defuzzify membership, non-membership and hesitancy values of IFS into a single crisp score. The causal diagram reveals that five indicators are classified as the cause group while five other indicators are classified as the effect group. The highest value of $(D + R)$ indicates that "cultivation" is the most important indicator of agriculture sustainability. The findings of this study could provide a valuable reference for developing policies related to SAP.

1. Introduction

Since the beginning of the first agricultural revolution 8,000–10,000 years ago, growing crops and breeding livestock have been the prime factors contributing to the loss and degradation of natural ecosystems (Millennium Ecosystem Assessment, 2005). The situation is getting worse with the long-term utilization of conventional agriculture and chemicals-based fertilization farming to increase productivity. Most of the agriculture-based countries have been facing a huge challenge with regard to the sustainable use of soils. Sustainability performance in agriculture such as soil erosion and degradation has threatened the availability and productivity of land for growing food (Reytar *et al.*, 2014). Extensive use of chemical-based fertilizers, particularly phosphorus and nitrogen, has led to soil and water pollution as well as the emissions of harmful greenhouse gases, especially nitrous oxide (Reytar *et al.*, 2014; Gupta *et al.*, 2015). Overuse of pesticide in preventing crop losses to insects and other pests has brought detrimental effects to water quality, wildlife, human health and other environmental factors. These detrimental effects very much depend on the toxicity of the component chemicals and the application conditions (Reytar *et al.*, 2014). The consequences of such negative environmental impacts have become one of the serious issues faced by many countries. It has been said that about 13% of global anthropogenic greenhouse gas emissions came from agricultural production such as the use of fertilizers, manure, ruminants, rice and on-farm energy use.

Nevertheless, these environmental impacts could be reduced by establishing sustainable agricultural practices (SAPs). The SAP is one of the most current empirically proven technologies that are considered crucial for efficient agriculture production (Kottegoda *et al.*, 2011). A way to curb the risk of human economy from overstraining the limit of the Earth's ecosystem is to approach the concept of a "green economy". This approach has contributed to a major increase in production and at the same time has aided in reducing environmental impacts. A careful intervention is essential in investigating the importance of multiple indicators that contribute to SAP.

It is hypothesized that these multiple indicators would contribute to SAP. Moreover, the relationship among multiple indicators is not fully investigated and it would be difficult to recognize the most dominant indicators of SAP. All these issues must be fairly deal with multicriteria approaches so as to practice sustainable agriculture where optimum benefit could be reaped from the environmental and economic aspects. However, selecting the important indicator in SAP is not a straightforward process as the selection is bounded to a set of multiple indicators, and at the same time, these indicators are dependent on each other. In other words, investigation of indicators in agriculture practices is a multicriteria decision-making (MCDM) approach where the right criteria are heavily dependent on the degree of importance out of many criteria. There are many MCDM methods available in the literature, and most of these methods considered the hierarchical structure of aim, criteria and alternatives. The analytic hierarchy process (AHP), for example, handles MCDM problems where the structure of alternatives, criteria and aim are considered concurrently. Unlike the AHP, the DEMATEL is one of the MCDM methods that only considered criteria and aim. With the aim to draw relationship among multiple criteria, the DEMATEL is germane to MCDM problems including the indicators in SAP.

However, indicators in SAP are sometimes vague and uncertain. For example, crop demand is unpredictable and its behavior is uncertain due to incomplete information. Therefore, many versions of fuzzy set such as intuitionistic fuzzy set (IFS) is fused into MCDM methods to handle fuzziness and uncertainty. The IF-DEMATEL method is one of the many fusions, where IFS is used in linguistic assessment of DEMATEL. Based on these premises, ten indicators related to the sustainable agriculture and environment, decision makers' (DMs) opinion or judgments, the DEMATEL method and IFS are the main frameworks of this study. Specifically, the purpose of this study is to identify the cause indicators and the effect indicators in managing SAP. In addition, the direction of relationships between indicators and degree of importance of indicators are also investigated.

2. Literature Review

Related research that used the DEMATEL, the fuzzy DEMATEL, the IF-DEMATEL and sustainable agriculture are the main highlight in this section. The Science and Human Affairs Program of the Battelle Memorial Institute of Geneva proposed the DEMATEL method between 1972 and 1976 (Gabus and Fontela, 1972). The method has been proven in resolving the complex and interrelated problems. The DEMATEL and its extensions are then successfully evolved and used in various fields for evaluation purposes such as auto spare parts industry (Wu and Tsai, 2011), medical tourism development (Chen, 2012), event quality (Yang *et al.*, 2013), coastal erosion (Najib *et al.*, 2017), green supply chain management (Wu and Chang, 2015) and knowledge management (Abdullah and Zulkifli, 2018). In another research, a study proposed by Lin *et al.* (2011) adopted the DEMATEL method to figure out the core competencies and causal effect of the integrated circuit design service company. This study interprets the interrelation among the core competencies using the DEMATEL method. Seven core competencies were selected and divided into causal and effect groups, allowing the readers to have a better understanding of the interactive relationship between them. A study proposed by Gandhi *et al.* (2015) has utilized the DEMATEL method to evaluate the influential factors in the implementation of successful green supply chain management. The DEMATEL method enables readers to study the interrelationship between the factors via a causal diagram. Bacudio *et al.* (2016) have suggested that the DEMATEL method is an effective tool used to analyze the barriers in implementing industrial symbiosis networks. The DEMATEL method is utilized to identify the cause–effect relationships among several factors in industrial networks.

The integration of fuzzy sets and DEMATEL also stimulated several application research. Chang *et al.* (2011) integrated the DEMATEL method with fuzzy theory in developing supplier selection criteria. The critical criteria to improve the performance in supplier performance were identified as well as a decision-making information for supply chain management supplier management was suggested using the DEMATEL method. The fuzzy theory was introduced for obtaining a more flexible analysis during the data collection. The results revealed that stable delivery of goods is the most influencing criteria as well as the highest correlation with other criteria. Integration with the DEMATEL is not only limited to the fuzzy sets. There are many cases where the DEMATEL is integrated with other MCDM methods. For example, Lin *et al.* (2016) integrated the DEMATEL method with VIKOR and ANP in improving the project risk management. In their study, a research framework was set up by the stimulus-organism-response framework by using DANP (DEMATEL-based ANP) and VIKOR methods to evaluate the relationship in project risk, project management and organizational performance.

As a generalization to fuzzy sets, the intuitionistic fuzzy sets (IFS) are also integrated with the DEMATEL. The IFS is applied in the DEMATEL method to minimize the linguistic imprecision and the uncertainty of human judgment. A study proposed by Vafadarnikjoo *et al.* (2016) utilized a hybrid approach of DEMATEL method based on IFS in prioritizing the selection criteria of bank branch locations. In order to satisfy the bank customers and stay competitive, optimally locating new bank branches is a strategic decision in the banking industry. In this research, six most widely used criteria were obtained from the literature to evaluate the Iranian bank branch location in the city of Rasht, Iran. In order to prioritize these criteria, the DEMATEL method was used to find out the interrelationships between criteria, and IFS was used to deal with the imprecision of subjective judgments of DMs. A study conducted by Govindan *et al.* (2015) developed green practices and performances in a green supply chain using the DEMATEL method based on IFS. By observing the causal diagram computed through the DEMATEL method, internal management support, green purchasing and internal management support were the most fundamental green supply chain practices that must be considered for improving both environmental and economic performances. A study proposed by Vafadarnikjoo *et al.* (2016) utilized intuitionistic fuzzy-based DEMATEL method in ranking the risk of construction projects. The IFS was applied in handling the uncertainty and ambiguity of human subjective judgments. Buyukozkan *et al.* (2017) suggested a new combination approach of IF-DEMATEL and IF-ANP in evaluating a customer relationship management (CRM) partner. Selecting the most suitable business partner is a strategic decision for outsourcing problems. Intuitionistic fuzzy analytic network process (IF-ANP) is utilized in this study for constructing and interpreting the criteria weights while IF-DEMATEL is used to figure out the interrelations between criteria under uncertainty environment. This finding has contributed to the existing literature by establishing a new combined IFS-based DEMATEL-ANP framework and constructing a novel evaluation model for a real industrial problem in improving the CRM partner selection process. A research proposed by Rahman *et al.* (2017) adopted intuitionistic fuzzy DEMATEL method in developing flood mitigation measures to minimize the impact of flood in Malaysia. Intuitionistic fuzzy DEMATEL method was utilized in dealing the ambiguity of input data set of flood mitigation measures and validating the results using sensitivity analysis.

A research study conducted by Veisi *et al.* (2016) utilized a decision analysis method to develop an ethical strategy of sustainable agriculture in Iran. There were nine criteria or indicators selected for sustainable agriculture development based on three ethical models, which are utilitarian-based, rights-based and virtue-based. The findings revealed that the indicators "supportive policies and self-reliance" and "resilience of agricultural systems and equity" were the most fundamental indicators for sustainable agriculture. In another study, Mohammed Shobri *et al.* (2016) conducted a study on identifying and improving good agriculture practice in Malaysia. Nutrient management, bio-diversity, energy, water management, pest management and soil health were selected as indicators in good agriculture practice based on the most frequently appeared indicators from the past research and study. It is believed that these six selected indicators need to be improved in order to have a better quality of life. The indicators are irrigation water, groundwater, crop planting pattern and weather conditions, which were the indicators for sustainable agricultural development (Chen *et al.*, 2018). It was concluded that groundwater plays a significant role in sustainable agricultural development in arid and semiarid areas.

3. Preliminaries

In this section, several mathematical definitions related to fuzzy sets and IFS are presented. These definitions are very much relevant with the proposed method that will be fully described in Section 4.

3.1. *Definitions of fuzzy sets and IFS*

As the exact data are literally difficult to determine due to impreciseness and vagueness of human judgements in many situations and conditions, the complexity of the problem also increases. Thus, the effective ways to deal with these uncertainties in data are using fuzzy set theory of linguistic terms and degrees of membership (Yager, 2009).

A fuzzy set is a class of objects with grades of membership and these grades show the degree of stability in which a certain element belongs or does not belongs to a fuzzy set (Yager, 2009). A fuzzy set A is defined as

$$A = \{(x, \mu_A(x)) | x \in A, \mu_A(x) \in [0, 1]\},$$

where $\mu_A(x)$ is a membership function belongs to the interval [0, 1]. The range between 0 and 1 is referred to as the membership grade or degree of membership.

The fuzzy set was initiated by Zadeh (1965), Since then, several new extensions and generalization have been proposed. One of the extensions to the fuzzy set is IFS theory. The IFS was proposed by Atanassov (1986). An IFS is one of the nonstandard fuzzy sets that have been found to be particularly useful in dealing with uncertainty. It is defined mathematically as follows.

Let X be a fixed set, an intuitionistic fuzzy set A in X is given as

$$A = \{\langle x, \mu_A(x), v_A(x)\rangle | x \in X\},$$

where $\mu_A(x) : X \to [0,1]$ and $v_A(x) : X \to [0,1]$ denote the membership function and non-membership function respectively, satisfying $0 \leq \mu_A(x) + v_A(x) \leq 1, \forall x \in X$.

The numbers $\mu_A(x)$ and $v_A(x)$ are degrees of membership and nonmembership of the element x to A respectively, for all $x \in X$. Additionally, $\pi_A(x) = 1-\mu_A(x)-v_A(x)$ is called the hesitation degree of $x \in A$, representing the degree of indeterminacy (uncertainty) or the degree of hesitancy of x to A with $0 \leq \pi_A(x) \leq 1, \forall x \in X$.

3.1.1. *Operational laws of intuitionistic fuzzy sets*

For any three IFNs $\alpha_1 = (\mu_{\alpha_1}, v_{\alpha_1})$, $\alpha_2 = (\mu_{\alpha_2}, v_{\alpha_2})$ and $\alpha = (\mu_\alpha, v_\alpha)$, the following operational laws are introduced (Atanassov, 1986; Xu and Yager, 2006):

(1) $\alpha_1 + \alpha_2 = (\mu_{\alpha_1} + \mu_{\alpha_2} - \mu_{\alpha_1}\mu_{\alpha_2}, v_{\alpha_1}v_{\alpha_2})$,
(2) $\alpha_1 \times \alpha_2 = (\mu_{\alpha_1}\mu_{\alpha_2}, v_{\alpha_1}v_{\alpha_2} - v_{\alpha_1}v_{\alpha_2})$,
(3) $\lambda_\alpha = (1-(1-\mu_\alpha)^\lambda, v_\alpha^\lambda), \lambda > 0$,
(4) $\alpha^\lambda = (\mu_\alpha^\lambda, 1-(1-v_\alpha)^\lambda), \lambda > 0$.

All these algebraic operations will be utilized in implementing the computational procedures of the case study.

The methodology of this research is described in the following section.

4. Methodology

This section is presented into two subsections. A brief profile of DMs and linguistic variables used in this study are described in first subsection. As the focal point of this study, the proposed modification of IF-DEMATEL method is presented afterward.

4.1. *Decision makers and linguistic variables*

Elicitation of DMs' opinions is the most commonly used method of data collection in decision-making approaches. This method of data collection could provide a solution particularly in the circumstances where information about data is limited and vague. In this research, agriculture officers from State Department of Agriculture, Terengganu and academicians attached to a public university in Malaysia were invited and assigned as the DMs. Two agriculture officers were assigned as the DMs for data collection as they are more knowledgeable and routinely engaged with farmers that practicing sustainable agriculture. Two academicians attached to a public university were chosen to be part of the DMs group owing to their knowledge in agriculture sustainability. All these lecturers are part of the teaching and research force for the programs of Bachelor of Agrotechnology (Crop Science) and Bachelor of Agrotechnology (Post Harvest Technology), where agriculture sustainability is one of the courses of the bachelor programs.

Group of DMs were asked to complete a simple survey forms through personal communication. The first part of the survey form evaluates the interrelationship between indicators that would affect agriculture sustainability, followed by the importance of each indicator. All these evaluations are made using the linguistic variables. The use of linguistic variable allows researchers to indicate the importance of the indicator compared to other indicators (pair-wise comparison) as well as the preference against the number of indicators, which could impact the SAP. For instance, the ratings of importance of indicators compared to other indicators could be expressed using linguistic variable such as "Very Important", "Important", "Quite important", etc. In addition, the values of linguistic variables could be quantified and extended to mathematical operators using IFS.

4.2. *Proposed IF-DEMATEL method*

The proposed algorithm of IF-DEMATEL in this study is mainly based on the work of Buyukozkan *et al.* (2017). However, some modifications are made as to ease the computation. The first modification is made on the step of defuzzification where the transformation equation is used. This modification leads to the values of $D + R$. Unlike the classical DEMATEL where the values of $D + R$ are positive, the proposed IF-DEMATEL may give negative values for $D + R$. The second modification is made on transforming IFS to a crisp set. An

IFS is transformed into a crisp set using a centroid method instead of coefficient risk preference, which normally used in most of the DEMATEL (Varghese and Kuriakose, 2012). Therefore, the equation is replaced with the centroid method to defuzzify the membership and nonmembership, hesitation values into a single score (Varghese and Kuriakose, 2012). The last modification happens at the step of constructing digraph. In the proposed method, the threshold equation proposed by Bacudio *et al.* (2016) replaces the simple average equation in the classical DEMATEL (Varghese and Kuriakose, 2012). This move would enhance the construction of digraph by setting up the inner dependence matrix. Detailed computational procedures of the proposed method are presented as follows.

Step 1: Define the indicators

Finite number of indicators $C_1, C_2, C_3, \ldots, C_n$ are defined.

Step 2: Define the linguistic variables "influence"

Step 3: Determine the weight of DMs

Each DM has a specific weight due to their varied working experience and knowledge.

Assuming that $\lambda_k = (\mu_k, v_k, \mu_k)$ is an IFS for relative importance weights of kth DM. The priority values of kth DM are computed as follows (Boran *et al.*, 2009):

$$\lambda_k = \frac{\left(\mu_k + \pi_k\left(\frac{\mu_k}{\mu_k + v_k}\right)\right)}{\sum_{k=1}^{l}\left(\mu_k + \pi_k\left(\frac{\mu_k}{\mu_k + v_k}\right)\right)} \tag{1}$$

where $\sum_{k=1}^{l} \lambda_k = 1$.

Step 4: Construct direct-relations matrix

DMs generate a $n \times n$ direct-relation matrix A^k by making a pairwise comparisons between indicators where n is the total number of indicators. Each direct-relation matrix A^k representing the opinion of each DM in evaluating the interrelationship between indicators. The direct-relation matrix is arranged in the way as shown in equation (2).

$$A^k = \left[a_{ij}^k\right]_{n\times n} = \begin{bmatrix} 0 & a_{12}^k & \cdots & a_{1n}^k \\ a_{n1}^k & 0 & \cdots & a_{2n}^k \\ \vdots & \vdots & \ddots & \vdots \\ a_{n1}^k & a_{n2}^k & \cdots & 0 \end{bmatrix}, \tag{2}$$

where a_{ij}^k is the rating by kth DM in the evaluation of interrelationship between indicators.

Step 5: Aggregate the DM's individual assessment

Each DM's individual opinions need to be aggregated to construct a collective intuitionistic fuzzy decision matrix. Let $a_{ij}^k = (\mu_{ij}^k, v_{ij}^k, \pi_{ij}^k)$ be the IFN given by kth DM on the assessment of indicator i on j where $\pi_{ij}^k = 1 - \mu_{ij}^k - v_{ij}^k$ (degree of hesitancy). Then, by using the intuitionistic fuzzy weighted average (IFWA) operator as shown below, the aggregated intuitionistic fuzzy ratings x_{ij} represents the influence level of indicator i on j are obtained.

$$\begin{aligned} a_{ij} &= IFWA(a_{ij}^1, a_{ij}^2, \ldots, a_{ij}^l) = \sum_{k=1}^{l} \lambda_k a_{ij}^k \\ &= \left(1 - \prod_{k=1}^{l} (1-\mu_{ij}^k)^{\lambda_k}, \prod_{i=1}^{l} (v_{ij}^k)^{\lambda_k}, \prod_{k=1}^{l} (1-\mu_{ij}^k)^{\lambda_k} \right. \\ &\quad \left. - \prod_{i=1}^{l} (v_{ij}^k)^{\lambda_k}\right) i,j \in \{1,2,3,\ldots n\}, \end{aligned} \tag{3}$$

where λ_k is the importance weight of kth DM, a_{ij}^k is the corresponding of intuitionistic fuzzy number of kth DM's opinion when comparing i to j.

Step 6: Construct the normalized aggregated direct-relation matrix

The normalized aggregated matrix of membership, nonmembership and hesitancy values are obtained using equation (4).

$$X = k \times A \tag{4}$$

where

$$k = \min\left(\frac{1}{\max\sum_{j=1}^{n} a_{ij}}, \frac{1}{\max\sum_{i=1}^{n} a_{ij}}\right), \; i, j \in \{1, 2, 3, \ldots n\}.$$

Step 7: Construct the total-relation matrix, T

This matrix can be obtained using equation (5).

$$T = X(I - X)^{-1}, \tag{5}$$

where I is the identity matrix.

From the total relation matrix (T) formula, we are able measure the combined impact of every indicator that is influenced by other indicators directly and indirectly.

Step 8: Transform both membership and non-membership values into a single score.

The centroid method is used to defuzzify both the membership and non-membership values in total relation matrices into a crisp number (see equation (6)).

$$p(x) = \frac{\mu - v + 1}{2}, \tag{6}$$

where p is a fuzzy number, μ is the membership values in total relation matrix, v is the non-membership values in total relation matrix.

Step 9: Construct the inner dependence matrix

The threshold value is computed using equation (7) so as to construct the inner dependence matrix.

$$\alpha = \frac{\sum_{j=1}^{n}\sum_{i=1}^{n} t_{ij}}{n^2}, \tag{7}$$

where α is the threshold value, t_{ij} is the total relation matrix of crisp value, n is the total number of indicators.

Step 10: Draw a digraph

A digraph of the causal–prominence relationship is constructed based on the matrix in Step 9.

Step 11: Obtain cause and effect diagram

The $(D + R)$ and $(D - R)$ values were computed by applying equations (8)–(10) with the aid of Microsoft Excel and MATLAB software. Summation of rows and columns of the total relation matrix produced vectors D and R, respectively.

$$T = [t_y]_{n\times n},\ \ i, j = 1, 2, \ldots, n, \tag{8}$$

$$R = \left[\sum_{i=1}^{n} t_y\right]_{1\times n} = [t_j]_{1\times n}, \tag{9}$$

$$D = \left[\sum_{j=1}^{n} t_y\right]_{n\times 1} = [t_i]_{n\times 1}. \tag{10}$$

The $(D + R)$, which is the horizontal axis vector named "Prominence", shows the importance of an indicator while $(D - R)$, which is the vertical axis called "Relation" is made. Based on the coordinate positions of $(D + R, D - R)$, indicators could be categorized into four types (Tsai *et al.*, 2015):

(1) **The $(D - R)$ is positive and $(D + R)$ is large:** The indicators are causes, which are also the driving factors for resolving problems.
(2) **The $(D - R)$ is positive and $(D + R)$ is small:** The indicators are independent and could influence only a few other indicators.
(3) **The $(D - R)$ is negative and $(D + R)$ is large:** The indicators are the core problems that must be solved but they are the effect type indicators, which could not be directly improved.
(4) **The $(D - R)$ is negative and $(D + R)$ is small:** The indicators are independent and could be influenced by only a few other indicators.

Basically, an indicator is considered as a member of the cause category if $(D-R)$ is positive. If $(D-R)$ is a negative value, the indicator is considered as effect group. Thus, the causal diagram is illustrated by mapping the data set of the $(D+R, D-R)$ in two planes, which provide us with the understanding about the behaviour and recognition of the entire system.

5. Results

In this section, the computational steps of IF-DEMATEL method using the linguistic data are presented.

Based on the literature review, ten indicators are chosen for this study. They are water tariff (C_1), cultivation (C_2), profitability (C_3), crop demand (C_4), water availability (C_5), water quality (C_6), soil texture (C_7), environmental condition (C_8), nutrient use efficiency (C_9) and production stability (C_{10}).

Based on the values of the inner dependence matrix, a digraph of the causal–prominence relationship are constructed. Figure 1 presents the directional relations of indicators in digraph.

The arrow shows the inter-relationships between the two indicators in implementing SAP. In the digraph, a dotted red line represents a uni-directional relationship while a solid blue line represents a bi-directional relationship.

The values of $(D+R)$ are called as prominence where the most important indicators can be identified based on their prominent values. Figure 2 shows the prominence values of each indicator.

The values of $(D+R)$ are represent the degree of influences of indicators. It can be seen that cultivation (C_2) has the highest value of $(D+R)$ followed by production stability (C_{10}). This indicates that cultivation (C_2) has the highest influence than other indicators and must be taken into consideration for SAPs. The next three highest degrees of influence are nutrient use efficiency (C_9), water availability (C_5) and profitability (C_3). It is also good to note that water tariff (C_1) is the least degree of influence of SAP indicator.

Values of $D+R$ and $D-R$ are plotted in a Cartesian plan. The $D+R$ axis divides the indicators into two groups, which are to be

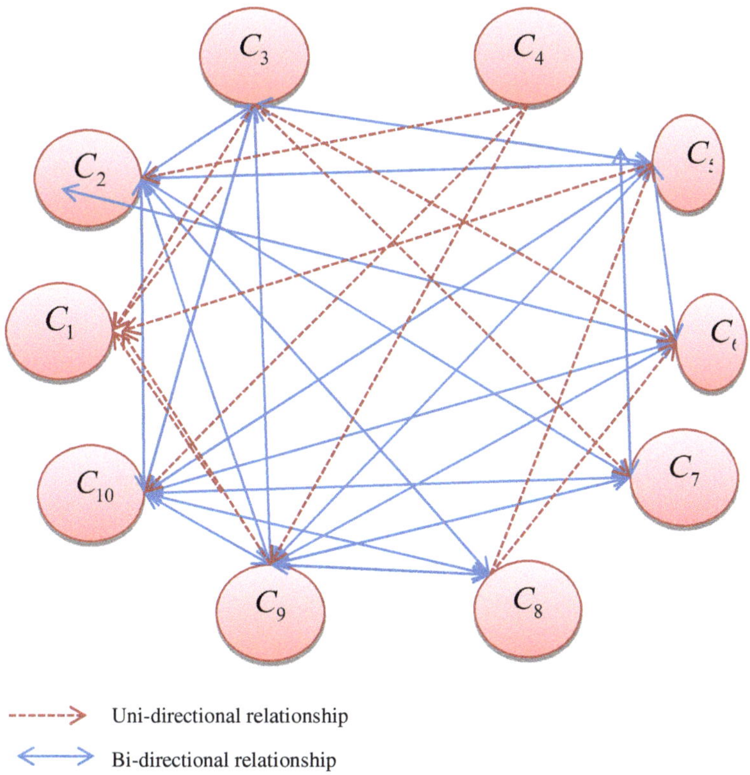

Figure 1: The digraph of the causal–prominence relationship.

called as cause group and effect group. Figure 3 shows the causal diagram.

It is shown that profitability (C_3), crop demand (C_4), soil texture (C_7), environmental condition (C_8) and nutrient use efficiency (C_9) are classified as the causal group. They should be given more priority as they direct impact the entire system and their performances highly affect the sustainability of agriculture. On the contrary, water tariff (C_1), cultivation (C_2), water availability (C_5), water quality (C_6) and production stability (C_{10}) are classified as the effect group. This suggests that they are easily influenced by other indicators. Indicators in the effect group are tended to be easily influenced by other indicators. This group of indicators does not have a direct impact on

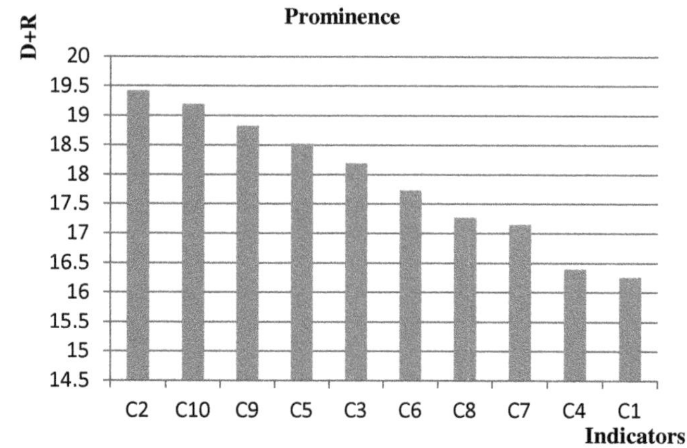

Figure 2: The prominent value of each indicator.

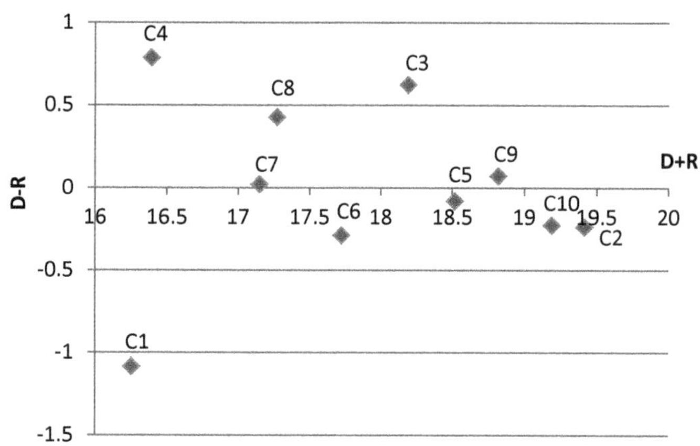

Figure 3: Causal diagram.

the system; however, it could provide a significant contribution as to cushion the affect.

6. Conclusions

Sustainable agriculture is a growing practice nowadays as it is very significant to the health and welfare of the people. SAP entails several indicators. However, the importance of an indicator over other indicators is not immediately known. Some might believe that, for example, the indicator water tariff is more important than the indicator environmental condition. Some might not agree with this opinion. Therefore, the importance of each indicator must be measured using a multi-indicator approach. Furthermore, it is believed that there is some extent of relationship between the indicators. The environmental condition, for example, may directly influence water tariff. Therefore, this paper aims to identify the cause and effect indicators in the relationship of SAP. To meet this objective, a multicriteria decision analysis, DEMATEL and IFS, is used in this analysis. The ambiguity and vague of criteria are dealt with IFS. The integrated IF-DEMATEL is meant to reduce the ambiguity of DM in data collection and also to evaluate the indicators. Ten indicators of SAP are evaluated using the 12-step computational procedures of the IF-DEMATEL. The results reveal that profitability, crop demand, soil texture, environmental condition and nutrient use efficiency are the causal group, which should be given more attention. On the other hand, water tariff, cultivation, water availability, water quality and production stability are the effect group that needs to be improved. It is good to note here that "cultivation" is the most prominent indicator in SAP. However, this research is far from complete and could be further improve in future research. Other multicriteria approaches plus interval value IFS is one of the possible methods that could be applied as interval values are said to be more comprehensive than nonintervals do. Moreover, a comparative analysis between DEMATEL based on set with single value and DEMATEL with interval values could be undertaken in future research.

Acknowledgments

The author would like to thank Miss Szeying Lau for her contributions in data collection and analysis.

References

Abdullah, L. and Zulkifli, N. (2018). A new DEMATEL method based on interval type-2 fuzzy sets for developing a causal relationship of knowledge management criteria, *Neural Comput Appl*, 1–17, Online first, DOI: .1007/s00521-017-3304-1.

Atanassov, K. (1986). Intuitionistic fuzzy sets. *Fuzzy Sets Syst* 20, 87–96.

Bacudio, L. R., Benjamin, M. F. D., Eusebio, R. C. P., Holaysan, S. A. K., Promentilla, M. A. B., Yu, K. D. S. and Aviso, K. B. (2016). Analyzing barriers to implementing industrial symbiosis networks using DEMATEL. *J Sustain Product Consump*, 7, 57–65.

Boran, E., Genc, S., Kurt, M. and Akay, D. (2009). A multi-criteria intuitionistic Fuzzy group decision making for supplier selection with TOPSIS method. *J Expert Syst Appl*, 36, 11363–11368.

Buyukozkan, G., Guleryuz, S. and Karpak, B. (2017). A new combined IF-DEMATEL and IF-ANP approach for CRM partner evaluation. *Int J Product Econ*, 191, 194–206.

Chang, B., Chang, C. and Wu, C. (2011). Fuzzy DEMATEL method for developing supplier selection criteria. *Expert Syst Appl*, 38, 1850–1858.

Chen, C. (2012). Using DEMATEL method for medical tourism development in Taiwan. *Am J Tour Res*, 1(1), 26–32.

Chen, H., Huo, Z., Dai, X., Ma, S., Xu, S. and Huang, G. (2018). Impact of agricultural water-saving practices on regional evapotranspiration: The role of groundwater in sustainable agriculture in arid and semi-arid areas, *Agri Forest Meteorol*, 263, 156–168.

Gabus, A. and Fontela, E. (1972). World Problems, An Invitation to Further Thought Within the Framework of DEMATEL. Battelle Geneva Research Centre, Geneva.

Ghandi, S., Mangla, S. K., Kumar, P. and Kumar, D. (2015). Evaluating factors in implementation of successful green management using DEMATEL: A case study. *J Int Strat Manage Rev*, 3, 96–109.

Govindan, K., Khodaverdi, R. and Vafadarnikjoo, A. (2015). Intuitionistic fuzzy based DEMATEL method for developing green practices and performances in a green supply chain. *Expert Syst Appl*, 42, 7207–7220.

Gupta, G., Parihar, S. S., Ahirwar, N. K., Snehi, S. K. and Singh, V. (2015). Plant growth promoting rhizobacteria (PGPR): Current and future prospects for development of sustainable agriculture. *J Microbiol Biochemi*, 7, 96–102.

Kottegoda, N., Munaweera, I., Madusanka, N. and Karunaratne, V. (2011). A green slow-release fertilizer composition based on urea-modified hydroxyapatite nanoparticles encapsulated wood. *J Curr Sci*, 101(1), 73–78.

Lin, C., Chen, Y. and Chuang, H. (2016). *Frontier Computing: Theory, Technologies and Applications*. Springer, Singapore.

Lin, Y., Yang, Y., Kang, J. and Yu, H. (2011). Using DEMATEL method to explore the core competences and causal effect of the IC design service company: An empirical case study. *Expert Syst Appl*, 38, 6262–6268.

Millennium Ecosystem Assessment (MEA) (2005). *Ecosystems and Human Well-Being: Synthesis*. Island Press, Washington, DC.

Mohammed Shobri, N. I., MdSakip, S. R. and Omar, S. S. (2016). Malaysian standards crop commodities in agricultural for sustainable living, *Procedia — Social and Behavioral Sciences*, 222, 485–492.

Najib, L., Ab Ghani, A. T., Abdullah, L. and Ahmad, M. F. (2017). An application of coastal erosion decision problem using interval type-2 fuzzy dematel method. *J Sustain Sci Manage*, 12(2), 204–217.

Rahman, N. A., Tarmudi, Z., Rossdy, M. and Muhiddin, F. A. (2017). Flood mitigation measures using intuitionistic fuzzy DEMATEL method. *Malay J Geosci*, 1(2), 1–5.

Reytar, K., Hanson, C. and Henninger, N. (2014). Indicators of sustainable agriculture: A scoping analysis. Working Paper of Installment 6 of Creating Sustainable Food Future. World Resources Institute, Washington, DC.

Tsai, B. S., Chien, M. F., Xue, Y., Li, L., Jiang, X., Chen, Q., Zhou, J. and Wang, L. (2015). Using the Fuzzy DEMATEL to determine environmental performance: A case of printed circuit board industry in Taiwan, *PLoS One*, 10(6), e0129153.

Vafadarnikjoo, A., Mobin, M. and Firouzabadi, S. M. A. K. (2016). An intuitionistic fuzzy based-DEMATEL to rank risks of construction projects. In *Proceedings of the 2016 International Conference on Industrial Engineering and Operations Management*. Detroit, Michigan, USA, pp. 1366–1377.

Varghese, A. and Kuriakose, S. (2012). Centroid of an intuitionistic fuzzy number. *Notes on Intuitionistic Fuzzy Sets*, 18(1), 19–24.

Veisi, H., Liaghat, H. and Alipour, A. (2016). Developing an ethics-based approach to indicators of sustainable agriculture using analytical hierarchy process (AHP). *Ecol Indicators*, 60, 644–654.

Wu, H. and Tsai, Y. (2011). A DEMATEL method to evaluate the causal relatins among the criteria in auto spare parts industry. *J Appl Math Comput*, 218, 2334–2342.

Wu, H. H. and Chang, S. Y. (2015). A case study of using DEMATEL method to identify critical factors in green supply chain management. *J Appl Math Comput*, 256, 394–403.

Xu, Z. S. (2007). Intuitionistic fuzzy aggregation operators. *IEEE Trans Fuzzy Syst*, 15(6), 1179–1187.

Xu, Z. S. and Yager, R. R. (2006). Some geometric aggregation operators based on intuitionistic fuzzy sets. *Int J General Syst*, 35(4), 417–433.

Yager, R. R. (2009). On decision-making with non-standard fuzzy subsets. *Int J Knowl Eng Soft Data Paradigms*, 1(1), 3–10.

Yang, W., Liu, W., Liu, H. and Lanasari, H. (2013). Evaluating influential factors in event quality using DEMATEL method. *An Int J Trade, Econ Fin*, 4(3), 92–97.

Zadeh, L. A. (1965). Fuzzy sets. *J Inform Control*, 8, 338–353.

Chapter 5

An Integrated Fuzzy Model for Solid Waste Management in Libya

Ibrahim Badi[*,§], Željko Stević[†,¶] and Siniša Sremac[‡,||]

[*]*Faculty of Engineering, Misurata University, 2429 Misurata, Libya*
[†]*Faculty of Transport and Traffic Engineering Doboj, University of East Sarajevo, Vojvode Mišića 52,74000 Doboj, Bosnia and Herzegovina*
[‡]*Faculty of Technical Sciences, University of Novi Sad, Trg Dositeja Obradovica 6, 21 000 Novi Sad, Serbia*
[§]*i.badi@eng.misuratau.edu.ly*
[¶]*zeljkostevic88@yahoo.com, zeljko.stevic@sf.ues.rs.ba*
[||]*sremacs@uns.ac.rs*

As in many developing countries, Libya is still inappropriately managing the solid waste. This has led to increase the amount of solid waste accumulated in the country. The objective of this study is to select an appropriate waste management technology for Libya. However, the selection of an appropriate waste management technology is a complex process, because it is necessary to consider a wide range of parameters. Best available techniques, technology quality method and multicriteria analysis were proposed as means of developing indices for evaluating municipal waste management systems. In this work, four treatment systems for waste management in Libya are evaluated using fuzzy analytic hierarchy process (AHP) and fuzzy technique for order preference by similarity to ideal solution (TOPSIS) in respect to four main criteria and 22 subcriteria. The treatment systems for waste management are anaerobic digestion, landfilling, incineration and compost. The selected criteria used in the evaluation of four treatment systems are environmental impacts, socio-cultural aspects, technical aspects and economic aspects. According to the evaluation, anaerobic digestion (AD) ranks the highest in classification in Libya. Compost ranks higher than landfilling and incineration. Sensitivity analysis, which consists of 20 different scenarios, shows that the proposed model is very stable. Furthermore, it should be noted that the rank of waste treatment systems can be changed according to the future technological developments.

1. Introduction

Today, in the conditions of rapid global changes, an adequate solid waste strategy, designed for underdeveloped and developing countries, needs to be proposed. The real challenge for these countries is to develop clear solid waste management plan, which is individually designed for each country, as it the same strategic option cannot be applicable for developed and economically advanced country as well as to the underdeveloped and developing countries.

Recent researches have shown that the selection of an adequate solid waste management strategy in underdeveloped and developing countries contributes to its optimization (Guerrero *et al.*, 2013; Sakurai, 2014). On the other hand, according to Hiraga (2007) the selection of optimal utility vehicles and other utility equipment contributes to the adequate and timely disposal of solid wastes. In the 21st century, the need for the implementation of the so-called 3R strategy (reduce, reuse and recycle) is growing, as it is necessary to reduce waste at the place of production, then to reuse it and the last option is to recycle (Novarlić *et al.*, 2017). That means there is a growing need for the recycling of waste, instead of burning it in the incineration plants or disposing of it in the open landfills.

The increasing volumes of waste generated and social environmental consciousness present prominent drivers for managers toward the achievement of a sustainable waste management system. Most often, to come up with the optimal solution among alternative scenarios, the decision makers require the consideration of a significant number of usually conflicting criteria. Therefore, in this chapter, a total number of 22 criteria were considered in order to make the decision makers more accurately evaluate the set strategies of waste management. This study evaluates four types of waste treatment options to overcome the pollution and solid waste environmental problems plaguing all cities in Libya and their citizens. This represents the main aim of the research, which can be achieved through the developed model in this chapter. The model consists of two very well used multicriteria decision-making (MCDM) methods in fuzzy form. Fuzzy AHP method is used for calculation weights of main criteria and their subcriteria, while the fuzzy TOPSIS method is used for the evaluation and selection of potential waste management strategies.

This chapter contains seven sections in total. In Section 1, the significance and motivation of research are briefly presented. Section 2 represents the current state in the field of waste management with a short overview of conducted studies in Libya. Section 3 shows municipal solid waste technologies and their explanation, while Section 4 represents the used methods. In Section 5, the case study is presented, while Section 6 represents sensitivity analysis and discussion. The chapter ends with conclusions and guidelines for future research.

2. Current State in Libya in the Field of Waste Management

Many studies concerning the current state of the municipal solid waste (MSW) in Libya have been conducted. Omran and Gebril (2018) studied the current situation of SWM practices in Ajdabiya city. The data was collected through a random survey. The study shows that the current services are inefficient. Also, Gebril *et al.* (2010) studied the current status of SWM practices in Benghazi city. The research concluded that the collection process is inefficient. Also, the only way for waste treatment is dumping into an open land. According to Hamad *et al.* (2014), the quantity of MSW generated in Libya is estimated at 3.2 million tons/year. Approximately 90% of waste end up in more than 2,300 open dumps. Bani Walid and Albayda cities has the same situation, where the lack of resources affects the waste management processes. Open dumping is currently the form of waste disposal in both cities. Also, poor collection and inadequate transportation of waste led to an accumulation of waste in the cities (Abdulnaser *et al.*, 2011; Omran *et al.*, 2017).

The absence of waste treatment systems led to the piling up of mountains of waste in the dumping sites. The authorities provide the MSW collection and transportation systems, the collectors, the equipment and the infrastructures such as temporary waste disposal. When the dumping sites are filled with waste, the common solution is to burn it.

The Waste Management Law was enacted in Libya in 1980 and has undergone major revisions in response to political, environmental and social changes, including law of environment (1982), law of atmosphere and air protection (1992) and law of transport of hazardous materials (2005) (Sawalem *et al.*, 2009).

The solution applied to 97% of waste produced is throwing it in open dumps (67%) or burning it in the open air in either public dumps or municipal, uncontrolled dumps (30%). Only 3% of the total quantity destined for recovery: 1% for recycling and 2% for composting, as illustrated in Figure 1. There are more than 2,300 open dumps in the country with an area of approximately 3,500 ha. Most of these dumps are nearly saturated (Hamad *et al.*, 2014).

The dumping sites can be in inappropriate locations and places such as lowlands, river valleys, old quarries, beaches, open drains and even along streets or directly into the sea. No prior treatment is applied before dumping. There is also no evaluation to the nuisance

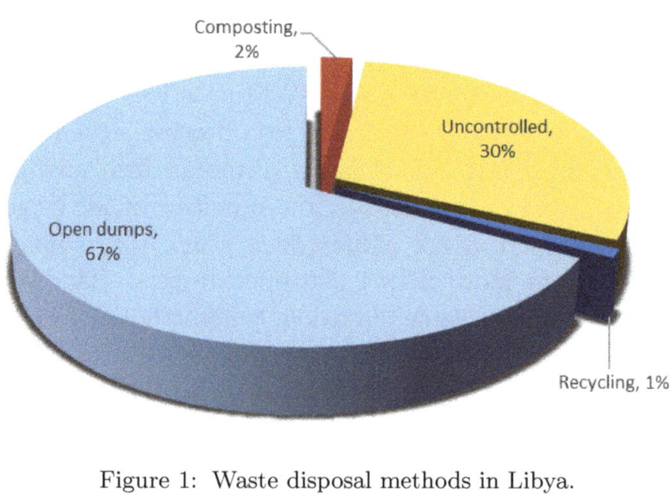

Figure 1: Waste disposal methods in Libya.

and harm it may cause to the environment. The solid waste is scattered at a large scale on any unused parcel of land, around buildings and in open market places. Shortage of resources and waste collection containers are among the important factors that are influencing the decision-making and operational processes of MSW (Omran *et al.*, 2017). Also, one of the main obstacles facing the proper management of the waste at Al-Abyear city is the limited documentation regarding the generation, handling and management of the solid waste (Gebril *et al.*, 2011). The lack of effective relationships between various stakeholders contributes to the unsuccessful good solid waste management practices in Al-Marj city (Hamad *et al.*, 2017). Furthermore, lack of manpower and vehicles are the main problems facing waste collection in Benghazi city. The wastes are just dumped anywhere, and there is no treatment provided to the waste after collection. Lack of suitable facilities and inadequate management and technical skills, improper bin collection, and route planning are among the issues resulting in poor collection and transportation of MSW. Therefore, this poor planning and environmental management is the main reason for solid waste pollution in Benghazi and its surrounding areas. There are different types of vehicles used to collect waste such as ordinary trucks, tractor–trailers, dumper–placers and tippers. The commonly used vehicle is the open-body trucks of 5–15-ton capacity. Tractor–trailers are used in smaller areas even if they were noisy and inefficient (Ali, 2013).

There is no monitoring or an evaluation procedure to the MSW activities provided by the government. Therefore, there is almost no performance or target that should be achieved by the MSW management. This affects the quality of the services provided by the workers. Also, the seasons had impacts on waste collections in households and schools (Alhderi *et al.*, 2015).

The percentage of MSW per capita was about 0.915 kg/day in 2006, while in 2010, it increased up to 1.1 kg/day (Elmnifi *et al.*, 2018). For instance, the average daily solid waste generation in Misurata city was found to be 1.48 kg/person. Open dumping is the most popular form of waste disposal for more than of 360 tons of waste daily generated in the city (Sawalem *et al.*, 2015). Also, the average generated quantity of waste in Tripoli city is 1,415 tons (Moftah *et al.*, 2016). The weight percentages of plastic waste and PVC in Tripoli were about 10.52% and 1.36%, respectively, with overall quantity generated is 57.80 tons, which includes 7.48 tons of PVC (Najjar *et al.*, 2015).

3. Municipal Solid Waste Technologies

Several types of recycling, energy recovery or waste neutralization technologies are used in a system of waste management. Each of them shows different technical and environmental characteristics. According to Pamučar *et al.* (2018a) Libya has a variety of energy sources, but also growing demands for electricity. In such a situation, adequate strategy of waste management can be useful.

Anaerobic digestion (AD) is a naturally occurring biological process that uses microorganisms to break down organic material in the absence of oxygen. AD is a process that makes any organic waste become biologically transformed into another form, in the absence of oxygen (Khalid *et al.*, 2011).

Landfilling is the ultimate disposal process for MSW management. A landfill is a facility in which solid waste is disposed of in a manner that limits their impact on the environment. Landfills consist of a complex system of interrelated components and subsystems that act together to break down and stabilize disposed waste over time (Ismail *et al.*, 2013).

Incineration is a thermal waste treatment process where raw or unprocessed waste can be used as feedstock. The process is done in the presence of oxygen, and maintaining it at high temperature for sufficient time to complete combustion to carbon dioxide and water (Jereme *et al.*, 2013).

Composting is a biological process in which the organic matter in the waste is converted into enriched inorganic nutrients (Brinton, 2003). The manure obtained has high nitrogen, phosphorus and potassium content.

The selection of proper and right disposal treatment will not only save money and time, but will also help in reducing the negative environmental impact. The right choice also protects human health and reduces risks to water, air and soil systems, as well as to plants and animals. At the same time, it causes no nuisance through odor or noise and does not disturb rural areas or locations of specific interest.

4. Methods

MCDM is one of the processes for finding the optimal alternative from all the feasible alternatives according to some criteria or attributes (Stević *et al.*, 2017). To adequately solve decision-making problems with vague or imprecise information according to Mardani *et al.* (2018), the fuzzy set theory (Radović *et al.*, 2018) and aggregation operator theory (Stević *et al.*, 2018a; Sremac *et al.*, 2018) have become powerful tools.

4.1. *Fuzzy sets*

The theory of fuzzy sets was first introduced by Zadeh (1965). Fuzzy theory is a very suitable tool for the treatment of uncertainty.

A fuzzy number $\tilde{A}$ on R to be a TFN if its membership function $\mu_{\tilde{A}}(x)$: $R \rightarrow (0,1)$ is equal to following equation (1):

$$M_A \sim (x) = \begin{cases} \frac{x-l}{m-l}, & l \leq x \leq m, \\ \frac{u-x}{u-m}, & m \leq x \leq u, \\ 0, & \text{otherwise}, \end{cases} \tag{1}$$

where l and u represent the lower and upper bounds of the fuzzy number $\tilde{A}$, and m is the crisp value for $\tilde{A}$.

The operations of TFN $\check{A}_1 = (l_1, m_1, u_1)$ and $\check{A}_2 = (l_2, m_2, u_2)$ are shown through following equations:

Addition:

$$\check{A}_1 + \check{A}_2 = (l_1, m_1, u_1) + (l_2, m_2, u_2) = (l_1 + l_2, m_1 + m_2, u_1 + u_2). \tag{2}$$

Multiplication:

$$\check{A}_1 x \check{A}_2 = (l_1, m_1, u_1) x (l_2, m_2, u_2) = (l_1 l_2, m_1 m_2, u_1 u_2). \tag{3}$$

Subtraction:

$$\check{A}_1 - \check{A}_2 = (l_1, m_1, u_1) - (l_2, m_2, u_2) = (l_1 - u_2, m_1 - m_2, u_{1-} l_2). \tag{4}$$

Division:

$$\frac{\breve{A}_1}{\breve{A}_2} = \frac{(l_1, m_1, u_1)}{(l_2, m_2, u_2)}$$
$$= \left(\frac{l_1}{u_2}, \frac{m_1}{m_2}, \frac{u_1}{l_2}\right) \text{ for } l_1 l_2 > 0; m_1 m_2 > 0; \; u_1 u_2 > 0. \tag{5}$$

Reciprocal:

$$\breve{A}^{-1} = (l_1, m_1, u_1)^{-1}$$
$$= \left(\frac{1}{u_1}, \frac{1}{m_1}, \frac{1}{l_1}\right) \text{ for } l_1 l_2 > 0; m_1 m_2 > 0; \; u_1 u_2 > 0. \tag{6}$$

4.2. *Fuzzy AHP method*

AHP method was proposed 40 years ago and, according to (Eshtaiwi *et al.*, 2018), is one of the most used MCDM techniques. Therefore, in the continuation of work only the basic items and ideas are given on which the method is based. According to the author of the method Saaty (1980), AHP presents measurement theory for pairs comparison, which is based on the opinion of experts. This method uses a nine-point scale called Saaty's scale. One of the very important items related to this method is that the problem is structured in a hierarchy that consists of multiple levels. The first level is goal, next levels are criteria and subcriteria and at the end are alternatives.

AHP is one of popular methods and it has the ability to identify and analyze the subjectivity of decision-makers in the estimation and evaluation process of hierarchy elements. A man is rarely consistent in assessing the value or relation of the quality elements in the hierarchy. AHP in a particular way solves that by measuring the ratio of consistency.

For each comparison matrix *Zk*, we determine consistency of value of experts. In order to check consistency (Saaty, 1980) has proposed consistency ratio *CR*. Some authors recommend values of *CR* depending on the size of the matrix, so in Lee *et al.* (2008), the highest allowed level of consistency is 0.05 for the matrix 3×3, 0.08 for the 4×4 matrix and for matrix 5×5 and more it is 0.1.

To check the CR in this research, we used Kwong's method (Kwong and Bai, 2003). A TFN, marked as $M = (l, u)$, can be defuzzified to a crisp number using following equation:

$$M_{-\text{crisp}} = \frac{(4m + l + u)}{6}. \tag{7}$$

TFN, which were used in this study, are labeled as (l_{ij}, m_{ij}, u_{ij}) where l_{ij} is the smallest possible value, m_{ij} is the value most nearest of crisp and u_{ij} is highest value.

In this research, the FAHP method by Chang (1996) is adopted. Some advantages of this method are effectively handling both qualitative and quantitative parameters and it is easy to implement and understand (Tüysüz, and Kahraman, 2006; Ertuğrul and Karakaşoğlu, 2008).

Chang's extended analysis consists of following steps.

Step 1: S_i with consideration to the ith criteria is expressed as

$$S_i = \sum_{j=1}^{n} M_{gi}^{j} \times \left[\sum_{\text{i}=1}^{\text{n}} \sum_{\text{j}=1}^{\text{m}} M_{gi}^{j} \right]^{-1}. \tag{8}$$

To calculate the next expression

$$\left[\sum_{i=1}^{\text{n}} \sum_{\text{j}=1}^{\text{m}} M_{gi}^{j} \right]^{-1} \tag{9}$$

it is needed to do supplementary fuzzy operations with m values

$$\sum_{j=1}^{m} M_{gi}^{j} = \left(\sum_{j=1}^{m} l_j, \sum_{j=1}^{m} m_j, \sum_{j=1}^{m} u_j \right), \tag{10}$$

$$\sum_{i=1}^{n} \sum_{j=1}^{n} M_{gi}^{j} = \left(\sum_{i=1}^{n} l_i, \sum_{i=1}^{n} m_i, \sum_{i=1}^{n} u_i \right). \tag{11}$$

After that, the inverse vector needs to be obtained:

$$\left[\sum_{i=1}^{n} \sum_{j=1}^{m} M_{gi}^{j} \right]^{-1} = \left[\frac{1}{\sum_{i=1}^{n} u_i}, \frac{1}{\sum_{i=1}^{n} m_i}, \frac{1}{\sum_{i=1}^{n} l_i} \right]. \tag{12}$$

Step 2: The degree of possibility of $S_b \geq S_a$ is expressed as

$$V(S_b \geq S_a) = \begin{cases} 1, & \text{if } m_b \geq m_a, \\ 0, & \text{if } l_a \geq u_b, \\ \frac{l_a - u_b}{(m_b - u_b) - (m_a - l_a)}, & \text{otherwise}, \end{cases} \tag{13}$$

where d ordinate of a largest cross-section in point D between μ_{Sa} and μ_{Sb} as shown in Figure 2.

Step 3: To find weight vector, the following equations are used:

$$V(S_i \geq S_1, S_2, \ldots, S_k) = \min V(S_i \geq S_k), = w'(S_i), \tag{14}$$

$$d'(A_i) = \min V(S_i \geq S_k), k \neq i, k = 1, 2, \ldots, n, \tag{15}$$

$$W' = (d'(A_1), d'(A_2), \ldots, d'(A_n))^T. \tag{16}$$

Step 4: The last step in this method is the normalization of previous values using next equation:

$$W = (d(A_1), d(A_2), \ldots, d(A_n))^T, \tag{17}$$

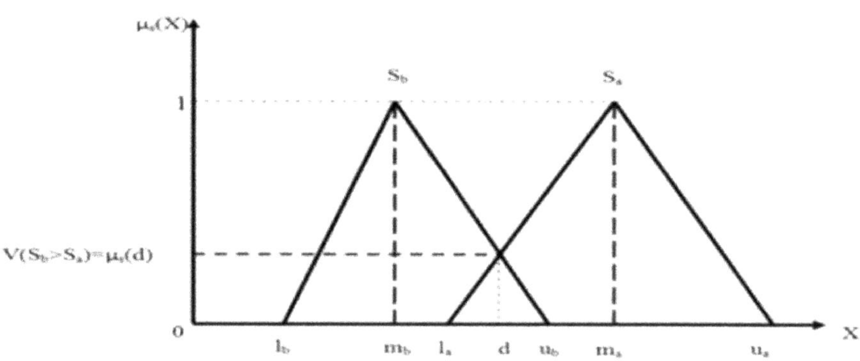

Figure 2: Intersection between S_a and S_b.

4.3. *Fuzzy TOPSIS method*

TOPSIS method in his conventional form represents a simple concept. This method has become popular and is often used in different areas of MCDM. However, regardless of that, this method is criticized in many situations because it does not have the ability to efficiently handle uncertainty in the moment when the DMs need exact results. For this reason, in this study, we use the extended fuzzy TOPSIS method, which allows proper handling of uncertainty and imprecision and it is completely appropriate for the ranking of potential solution.

TOPSIS was first proposed by Hwang and Yoon (1981) and a fuzzy TOPSIS method was later introduced by Chen and Hwang (1992).

The algorithm of the FTOPSIS method can be expressed as follows (Chen, 2000):

Step 1: After forming a team of decision-makers, the definition criteria are then selected for evaluation.

Step 2: Select the suitable linguistic variables for the weight of the criteria and the linguistic ratings for alternatives with respect to criteria.

Step 3: Aggregate the weight of criteria to get the aggregated fuzzy weight $\tilde{w}_{ij}$ of criterion C_j, and pool the decision makers' opinions to get the aggregated fuzzy rating $\tilde{x}_{ij}$ of alternative A_i under criterion C_j

$$\tilde{R}_k = (a_k, b_k, c_k), \quad k = 1, 2, 3, \ldots K, \tag{18}$$

then the aggregated fuzzy rating can be calculated as

$$R = (a, b, c), \quad k = 1, 2, 3, \ldots K. \tag{19}$$

$$a = \min_k(a_k), b = \frac{1}{K}\sum_{k=1}^{K} b_k, \quad c = \max_k(c_k). \tag{20}$$

Step 4: The fuzzy decision matrix is formed and the normalized fuzzy decision matrix is formed:

$$\tilde{R}_k = [r_{ij}]_{\text{mxn}} i = 1, 2, 3, \ldots, m;\ j = 1, 2, 3, \ldots, n \tag{21}$$

where B and C are the set of benefit criteria and cost criteria, respectively, and

$$r_{ij} = \left(\frac{a_{ij}}{c_j^*}, \frac{b_{ij}}{c_j^*}, \frac{c_{ij}}{c_j^*}\right), j \in B, \tag{22}$$

$$r_{ij} = \left(\frac{a_{\bar{j}}}{c_{ij}}, \frac{a_{\bar{j}}}{b_{ij}}, \frac{a_{\bar{j}}}{a_{ij}}\right), j \in C, \tag{23}$$

$c_j^* = \max_i c_{ij}$ if $j \in B$;
$a_{\bar{j}} = \min_i a_{ij}$ if $j \in C$.

Step 5: Considering the different weights of each criterion, we can construct the weighted normalized fuzzy decision matrix as

$$\tilde{V} = [\tilde{v}_{ij}]_{\text{mxn}}\, i = 1, 2, \ldots, m;\ j = 1, 2, \ldots, n, \tag{24}$$

$$\tilde{v}_{ij} = r_{ij} W, \tag{25}$$

where W is the weighted vector of evaluating criteria.

Step 6: Determine the fuzzy positive ideal solution (FPIS) and fuzzy negative ideal solution (FNIS) where according Yu *et al.* (2013):

$$A^* = (\tilde{v}_1^*, \tilde{v}_2^*, \ldots, \tilde{v}_n^*) = (\max_j v_{ij} | i \epsilon B), .$$

$$(\min_j v_{ij} | i \epsilon C), \quad i = 1, 2 \ldots m;\ j = 1, 2 \ldots n, \tag{26}$$

$$A^- = (\tilde{v}_1^-, \tilde{v}_2^-, \ldots, \tilde{v}_n^-) = (\min_j v_{ij} | i \epsilon B),$$

$$(\max_j v_{ij}, i \epsilon C), \quad i = 1, 2 \ldots m;\ j = 1, 2 \ldots n, \tag{27}$$

where B and C are the set of benefit criteria and cost criteria, respectively.

Step 7: Calculate the distance of each alternative from FPIS and FNIS, respectively. The distance of each alternative from A^* and A^- can be calculated as

$$d_i^* = \sum_{j=1}^{n} d(\tilde{v}_{ij}, \tilde{v}_j^*), \quad i = 1, 2, \ldots m, \tag{28}$$

$$d_i^- = \sum_{j=1}^{n} d(\tilde{v}_{ij}, \tilde{v}_j^-), \quad i = 1, 2, \ldots m, \tag{29}$$

where $d(.;.)$ is the distance measurement between two fuzzy numbers.

Step 8: Calculate the closeness coefficient of each alternative.

A closeness coefficient is defined to determine the ranking order of all alternatives once the d_i^* and d_i^- of each alternative $A_i (i = 1; 2; m)$ has been calculated. The closeness coefficient of each alternative is calculated as

$$CC_i = \frac{d_i^-}{d_i^* + d_i^-}, \quad i = 1, 2, \ldots, m \tag{30}$$

Step 9: According to the closeness coefficient, the ranking order of all alternatives can be determined.

5. The Case Study

Of late, the area of MCDM is rapidly developing, thanks to large number of researches dealing with the adoption of certain decisions based on the applied methods that belong to the specified field (Pamučar *et al.*, 2018b). MCDM methods belong to operational research and the present area of waste disposal, which is very fast developing (Stević *et al.*, 2016). Multicriteria methods for decision-making are used to resolve a large number of problems in all spheres of business, primarily due to a large number of methods that have been developed (Badi *et al.*, 2018), particularly within the last decade. The combination of these methods with fuzzy logic gives excellent results (Stević *et al.*, 2018b) because classical methods such as AHP cannot with such precision perform the required quantification and this is where fuzzy logic shows all its advantages (Rezaei *et al.*, 2013; Puška *et al.*, 2018).

Management of waste has become an issue of critical importance during last decades mostly due to the complexity of waste streams and the steadily increasing produced volumes (Achillas *et al.*, 2013). In last three decades, problems in area of waste management have been solved through the applications of MCDA techniques. According Babalola (2015), this usually involves the integration of environmental, political, social, cultural and economic parameters, which involve the preferences of stakeholders while considering the challenges of monetizing essentially non-monetary factors. The AHP is one of the extensively used MCDM methods and has successfully been used area of waste management (Milutinović *et al.*, 2014; Nouri *et al.*, 2016; Herva and Roca, 2013; Mirzazadeh *et al.*, 2017). According to Soltani *et al.* (2015) AHP has been the predominant candidate for combining with other methods and in this study we use integrated fuzzy AHP and fuzzy TOPSIS method.

The goal of this study is to determine the most suitable waste treatment option for treating the amount of MSW generated in Libya. The only treatment provided to the waste after collection is the composting of the waste. However, in Libya there are five MSW plants and were established in several cities. These plants are outdated and suffer from many obstacles, because of the lack of maintenance and there is a need to be updated or replaced to work efficiently. For instance, the compost plant in Misurata currently has a capacity of only 60 tons, where it had been 120 tons/day when was opened in 1982. Developing an adequate plan for waste management in Libya is a real challenge. This inappropriate waste management practices may cause degradation of valuable land resources and create long-term environmental and serious human health problems as a result of fires, explosions, and contamination of air, and might affect water as well. Definitely, the waste polluting the environment and threatening human health requires public education and awareness.

The suggested treatment options in this work are: AD (mentioned A_1 in the calculations), landfilling (A_2), incineration (A_3), and composting (A_4). Figure 3 shows the hierarchical structure of the suggested model.

Figure 3 shows a proposed model for waste management in Libya. The criteria used in the evaluation of four treatment systems are environmental impacts, socio-cultural aspects, technical aspects and economic aspects. The first and fourth criteria consist of six subcriteria, while the second and third criteria have five subcriteria.

- **Environment:** This deals with any change, positive or negative, to land, ecosystems and human health as a result of action resulting from waste mismanagement. Disposing of waste has its own specific impacts and can cause severe problems to public health.
- **Socio-cultural:** Waste generation is usually determined by socio-economic characteristics and people's attitudes. It is known that campaigns, educational measures and public awareness positively influence people's approach to waste.
- **Technical:** Their importance is usually in view of the possibility of subsequent increases in the daily tonnage of waste that the facility will be required to manage, for added processing capabilities. These criteria also determine what equipment and training will be necessary to perform the waste management responsibilities.
- **Economic:** It is a significant part of tactical planning and investment programming for any waste disposal facility. Estimating the preliminary investment capital requires long-term operating and maintenance costs in relation to the different waste management activities.
- **Sub-criteria:** C11 — Air and Water Pollution, C12 — Land Use, Requirement, and Contamination, C16 — Disamenity, such as Noise and Dust, C25 — Vulnerability of the Area, C41 — Capital and Construction Cost, C42 — Operating and Maintenance Cost are cost criteria, while other are beneficial.

Upon establishment of criteria, the expert team comprised of five members and compared them on the base of triangular fuzzy scale (27). So, this is a group decision, where decision makers influence the value of the selected criteria. In Table 1, values of expert comparison of 22 criteria grouped into levels are shown.

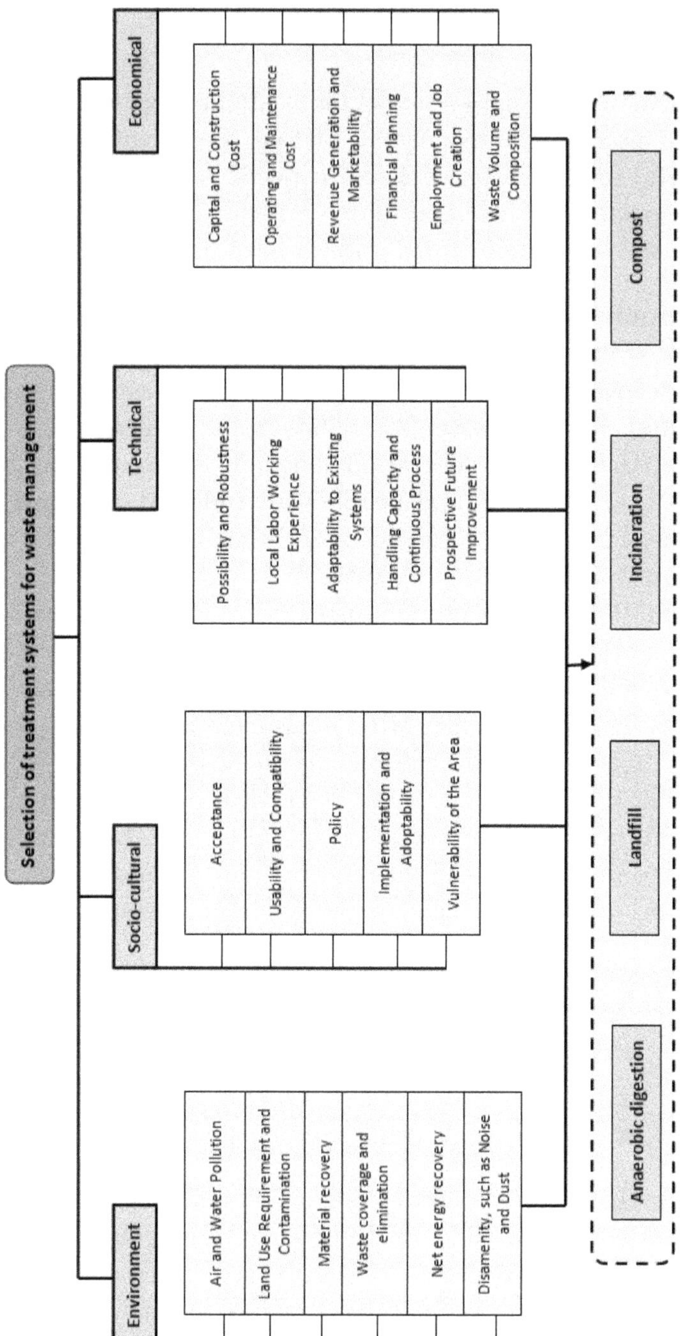

Figure 3: Proposed model for Waste Management in Libya.

Table 1: Comparison main criteria by five experts.

		C_1	C_2	C_3	C_4
C_1	E_1	(1,1,1)	(1,3/2,2)	(2,5/2,3)	(5/2,3,7/2)
	E_2	(1,1,1)	(1,3/2,2)	(2,5/2,3)	(5/2,3,7/2)
	E_3	(1,1,1)	(1/2,1,3/2)	(3/2,2,5/2)	(2,5/2,3)
	E_4	(1,1,1)	(1/2,1,3/2)	(3/2,2,5/2)	(2,5/2,3)
	E_5	(1,1,1)	(1,3/2,2)	(2,5/2,3)	(5/2,3,7/2)
C_2	E_1	(1/2,2/3,1)	(1,1,1)	(1/2,1,3/2)	(1/2,1,3/2)
	E_2	(1/2,2/3,1)	(1,1,1)	(1/2,1,3/2)	(1/2,1,3/2)
	E_3	(2/3,1,2)	(1,1,1)	(1,3/2,2)	(1/2,1,3/2)
	E_4	(2/3,1,2)	(1,1,1)	(1,3/2,2)	(1/2,1,3/2)
	E_5	(1/2,2/3,1)	(1,1,1)	(1/2,1,3/2)	(1/2,1,3/2)
C_3	E_1	(1/3,2/5,1/2)	(2/3,1,2)	(1,1,1)	(1/2,1,3/2)
	E_2	(1/3,2/5,1/2)	(2/3,1,2)	(1,1,1)	(1,3/2,2)
	E_3	(2/5,1/2,2/3)	(1/2,2/3,1)	(1,1,1)	(3/2,2,5/2)
	E_4	(2/5,1/2,2/3)	(1/2,2/3,1)	(1,1,1)	(1,1,1)
	E_5	(1/3,2/5,1/2)	(2/3,1,2)	(1,1,1)	(1,3/2,2)
C_4	E_1	(2/7,1/3,2/5)	(2/3,1,2)	(2/3,1,2)	(1,1,1)
	E_2	(2/7,1/3,2/5)	(2/3,1,2)	(1/2,2/3,1)	(1,1,1)
	E_3	(1/3,2/5,1/2)	(2/3,1,2)	(2/5,1/2,2/3)	(1,1,1)
	E_4	(1/3,2/5,1/2)	(2/3,1,2)	(1,1,1)	(1,1,1)
	E_5	(2/7,1/3,2/5)	(2/3,1,2)	(1/2,2/3,1)	(1,1,1)

Table 2: Fuzzy important weight of the criteria.

	C_1			C_2			C_3			C_4		
C_1	1.000	1.000	1.000	0.758	1.275	1.783	1.783	2.287	2.789	2.287	2.789	3.291
C_2	0.561	0.784	1.320	1.000	1.000	1.000	0.660	1.176	1.683	0.574	1.084	1.589
C_3	0.359	0.437	0.561	0.594	0.850	1.516	1.000	1.000	1.000	1.046	1.351	1.719
C_4	0.304	0.359	0.437	0.667	1.000	2.000	0.582	0.740	1.059	1.000	1.000	1.000

Lee (2009) recommended the calculation of fuzzy weight of the criteria using geometric mean of the responses of the DMs, which is shown in Table 2.

An example of the calculation procedure in the first step is shown as follows:

$$C_1 = (1 + 0.758 + 1.783 + 2.287; 1 + 1.275 + 2.287 + 2.789; 1 + 1.783 + 2.789 + 3.291) = (5.827; 7.351; 8.862), \text{ etc.}$$

The $\sum_{i=1}^{n}\sum_{j=1}^{n} M_{gi}^{j}$ value is

$$(5.827; 7.351; 8.862) + (2.795; 4.045; 5.591) + (2.998; 3.639; 4.795) + (2.552; 3.099; 4.497) = (14.173; 18.133; 23.746)$$

Then, $S_i = \sum_{j=1}^{n} M_{gi}^{j} \times \left[\sum_{\mathrm{i=1}}^{\mathrm{n}}\sum_{\mathrm{j=1}}^{\mathrm{m}} M_{gi}^{j}\right]^{-1}$:

$$S_1 = (5.827; 7.351; 8.862)x(1/23.746; 1/18.133; 1/14.173) = (0.245; 0.405; 0.625), \text{etc. (see Table 3)}.$$

The V values are calculated using second step, i.e., equation (13) (Table 4).

After third and fourth steps of fuzzy AHP method, weight values are obtained and shown in Table 5. After defuzzification using equation (7), by applying the steps of conventional AHP method, we obtain CR = 0.065, which means that the consistency ratio is 0.065, which is less than the maximum permitted limit of 0.08 for this matrix size. After application of the described steps of FAHP method for sub-criteria are obtained all weights, which are also shown in Table 5.

Table 3: Values obtained after first step.

	l	m	u	$\frac{1}{\sum_{i=1}^{n} u_i}$	$\frac{1}{\sum_{i=1}^{n} m_i}$	$\frac{1}{\sum_{i=1}^{n} l_i}$
C_1	5.827	7.351	8.862	0.245	0.405	0.625
C_2	2.795	4.045	5.591	0.118	0.223	0.395
C_3	2.998	3.639	4.795	0.126	0.201	0.338
C_4	2.552	3.099	4.497	0.107	0.171	0.317
Σ	14.173	18.133	23.746			

Table 4: Values obtained after second step.

$V(S_1 \geq S_2)$	1.000	$V(S_2 \geq S_1)$	0.450
$V(S_1 \geq S_3)$	1.000	$V(S_2 \geq S_3)$	1.000
$V(S_1 \geq S_4)$	1.000	$V(S_2 \geq S_4)$	1.000
$V(S_3 \geq S_1)$	0.312	$V(S_4 \geq S_1)$	0.235
$V(S_3 \geq S_2)$	0.908	$V(S_4 \geq S_2)$	0.793
$V(S_3 \geq S_4)$	1.000	$V(S_4 \geq S_3)$	0.865

Table 5: Results obtained using fuzzy AHP method.

	Crit.	*wj*	*CR*		Subcriteria	Local *wi*	Global *wi*	*CR*
				C_{11}	Air and Water Pollution	0.476	0.238	
				C_{12}	Land Use, Requirement, and Contamination	0.195	0.098	
C_1	Environmental	0.501		C_{13}	Material Recovery	0.111	0.056	0.029
				C_{14}	Waste Coverage and Elimination	0.083	0.042	
				C_{15}	Net Energy Recovery	0.010	0.005	
				C_{16}	Disamenity, such as Noise and Dust	0.125	0.062	
				C_{21}	Acceptance	0.808	0.182	
				C_{22}	Usability and Compatibility	0.044	0.010	
C_2	Socio-Cultural	0.225		C_{23}	Policy	0.027	0.006	
				C_{24}	Implementation and Adoptability	0.000	0.000	0.021
				C_{25}	Vulnerability of the Area	0.120	0.027	
			0.065	C_{31}	Possibility and Robustness	0.316	0.049	
				C_{32}	Local Labor Working Experience	0.156	0.024	
C_3	Technical			C_{33}	Adaptability to Existing Systems	0.160	0.025	0.055
		0.156		C_{34}	Handling Capacity and Continuous Process	0.265	0.041	
				C_{35}	Prospective Future Improvement	0.103	0.016	
				C_{41}	Capital and Construction Cost	0.582	0.068	
				C_{42}	Operating and Maintenance Cost	0.000	0.000	
C_4	Economical	0.118		C_{43}	Revenue Generation and Marketability	0.000	0.000	0.037
				C_{44}	Financial Planning	0.252	0.030	
				C_{45}	Employment and Job Creation	0.166	0.020	
				C_{46}	Waste Volume and Composition	0.000	0.000	

According to the results presented in Table 5, each individual subcriterion was ranked under its main criteria. The most important criterion in model is C_{11} with a value of 0.298, which is dominant.

The second most important criterion is C_{21} (Acceptance) with a value 0.182, while other criteria have much less values. All consistency ratio is less than 0.1 and is acceptable.

However, the initial suggested criteria were modified; four subcriteria, which have got local weight zero, were omitted from the initial group. Finally, 18 sub-criteria across four main criteria were used as measures for decision-making process using fuzzy TOPSIS method.

Table 6 shows the evaluation of alternatives by five experts using the linguistic variables. Based on the characteristics of the alternatives and the expert opinion Table 6 was formed.

By applying the third and fourth steps of the FTOPSIS method, we get the values presented in Table 7, which represent a normalized fuzzy decision matrix.

Table 8 shows the final results and alternatives ranking: $A_1 > A_4 > A_3 > A_2$.

6. Sensitivity Analysis and Discussion

The sensitivity analysis includes the experiment of 20 sets where the values of criteria are changed. The first 18 sets means increasing each criterion separately by 17% starting from the first one to the last. The set number 19 means equal weighting of all the criteria, while in the set number 20, criteria C_{15}, C_{22} and C_{23} increase by 10%.

After the formation of sets and the analysis shown in Figures 4 and 5, it is evident that the first alternative-AD is ranked as the most acceptable solution in 18 out of a total of 20 cases; therefore, it holds the first position. In the eighth set the change of the criterion C_{22} (Usability and Compatibility) affect the change of preferred alternative, and then the fourth alternative becomes number one. In set number 20 criteria C_{15}, C_{22} and C_{23} increase by 10% and alternative four is also most suitable.

The second alternative — Landfill — is in sets 8, 9, 11, 14 and 17 on third place, while in rest of the sets, it is on last place and represents the most worst solution. The third alternative — Incineration — in 15 sets is on third place and rest on the last, while alternative four is in 18 sets at the second place.

Table 6: Rating of the alternatives in linguistic terms.

Expert	Al.	C_{11}	C_{12}	C_{13}	C_{14}	C_{15}	C_{16}	C_{21}	C_{22}	C_{23}	C_{25}	C_{31}	C_{32}	C_{33}	C_{34}	C_{35}	C_{41}	C_{44}	C_{45}
E_1	A_1	VG	G	MG	MG	MG	VG	MG	MP	G	G	MG	MG	MG	VG	VG	G	G	MG
	A_2	MP	P	P	VP	VP	F	P	MG	MG	MG	F	F	MP	G	F	MG	F	MP
	A_3	P	G	G	VG	VG	MP	VP	F	P	MP	F	MP	MP	F	MG	F	P	G
	A_4	MG	VG	VG	G	G	MG	VG	G	F	P	G	G	G	F	F	F	P	F
E_2	A_1	G	G	G	G	MG	VG	G	P	G	VG	MG	MG	MG	G	G	G	G	MG
	A_2	F	MP	P	VP	VP	MP	P	G	MG	MG	MG	F	MP	VG	F	MG	F	F
	A_3	P	VG	G	VG	G	MP	VP	G	P	F	F	MP	MP	F	G	F	P	G
	A_4	MG	VG	G	VG	G	MG	G	G	F	F	G	VG	G	F	F	MG	MG	F
E_3	A_1	G	VG	VG	G	MG	G	MG	F	MG	G	MG	G	G	VG	VG	G	G	MG
	A_2	MP	P	P	VP	VP	F	P	F	F	MG	F	F	MP	G	MP	F	F	MP
	A_3	MP	MG	G	VG	VG	F	VP	MP	MP	MP	F	P	MP	MG	MG	F	P	G
	A_4	MG	MG	VG	MG	G	MG	MG	G	F	P	G	G	G	MG	F	F	P	F
E_4	A_1	VG	VG	MG	VG	VG	VG	MG	MP	G	G	MG	MG	MG	VG	VG	G	G	MG
	A_2	MP	MP	MP	VP	P	P	P	MG	F	MG	F	F	MP	MG	F	MG	F	MP
	A_3	G	G	MG	VG	VG	MP	VP	F	P	P	P	MP	P	F	MG	F	P	P
	A_4	G	G	VG	G	VG	MG	G	G	F	P	G	F	F	F	F	G	P	F
E_5	A_1	VG	G	MG	MG	MG	VG	G	F	G	G	MG	G	G	VG	VG	G	G	G
	A_2	P	MP	MP	P	VP	F	P	G	MG	MG	F	P	MP	G	F	MG	F	P
	A_3	MP	MG	MG	MG	G	F	P	MP	MP	F	F	F	F	F	G	F	VP	G
	A_4	G	G	VG	VG	G	G	VG	G	MG	P	MG	MG	MG	F	MP	F	P	F

Table 7: Normalized fuzzy decision matrix.

	C_{11}	C_{12}	C_{13}	C_{14}	C_{15}	C_{16}
A_1	(0,0,0)	(0,0,0)	(0.5,0.8,1)	(0.5,0.84,1)	(0.5,0.76,1)	(0,0,0)
A_2	(0,0,0)	(0,0,0)	(0,0.18,0.5)	(0,0.02,0.3)	(0,0,0.1)	(0,0,0)
A_3	(0,0,0)	(0,0,0)	(0.5,0.82,1)	(0.5,0.94,1)	(0.7,0.96,1)	(0,0,0)
A_4	(0,0,0)	(0,0,0)	(0.7,0.98,1)	(0.5,0.9,1)	(0.7,0.92,1)	(0,0,0)
	C_{21}	C_{22}	C_{23}	C_{25}	C_{31}	C_{32}
A_1	(0.5,0.78,1)	(0,0.34,0.7)	(0.5,0.86,1)	(0,0,0)	(0.5,0.7,0.9)	(0.5,0.78,1)
A_2	(0,0.1,0.3)	(0.3,0.74,1)	(0.3,0.62,0.9)	(0,0,0)	(0.3,0.54,0.9)	(0,0.42,0.7)
A_3	(0,0.02,0.3)	(0.1,0.5,1)	(0,0.18,0.5)	(0,0,0)	(0,0.42,0.7)	(0,0.3,0.7)
A_4	(0.5,0.9,1)	(0.7,0.9,1)	(0.3,0.54,0.9)	(0,0,0)	(0.5,0.86,1)	(0.3,0.8,1)
	C_{33}	C_{34}	C_{35}	C_{41}	C_{44}	C_{45}
A_1	(0.5,0.78,1)	(0.7,0.98,1)	(0.7,0.98,1)	(0.3,0.33,0.43)	(0.7,0.9,1)	(0.5,0.74,1)
A_2	(0.1,0.3,0.5)	(0.5,0.88,1)	(0.1,0.46,0.7)	(0.33,0.45,1)	(0.3,0.5,0.7)	(0,0.3,0.7)
A_3	(0,0.3,0.7)	(0.3,0.54,0.9)	(0.5,0.78,1)	(0.43,0.60,1)	(0,0.08,0.3)	(0,0.74,1)
A_4	(0.3,0.78,1)	(0.3,0.54,0.9)	(0.1,0.46,0.7)	(0.3,0.48,1)	(0,0.22,0.9)	(0.3,0.5,0.7)

Table 8: Closeness coefficient of alternatives and their ranking.

	d_i*	d_i^-	CC_i	Rank
A_1	0,157	0,432	0,733	1
A_2	0,393	0,207	0,345	4
A_3	0,369	0,243	0,397	3
A_4	0,208	0,413	0,665	2

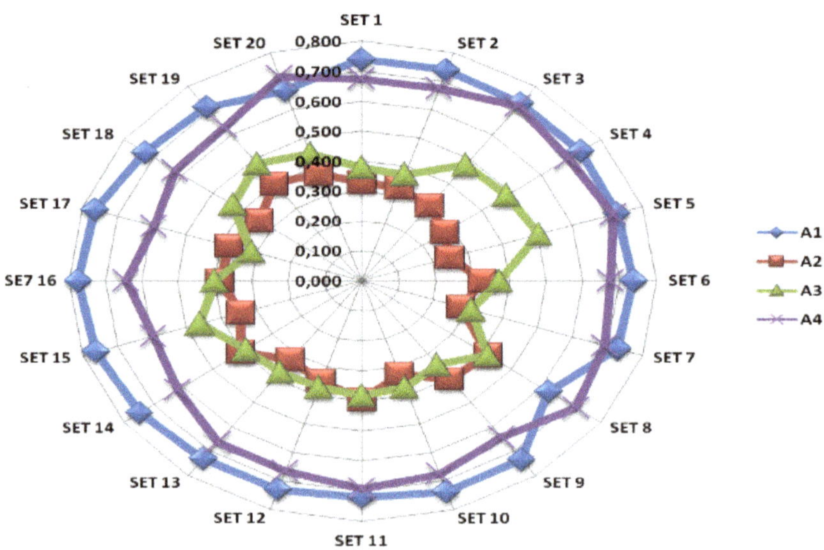

Figure 4: Results of sensitivity analysis.

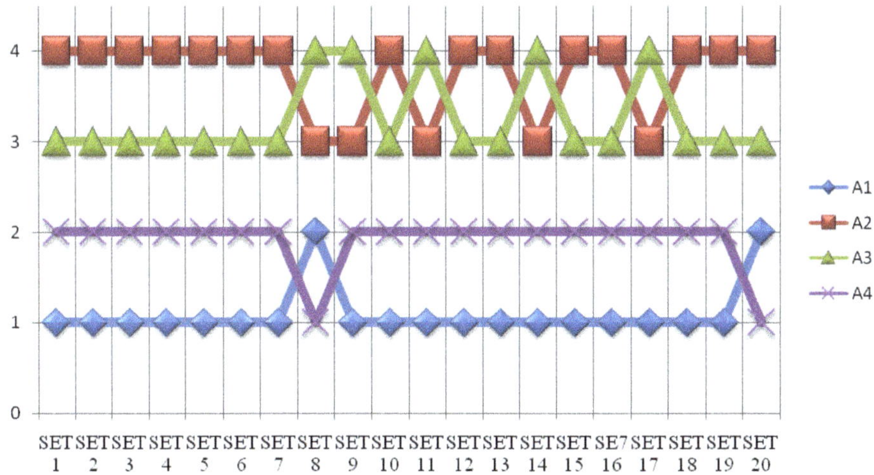

Figure 5: Ranking alternatives in 20 different sets of sensitivity analysis.

It is important to note that the first alternative in those two cases where it is not the most appropriate solution still holds second position, which speaks volumes about the qualities of the same. Even in the situation when all criteria are equally important with the same values, this alternative is the best solution.

7. Conclusion

Fuzzy AHP and fuzzy TOPSIS were applied to compare four municipal waste treatment options in Libya. These options are anaerobic digestion, landfilling, incineration and composting. The results of the model are arrived at through treatment options ranking. Four main criteria and 22 subcriteria were considered. The environmental criteria had the highest impact on the decision-making for selecting the treatment options. The selected treatment option was anaerobic digestion, followed by composting in the second position. Sensitivity analysis for 20 sets were conducted. The first alternative, anaerobic digestion, ranked as the most acceptable solution in 18 out of a total of 20 cases.

Furthermore, it should be noted that the rank of waste treatment systems can be changed according to the future technological developments. The model presented in this work can be used as a strategic decision supporting tool for decision makers, politicians and planners. The technology options performance based on the described criteria is a robust one similar to the synthesis results.

References

Abdelnaser, O., Alsadey, S. and Gavrilescu, M. (2011). Municipal solid waste management in Bani Walid City, Libya: Practices and challenges. *J Environ Manage Tour*, 2(2), 228–237.

Achillas, C., Moussiopoulos, N., Karagiannidis, A., Banias, G. and Perkoulidis, G. (2013). The use of multi-criteria decision analysis to tackle waste management problems: A literature review. *Waste Manage Research*, 31(2), 115–129.

Alhderi, N. A. A., Purwanto, P. and Sunoko, H. R. (2015). Impacts of four seasons in municipal solid waste in Al-khums city. *Sci J Environ Eng Res*, 15, 1–5.

Ali, A. O. G. (2013). Solid waste pollution and the importance of environmental planning in managing and preserving the public environment in Benghazi city and its surrounding areas. *Int J Environ, Chem, Ecol, Geol Geophys Eng*, 7(12), 924–930.

Babalola, M. A. (2015). A multi-criteria decision analysis of waste treatment options for food and biodegradable waste management in Japan. *Environments*, 2(4), 471–488.

Badi, I., Ballem, M. A. and Shetwan, A. G. (2018). Site selection of desalination plant in LIBYA by using COmbinative Distance-based ASsessment (CODAS) METHOD. *Int. J. Qual. Res*, 12.

Brinton, W. F. (2003). Interpretation of waste and compost tests. *J Woods End Res Lab*, 1(4), 1–6.

Chang, D. Y. (1996). Applications of the extent analysis method on fuzzy AHP. *Eur J Oper Res*, 95(3), 649–655.

Chen, C. T. (2000). Extensions of the TOPSIS for group decision-making under fuzzy environment. *Fuzzy Sets Syst*, 114(1), 1–9.

Chen, S. J. and Hwang, C. L. (1992). Fuzzy multiple attribute decision making methods. *Fuzzy Multiple Attribute Decision Making*. Springer Berlin Heidelberg, pp. 289–486.

Elmnifi, M., Alshilmany, M. and Abdraba, M. (2018). Potential of municipal solid waste in Libya for energy utilization. *Open J Mech Eng*, 2(2), 1–5.

Ertuğrul, İ. and Karakaşoğlu, N. (2008). Comparison of fuzzy AHP and fuzzy TOPSIS methods for facility location selection. *Int J Adv Manufact Technol*, 39(7), 783–795.

Eshtaiwi, M., Badi, I., Abdulshahed, A. and Erkan, T. E. (2018). Determination of key performance indicators for measuring airport success: A case study in Libya. *J Air Transp Manage*, 68, 28–34.

Gebril, A. O., Omran, A. and Pakir, H. K. (2011). solid waste management in al-abyar city, Libya. In *11th International Multidisciplinary Scientific GeoConference SGEM 2011*, Vol. 3, pp. 617–622.

Gebril, A. O., Omran, A., Pakir, A. H. K. and Aziz, H. A. (2010). Municipal solid waste management in Benghazi (Libya): Current practices and challenges. *Environ Eng Manag J (EEMJ)*, 9(9), 1289–1296.

Guerrero, L. A., Maas, G. and Hogland, W. (2013). Solid waste management challenges for cities in developing countries. *Waste Manage*, 33(1), 220–232.

Hamad, J. R. J., Hanafiah, M. M. and Abdullah, S. (2017). Problems and current practices of solid waste management in the city of Al-Marj. *Clean WAS*, 1(1), 01–05.

Hamad, T. A., Agll, A. A., Hamad, Y. M. and Sheffield, J. W. (2014). Solid waste as renewable source of energy: Current and future possibility in Libya. *Case Stud Thermal Eng*, 4, 144–152.

Herva, M. and Roca, E. (2013). Ranking municipal solid waste treatment alternatives based on ecological footprint and multi-criteria analysis. *Ecol Indicat*, 25, 77–84.

Hiraga, R. (2007). *Waste anagement and 3R Policies in Japan.* Clean Japan Center, Tokyo.

Hwang, C. L. and Yoon, K. (1918). Methods for multiple attribute decision making. *Multiple Attribute Decision Making.* Springer Berlin Heidelberg, pp. 58–191.

Ismail, S. N. S. and Manaf, L. A. (2013). The challenge of future landfill: A case study of Malaysia. *J Toxicol Environ Health Sci*, 5(6), 86–96.

Jereme, I. A., Siwar, C. and Bhuiyan, M. A. H. (2013). Incineration and its implications: The need for a sustainable waste management system in Malaysia. *Int J Environ Sci*, 4(3), 367–381.

Khalid, A., Arshad, M., Anjum, M., Mahmood, T. and Dawson, L. (2011). The anaerobic digestion of solid organic waste. *Waste Manage*, 31(8), 1737–1744.

Kwong, C. K. and Bai, H. (2003). Determining the importance weights for the customer requirements in QFD using a fuzzy AHP with an extent analysis approach. *IIE Trans*, 35(7), 619–626.

Lee, A. H. (2009). A fuzzy supplier selection model with the consideration of benefits, opportunities, costs and risks. *Expert Syst Appl*, 36(2), 2879–2893.

Lee, A. H., Chen, W. C. and Chang, C. J. (2008). A fuzzy AHP and BSC approach for evaluating performance of IT department in the manufacturing industry in Taiwan. *Expert Syst Appl*, 34(1), 96–107.

Mardani, A., Nilashi, M., Zavadskas, E. K., Awang, S. R., Zare, H. and Jamal, N. M. (2018). Decision making methods based on fuzzy aggregation operators: Three decades review from 1986 to 2017. *Int J Inform Technol Decis Making*, 17(02), 391–466.

Milutinović, B., Stefanović, G., Dassisti, M., Marković, D. and Vučković, G. (2014). Multi-criteria analysis as a tool for sustainability assessment of a waste management model. *Energy*, 74, 190–201.

Mirzazadeh, F., Hadinejad, F. and Roshan, N. A. (2017). Investigating utility level of waste disposal methods using multicriteria decision-making techniques (case study: Mazandaran-Iran). *J Mater Cycles Waste Manage*, 20(1), 1–11.

Moftah, W. A., Marković, D., Moftah, O. A. and Nesseef, L. (2016). Characterization of household solid waste and management in tripoli City-Libya. *Open J Ecol*, 6(07), 435–449.
Najjar, A. K., Elmelah, M. A., Bannani, F. K. and Ghania, F. M. (2015). Estimation of polyvinylchloride (PVC) in household solid waste at Tripoli-Libya: A case study. *Sci Rev Chem Commun*, 5(2), 51–56.
Nouri, D., Sabour, M. R. and GhanbarzadehLak, M. (2016). Industrial solid waste management through the application of multi-criteria decision-making analysis: A case study of Shamsabad industrial complexes. *J Mater Cycles Waste Manage*, 20(1), 1–16.
Novarlić, B., Stević, Ž., Durić, P. and Vasiljević, M. (2017). Efficiency in organizing transport routes as part of the city waste management: Proposal for innovative way of transport. *Int J Quality Res*, 11(3).
Omran, A. and Gebril, A. O. (2018). Municipal Solid waste management practices in the central part of Libya. *The Impact of Climate Change on Our Life*. Springer, Singapore, pp. 149–165.
Omran, A., Altawati M. and Davis, G. (2017). Identifying municipal solid waste management opportunities in Al-Bayda City, Libya. *Environ, Develop Sustain*, 95(3), 1–17.
Pamučar, D., Badi, I., Sanja, K. and Obradović, R. (2018a). A novel approach for the selection of power-generation technology using a linguistic neutrosophic CODAS method: A case study in Libya. *Energies*, 11(9), 2489.
Pamučar, D., Stević, Ž. and Sremac, S. (2018b). A new model for determining weight coefficients of criteria in MCDM models: Full consistency method (FUCOM). *Symmetry*, 10(9), 393.
Puška, A., Kozarević, S., Stević, Ž. and Stovrag, J. (2018). A new way of applying interval fuzzy logic in group decision making for supplier selection. *Econ Comput Econ Cybernet Stud Res,* 52(2).
Radović, D., Stević, Ž., Pamučar, D., Zavadskas, E., Badi, I., Antuchevičiene, J. and Turskis, Z. (2018). Measuring performance in transportation companies in developing countries: A novel rough ARAS model. *Symmetry*, 10(10), 434.
Rezaei, P., Rezaie, K., Nazari-Shirkouhi, S. and Tajabadi, M. R. (2013). Application of fuzzy multi-criteria decision making analysis for evaluating and selecting the best location for construction of underground dam. *Acta Polytechnica Hungarica,* 10(7), 187–205.
Saaty, T. L. (1980). *The Analytic Hierarchy Process.* Mc Graw-Hill, New York.
Sakurai, K. (2014). A time and motion study of municipal solid waste collection system. *Jap Int Cooperation Agency.*
Sawalem, M., Badi, I. and Aljamel, S. (2015). Evaluation of solid wastes for utilization in biogas plants in Libya: A case study. *Int J Eng Sci Res Technol,* 4(11), 577–583.
Sawalem, M., Selic, E. and Herbell, J. D. (2009). Hospital waste management in Libya: A case study. *Waste Manage*, 29(4), 1370–1375.

Soltani, A., Hewage, K., Reza, B. and Sadiq, R. (2015). Multiple stakeholders in multi-criteria decision-making in the context of municipal solid waste management: a review. *Waste Manage*, 35, 318–328.

Sremac, S., Stević, Ž., Pamučar, D., Arsić, M. and Matić, B. (2018). Evaluation of a third-party logistics (3PL) provider using a rough SWARA–WASPAS model based on a new rough Dombi aggregator. *Symmetry*, 10(8), 305.

Stević, Ž., Pamučar, D., Subotić, M., Antuchevičiene, J. and Zavadskas, E. (2018a). The location selection for roundabout construction using rough BWM-Rough WASPAS approach based on a new rough hamy aggregator. *Sustainability*, 10(8), 2817.

Stević, Ž., Vasiljević, M., Zavadskas, E. K., Sremac, S. and Turskis, Z. (2018b). Selection of carpenter manufacturer using fuzzy EDAS method. *Eng Econom*, 29(3), 281–290.

Stević, Ž., Pamučar, D., Vasiljević, M., Stojić, G. and Korica, S. (2017). Novel integrated multi-criteria model for supplier selection: Case study construction company. *Symmetry*, 9(11), 279.

Stević, Ž., Tanackov, I., Vasiljević, M., Novarlić, B. and Stojić, G. (2016). An integrated fuzzy AHP and TOPSIS model for supplier evaluation. *Serbian J Manage*, 11(1), 15–27.

Tüysüz, F. and Kahraman, C. (2006). Project risk evaluation using a fuzzy analytic hierarchy process: An application to information technology projects. *Int J Intell Syst*, 21(6), 559–584.

Yu, X., Guo, S., Guo, J. and Huang, X. (2011). Rank B2C e-commerce websites in e-alliance based on AHP and fuzzy TOPSIS. *Expert Syst Appl*, 38(4), 3550–3557.

Zadeh, L. A. (1965). Information and control. *Fuzzy Sets*, 8(3), 338–353.

Chapter 6

Development of Hybrid Tool using AHP and BSC Models for Analyzing Performance of Waste Disposal Management System of a Municipality

Kanika Prasad*, Rishi Dwivedi†,‡ and Ayushi Sharma†

**National Institute of Technology, Department of Manufacturing Engineering, Jamshedpur 831014, India*

†Xavier Institute of Social Service, Department of Finance, Ranchi 834001, India

‡rishidwivedi12@yahoo.co.in

Waste management is important as well as challenging for humankind and many market segments. It is linked to diverse worldwide challenges, like health, climate change, poverty alleviation, food and resource security, and sustainable manufacturing and utilization. The sustainable development objective for an organization cannot be achieved without addressing waste management as a priority. Therefore, waste disposal in an accountable, secure and principled manner is enormously significant for waste management. Moreover, efficient waste management system benefits an enterprise through cost savings and promoting environment-friendly practices. India faces foremost ecological threats related to waste generation and derisory waste collection, transport, handling and dumping. Existing systems in India can barely effectively deal with the amount of waste generated by a growing metropolitan populace. The challenges and blockade are noteworthy, but so are the prospects. Hence, it is imperative to understand the key performance drivers of waste disposal management system of a municipality. Balanced Scorecard (BSC) model is a modern strategic management technique that can help managers to identify an organization's Key Performance Indicators (KPIs) on four interrelated perspectives, i.e., financial, customer, internal business processes and learning and growth. Therefore, this chapter first develops a BSC model for a Municipal Corporation in India to facilitate enterprise view of an organization's overall performance to the policymakers. Next, a framework based on Analytical Hierarchy Process (AHP) method is designed to measure the performance of the Municipal Corporation over a period of time with respect to KPIs in the BSC model in order to validate the efficacy of strategies employed.

1. Introduction

In today's competitive world, there is a scramble for limited resources and energy is getting more costly. Therefore, all around the globe, soil, wind and water effluents present a threat to continual progress. Waste management is strongly connected to this problem and devoted to handling waste production, characterization, minimization, compilation, separation, treatment and disposal. Appropriate waste disposal is mandatory as certain types of wastes can be dangerous and can pollute the atmosphere if not managed suitably. Further, problems related to waste disposal are aggravated by shifting patterns of spending, business expansion and citification. This consecutively means that conventional methods for waste removal and reprocessing are no more appropriate. The progress to a new sustainable world needs superior processing to handle dissipated materials. A conservative reductionist technique is untenable as effortless removal and long-term view are necessary. Additionally, disposing of waste in an accountable, secure and principled manner is tremendously vital when it comes to waste supervision. A sturdy focus on reprocessing can diminish the undesirable effects of waste accumulation on the surroundings. Commitment to waste reduction and recycling will diminish the quantity of materials being sent to dump yards and incinerators, and the emission of greenhouse gas and other forms of contaminants will be minimized.

With speedy urbanization, India is witnessing a huge waste disposal and administration problem. Based on report published by *Down To Earth* in 2018, in India, over 377 million urban people live in 7,935 towns and cities and produce 62 million tonnes (MT) of municipal solid waste per annum. Only 43 MT of the waste is accumulated, whereas 11.9 MT is treated and 31 MT is dumped in landfill sites. Solid waste administration is one of the fundamental services offered by municipal authorities in the country to maintain towns dirt-free. The key to proficient waste management is to guarantee efficient separation of waste at its source and to make sure that the waste goes through diverse streams of reprocessing and recovery of resources. Subsequently, condensed final dregs are dumped in sanitary landfills. Almost all municipal establishments dump solid waste at a junkyard inside or outside the town unsystematically. Therefore, it is important to understand the key indicators of performance for a municipality in order to achieve sustainability. Furthermore, around 100 cities are identified to be developed as smart cities. Hence, municipalities have to redraw their enduring vision in solid waste supervision and amend their policies according to transforming standards of living. Balanced Scorecard (BSC) is an exceptionally prominent management technique that has found great acceptance by enterprises around the globe. It has evolved to become one of the most efficient strategy implementation frameworks. BSC model assists organizations to articulate their policy in actionable terms. It provides a road map for marshalling and supporting executives and employees and making strategy a continual process. BSC is a multifaceted method to analyze performance of an organization by incorporating both financial and nonfinancial parameters. It stabilizes four performance perspectives, i.e., financial, customer, process, and learning and growth. With the BSC technique, enterprises can effortlessly establish the cause–effect connection between actions and outcomes, analyze the effect a new software implementation will have on financial performance or the impact a new machinery installment will have on hastening product improvement. No other management technique exemplifies linking of diverse parameters with such precision. Analytical hierarchy process (AHP) is a multicriteria model that presents a method for evaluating alternatives by structuring the criteria into a hierarchy, providing for pairwise assessment of criteria at the smallest echelon of the hierarchy to be entered by the client/user and synthesizing the results into a solitary numerical value. It helps to institute the relation among the goal, criteria, subcriteria and alternatives.

AHP is employed to measure intangibles while utilizing human opinion. It has application in group decision-making and is used all over the globe in an extensive range of decision situations, like government, business, industry, health care, shipbuilding and education.

Hence, this chapter intends to first apply the BSC model in a municipal corporation of a city of India to demonstrate its efficiency in resolving different tactical, operational and strategic management problems. AHP technique is then implemented in this research work for quantitative evaluation of significance of all the identified key performance indicators (KPIs) in an already applied BSC model. Subsequently, a performance management framework is developed to examine the performance of said municipal corporation over two time periods while using the weightages of the chosen KPIs as obtained through AHP technique.

2. Literature Review

Leung *et al.* (2006) showed that the AHP and its variant, the analytic network process (ANP) models, could be tailor-made for specific situations. Numerical examples were also included for better understanding of the methodology. Gumus (2009) evaluated hazardous waste transportation enterprises while using two-step fuzzy-AHP and technique for order of preference by similarity to ideal solution (TOPSIS) methodologies. Govindan *et al.* (2014) identified barriers to the implementation of a green supply chain management based on procurement effectiveness. Influential barriers/priorities were identified through the AHP method. Gupta *et al.* (2015) presented an AHP model of manufacturing sustainability through different manufacturing practices, which provided a better environment in the organizations. Sedighi *et al.* (2015) identified and prioritized critical success factors for the implementation of knowledge management system in the Iranian energy sector while employing the AHP model. Veisi *et al.* (2016) determined the strategies and ethics of sustainability in agriculture and food systems in Iran through implementation of AHP method. Ho and Ma (2018) reviewed the literature on integrated AHP approaches and its applications published between

2007 and 2016, and compared those studies with papers published during the previous decade, i.e., 1997–2006. Roodposhti *et al.* (2018) integrated AHP and numerical taxonomy techniques for determining the weights of criteria by decision-makers and subsequently rankings of the firms were carried out using those estimated weights.

Bhagwat and Sharma (2007) applied the AHP technique to prioritize and formulate viable performance measurement strategies of an organization in a volatile and complex global decision environment from different BSC perspectives. A demonstration of the application of the proposed methodology in a real-time problem was also presented. Cebeci (2009) implemented BSC and fuzzy AHP method to compare various enterprise resource planning (ERP) system solutions for a textile manufacturing company. Huang (2009) integrated BSC and AHP methodologies to develop an intellectual BSC knowledge-based system. This system could automate the strategic planning process for a business. Kurien and Qureshi (2012) integrated the AHP method with modified BSC model to facilitate effective green supply chain performance measurement. Modak *et al.* (2017) developed an effective performance evaluation framework based on BSC and fuzzy AHP techniques to examine the suitability of an organization's strategic decision of outsourcing in alignment with the organizational performance for the Indian coal mining organization. Singh *et al.* (2018) proposed a sustainability evaluation method for measuring performance of manufacturing SMEs using integrated fuzzy AHP and fuzzy inference system. A BSC framework was applied to categorize the performance indicators among four different aspects.

Krivokapiü and Jovanoviü (2009) focused on an analysis of environmental management system integration into the classical model of the BSC to improve environmental performance through the implementation of strategy. Ravi *et al.* (2009) developed a framework based on the combination of BSC and ANP approaches for the selection of an alternative for the reverse logistics operations for end-of-life computers. Wu *et al.* (2009) designed a fuzzy multiple criteria decision-making (FMCDM) evaluation model for measuring banking performance employing the BSC framework. Yüksel and Dagdeviren (2010) integrated the BSC approach with fuzzy ANP methodology to determine the performance level of a business based on its vision and strategies. The proposed model showed that the performance indicators with different structures included in BSC method could be consolidated with the help of fuzzy ANP technique. Staš *et al.* (2015) developed a framework for the evaluation of green transport performance and support of an effective implementation of green transport strategies using ANP and BSC methodologies. The verification of the designed conceptual framework was performed on a real supply chain of the European automotive industry. Monteiro and Ribeiro (2017) presented different possibilities for the integration of environmental issues into the BSC model, not only in the private sector but also in the public sector. The work emphasized the need to adapt the original model of the BSC to public entities. Duman *et al.* (2018) proposed a BSC-based approach while integrating Decision-Making Trial and Evaluation Laboratory (DEMATEL) and ANP methodologies for performance evaluation of a US-based food franchise. Farokhi and Roghanian (2018) designed a quantitative methodology for setting targets in the framework of BSC model to achieve vision and goals of an organization. The desirability function-based approach of response surface methodology was used to assess the significant relationships that should be included in the strategy map and the optimal values of performance measures. Rodrigues *et al.* (2018) developed an assessment model-based Multi-Criteria Decision Aid-Constructivist (MCDA-C) methodology, which allowed comparison of the proposed strategic objectives for solid waste management and the available performance information.

It can be concluded from the review of past research works that both AHP and BSC models have been implemented independently to assess and streamline the various organization's strategies. Further, there are substantial published research papers on the integrated applications of AHP and BSC models to devise competitive management schemes and guidelines for the enterprises. But, the earlier papers have not explored the applicability of integrated AHP–BSC technique to analyze the performance of a government municipal corporation. Therefore, this paper proposes the development of a combined AHP–BSC model for a municipal corporation, which works in coordination with the state government in India for the effective execution of the various plans and programs. The detailed implementation procedure of the integrated model in that municipal corporation is also illustrated. The results derived from the implementation of the proposed combined AHP–BSC technique would not only assist the administrators of the municipal corporation to augment the productivity and performance the enterprise in order to achieve sustainable competitive edge, but also help them to determine the comprehensive advancement of the municipality during a time period in quantitative terms.

3. BSC Model

The BSC technique, proposed by Kaplan and Norton (2001), is a modern-age management technique designed to overcome the limitations of traditional performance management (TPM) system, which encompasses only financial measures. It is one of the most appropriate performance management tool and strategic control technique that has evolved over the years. Its main focus is to present management with a brief summary of the significant success factors of a business and also to ease the arrangement of business functioning in sync with the global strategy. A wide number of enterprises across various sectors have incorporated the BSC model while catering to a variety of management needs. The BSC model helps to convert the vision into the objectives and these objectives are then converted into a structure of performance measurement that efficiently delivers an impressive, forward looking, strategic focus to the organization for both long-term and short-term objectives. The BSC technique facilitates the inducement of new and better processes, which in turn help in achieving long-term sustainability. This model is designed on four interdependent perspectives, i.e., financial, customer, internal business processes, and learning and growth. These different perspectives have different KPIs that are vital for different stakeholders of business (shareholders, customers and employees). The financial perspective stresses on the long-term objectives of the organization in terms of earning growth and productivity improvement; this perspective should be the end objective for the rest of the perspectives. The customer perspective stresses on the lifetime relation and service delivery with customer, whereas the internal process perspective emphasizes on the usage of customers' information to deal with new products and services according to the needs of the clients. In addition, the learning and growth perspective is the base of the BSC methodology, and emphasizes on the motivation, training, satisfaction, retention and capacity that employees require for developing innovative strategies. A performance evaluation based on identified KPIs under these four perspectives of BSC model assists the management in better decision-making while instituting a balance between short-term and long-term objectives. This model thus helps in finding out the answers of some questions like

- How well one should present them in front of the shareholders?
- How to manage customers in most efficient way?
- What are the processes that enterprise should focus on in order satisfy diverse stakeholders?
- What is being done to ensure that the firm adapts and innovates to fulfill its vision?

Thus, in a nutshell, it can be observed that the main objective of the BSC model is to find and define a close set of measures that help the management to understand the position of the organization so as to regulate and guide it.

4. Designing of a BSC Model for the Municipal Corporation

For long-term stability and validation, it is very essential and important for an entity to design a suitable BSC model based on its extraordinary sets of constraints. Further, while developing the model, proper care should be provided for an equitable depiction of both financial as well as operational measures in it. Here, a team consisting of the mayor and deputy mayor along with the subject experts is consulted for developing the BSC model for one of the municipal corporations in the Jharkhand state of India taking into consideration the limitations, preferences and geographical area. The developed

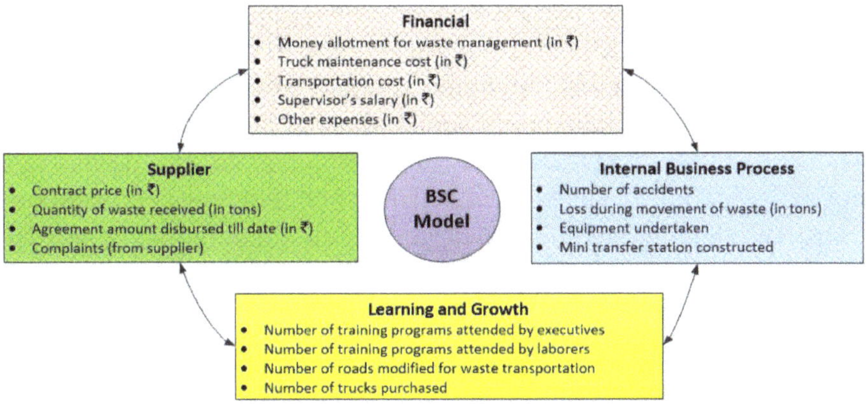

Figure 1: Developed BSC model for the municipal corporation.

BSC model for the municipal corporation is shown in Figure 1. In the designed BSC model, customer perspective is interchanged with supplier perspective as municipal corporation has to mainly deal with the supplier.

Each recognized performance parameter depicts a specific objective or strength of the municipal corporation. For instance, contract price depicts how well the waste management is carried out by the authorities while covering all the necessary requirements within the estimated budget. It also presents an idea about the viability of the considered waste management project. The quantity of waste received shows how much tons of waste are lifted and transported to dump yard for ultimate recycling. The agreement amount disbursed till date means the efficiency of the municipality in terms of fund allocation, whereas complaints from suppliers provides an indication of the level of satisfaction or dissatisfaction of teams working with them. Similarly, the performance measures of number of training programs attended by executives and laborers shows municipal corporation's efficacy in modernizing skill sets of its employees. Besides, number of roads modified for waste transportation indicate the flexibility of mobility of waste transportation system. It shows how efficient the municipal corporation is in proper disposal of municipal waste. New trucks purchased show coverage competency of the waste collection system of the municipality. Apart from all these money allotment for waste management, truck maintenance cost, transportation cost, supervisor's salary and other expenses show the financial effectiveness of the said enterprise. Moreover, number of accidents, loss during movement of waste, equipment undertaken and construction of Mini Transfer Station (MTS) provide an overview of the key parameters from the internal business perspective of the municipal corporation. Thus, BSC model helps the management to adopt better strategies for future and also aids in monitoring the growth of the organization.

5. AHP Model

AHP is an implicit technique that works on complex and multi-criteria decision-making (MCDM) problem for a preferable decision-making through prioritizing certain factors and evaluating alternatives on those weighted criteria. Saaty (1990) developed this model to measure managerial judgments of the relative significance of each of numerous conflicting criteria employed in the decision-making process. It is a tool that takes into consideration both qualitative as well as quantitative data. It also helps in checking the consistency of the output attained and thus the main feature of this technique is that it reduces bias in the output as well as in decision-making. The following steps of AHP method have been used to evaluate the comparative importance of the selected attributes. In the first step the ultimate objective, criteria and decision options are listed. Next, a pairwise comparison matrix, say matrix A, is developed,where A is a matrix of $m \times m$ dimension, and m is the criteria that have been taken into consideration. Each entry a_{jk} of the matrix A represents the usefulness of the jth criterion that is relative to the kth criterion. If $a_{jk} > 1$, then the kth criterion is less important than the jth criterion, while if $a_{jk} < 1$, then the kth criterion is more significant than the jth criterion. If both the criteria have equal weight, then in that case the entry a_{jk} is 1. The product of criterion a_{jk} and a_{kj} should always be equal to 1. Saaty's (1990) numerical scale of 1–9 is used, as shown in Table 1, for measuring the relative importance among two criteria, and an assumption is made that jth criterion is more important than kth criterion or both hold an equal position.

Table 1: Relative scale for pairwise comparison of criteria in AHP method.

Intensity of importance	Definition
1	Equal importance
3	Somewhat more important
5	Much more important
7	Very much more important
9	Absolute much more important
2,4,6,8	Intermediate Values

Besides, there might be situations where ratings may show slight inconsistency although it does not affect the working of AHP.

The subsequent step in the AHP methodology after the development of pairwise comparison matrix is to extract the relative significance implied by the previous comparisons. That is, how important are the identified criteria when they are considered in terms of the chosen objective? Saaty (1990) asserts that to respond to this question one has to calculate the right principal eigenvector of the previous matrix. Given a pairwise comparison matrix with relative importance, the corresponding maximum left eigenvector is computed by using the geometric mean of each row. That is, the elements in each row are multiplied with each other and then the mth root is taken (where m is the number of elements in the row). After that the numbers are normalized by dividing them with their summation. The judgment carried out in AHP method through pairwise comparison is considered to be sufficiently consistent if the corresponding consistency ratio (CR) is less than 10%. For estimation of CR coefficient, the consistency index (CI) needs to be calculated. This is carried out through adding the columns in the pairwise comparison matrix and multiplying the resulting vector by the vector of priorities (i.e., the estimated eigenvector) obtained earlier. This results in an approximation of the maximum eigenvalue, represented by $\lambda_{\max}$. Afterwards, the CI value is computed through employing the formula; $\mathrm{CI} = (\lambda_{\max} - m)/(m - 1)$. Then, the CR is estimated by dividing the CI value by the random consistency index (RCI).

5.1. *Estimation of importance of BSC performance parameters using AHP*

The AHP method is adopted here to estimate the weight for each identified performance measure of BSC model in considered municipal corporation. Here, the objective of the said enterprise is to achieve sustainable operational performance. For achieving long-term existence, 17 performance parameters for the four perspectives of BSC model are identified for municipal corporation. Then, following the steps of the AHP approach as discussed in previous section, at first pairwise comparison matrix is developed for the 17 selected KPIs. Then, eigenvector is calculated in order to finally estimate the significance of chosen BSC parameters. Table 2 shows the computed normalized weights of different BSC performance parameters, as estimated by employing the AHP approach.

In order to check the consistency of these estimated weights of diverse KPIs of municipal corporation, CR is calculated. It is observed that CR for the judgments carried out in above mentioned problem is 9.8%, which signifies adequate consistency.

Table 2: Normalized weights of BSC performance measures.

Performance measure	Normalized weight (%)
Contract price	4.669
Quantity of waste received	10.171
Agreement amount disbursed till date	3.984
Complaints (from supplier)	2.154
Money allotment for waste management	2.729
Truck maintenance cost	8.849
Transportation cost	7.517
Other expenses	2.982
Supervisor's salary	8.347
Number of accidents	2.045
Loss during movement of waste	14.907
Equipment undertaken	3.770
MTS constructed	3.346
Number of training programs attended by executives	10.489
Number of training programs attended by laborers	4.623
Number of roads modified for waste transportation	4.981
Number of trucks purchased	4.438

6. Designing of a Performance Index to Evaluate the Overall Performance of the Municipal Corporation

After the determination of relative importance of all KPIs of the BSC model, an index is established to analyze and evaluate the comprehensive performance of municipal corporation (Liberatore and Miller, 1998). Thus, for the calculation of index and evaluation of the progress of the chosen enterprise with consideration of the parameter of the BSC model, data for two distinctive years is required for 17 KPIs. The primary set of data linked to the 17 KPIs for the initial period is set as the baseline value, which in this case is accumulated from municipal corporation's record books for the financial year 2016–17. Besides, current period value exhibits the second data set for those parameters and is gathered from the record books of municipal corporation for the financial year 2017–18. An initial value of 100 is allocated to all KPIs and consequently, their weighted points are estimated using the following equation:

$$\text{Weighted point} = \text{Normalized performance measure weight (NPMW)} \times \text{Initial point} \tag{1}$$

After that, the weighted points estimated for each of the 17 KPIs are summed up to determine the performance index for the initial period. A detailed estimation of performance index of municipal corporation for the initial period is shown in Table 3.

With a view to analyze and evaluate the progress of the organization over the considered time period, the performance index for the next period is also needed. Thus, the preliminary stride in computing the performance index for the current quarter is to calculate the current period points, employing either Equation (2) (for beneficial performance parameters) or Equation (3) (for non-beneficial performance parameters).

$$\text{Current period point} = (\text{Current quarter value}/\text{Baseline value}) \times 100 \tag{2}$$

Table 3: Performance index for the initial period.

Performance measure	Baseline value	Initial point	Weighted point
Contract price	1900	100	4.669
Quantity of waste received	330	100	10.171
Agreement amount disbursed till date	12	100	3.984
Complaints (from supplier)	20	100	2.154
Money allotment for waste management	8900000	100	2.729
Truck maintenance cost	9000	100	8.849
Transportation cost	600	100	7.517
Other expenses	25	100	2.982
Supervisor's salary	8000	100	8.347
Number of accidents	3	100	2.045
Loss during movement of waste	50	100	14.907
Equipment undertaken	80	100	3.770
MTS constructed	2	100	3.346
Number of training programs attended by executives	1	100	10.489
Number of training programs attended by laborers	1	100	4.623
Number of roads modified for waste transportation	3	100	4.981
Number of trucks purchased	150	100	4.438
Performance index value	—	—	100.000

$$\text{Current period point} = 200 - ((\text{Current quarter value}/\text{Baseline value}) \times 100) \quad (3)$$

The assessment of the current period weighted point for the individual performance parameter is now in similar way estimated using Equation (4).

$$\text{Current period weighted point} = \text{NPMW} \times \text{Current period point} \quad (4)$$

Moreover, to calculate the performance index for the current period, all the current period weighted points for all the individual performance parameters are added. A detailed estimation of performance index of municipal corporation for the current period as computed using integrated AHP and BSC models is shown in

Table 4: Current period performance index.

Performance measure	Current year value	Current period point	Current period weighted point
Contract price	1950	97.368	4.546
Quantity of waste received	297	110.000	11.188
Agreement amount disbursed till date	8	66.667	2.656
Complaints (from supplier)	25	75.000	1.615
Money allotment for waste management	9000000	98.876	2.698
Truck maintenance cost	10000	88.889	7.866
Transportation cost	730	78.333	5.888
Other expenses	30	80.000	2.385
Supervisor's salary	10000	75.000	6.260
Number of accidents	2	133.333	2.727
Loss during movement of waste	80	160.000	23.851
Equipment undertaken	115	143.750	5.420
MTS constructed	5	250.000	8.366
Number of training programs attended by executives	2	200.000	20.978
Number of training programs attended by laborers	3	300.000	13.868
Number of roads modified for waste transportation	4	133.333	6.641
Number of trucks purchased	180	120.000	5.326
Performance index value	—	—	132.279

Table 4. An evaluation of the performance index values for the current period with those for the initial period, derived from Tables 4 and 3, respectively, helps in measuring the holistic progress of the municipal corporation over the two years and thus enabling the policymakers to realize the effectiveness of the various programs implemented for waste management.

The overall performance of the municipal corporation has improved by 32.279 points. This signifies that the said enterprise is moving in a positive direction with respect to identified KPIs. In other words, the increment in performance index value over the two time periods shows the overall efficiency of waste management system employed by the policymakers. Moreover, the maximum individual improvement in the index value over two chosen years is observed in a number of training programs attended by executives, which shows that the executives of this municipality are updating themselves in order to be in sync with the dynamic business environment. Further, it is noticed that the least improvement with respect to index value is observed in number of accidents parameter. It implies that on that criteria the enterprise is touching the threshold limit and its management needs to take some preventive measure in accident management before things go into the negative region. The municipal corporation has performed worst on supervisor's salary as derived from the comparison of index values for two time periods, and therefore some cost control techniques need to be adopted.

7. Conclusion

With ever-increasing urbanization and elevated purchasing power to consume more, all nations in the world are encountering the mounting problem of waste management. Countries with superior technological and economic competency have been able to turn their garbage into capital, avoiding environmental pollution, while numerous developing nations have been struggling to find effective solutions to the problem. The benefits of innovative garbage disposal are extensively discussed all over the globe but not effectively implemented in countries like India. Therefore, there is an immediate need for the development of an innovative tool, which not only helps in providing management-oriented information but also assists in instituting establishment of a sound strategy development system at the enterprise level. This chapter demonstrates the applicability of hybrid AHP–BSC model in providing information that can act as a supporting tool in regulating and administering garbage, including planning for waste management and waste avoidance, minimization and reuse, regulation of waste transport, storage, treatment, resource recovery and disposal, and managing the impacts of waste management activities. Besides, the results derived from the implementation of proposed technique aid in monitoring the performance of the municipal corporation, which can be consequently utilized as a driver for improving collection, management and disposal of urban waste. This improves recycling, reusing and reducing of waste to minimize the amount of garbage that is sent to the landfills, protects the dwindling resources, reduces emissions, pollution and contamination, which is ultimately required for achieving sustainability. Further, it can be comprehended from the results obtained from the application of the proposed performance measurement tool in municipal corporation for the considered time periods that the said enterprise is moving in the right direction with respect to all selected performance parameters, except contract price, agreement amount disbursed till date, complaints (from supplier), money allotment for waste management, truck maintenance cost, transportation cost, other expenses and supervisor's salary. Even though the integrated AHP–BSC model is designed and developed for a municipal corporation, it can also be employed to other enterprises with slight modifications.

References

Bhagwat, R. and Sharma, M. K. (2007). Performance measurement of supply chain management using the analytical hierarchy process. *PP&C*, 18(8), 666–680.

Cebeci, U. (2009). Fuzzy AHP-based decision support system for selecting ERP systems in textile industry by using balanced. *Expert Syst Appl*, 36(5), 8900–8909.

Duman, G. M., Taskaynatan, M., Kongar, E. and Rosentrater, K. A. (2018). Integrating environmental and social sustainability into performance evaluation: A balanced scorecard-based grey-DANP approach for the food industry. *Front Nutr*, 5, 65.

Farokhi, S. and Roghanian, E. (2018). Determining quantitative targets for performance measures in the balanced scorecard method using response surface methodology. *Manage Decision*, 56(9), 2006–2037.

Govindan, K., Kaliyan, M., Kannan, D. and Haq, A. N. (2014). Barriers analysis for green supply chain management implementation in Indian industries using analytic hierarchy process. *Intl. J. Prod. Econ*, 147(B), 555–568.

Gumus, A. T. (2009). Evaluation of hazardous waste transportation firms by using a two-step fuzzy-AHP and TOPSIS methodology. *Expert Syst Appl*, 36(2), 4067–4074.

Gupta, S., Dangayach, G. S., Singh, A. K. and Rao, P. N. (2015). Analytic hierarchy process (AHP) model for evaluating sustainable manufacturing practices in Indian electrical panel industries. *Procedia Soc Behav Sci*, 189, 208–216.

Ho, W. and Ma, X. (2018). The state-of-the-art integrations and applications of the analytic hierarchy process. *Eur J Oper Res*, 267(2), 399–414.

Huang, H-C. (2009). Designing a knowledge-based system for strategic planning: A balanced scorecard perspective. *Expert Syst Appl*, 36(1), 209–218.

Kaplan, R. S. and Norton, D. P. (2001). Transforming the balanced scorecard from performance measurement to strategic Management: Part II. *Account Horiz*, 15(2), 147–160.

Krivokapiü, Z. and Jovanoviü, J. (2009). Using balanced scorecard to improve environmental management system. *J Mech E*, 55(4), 262–271.

Kurien, G. P. and Qureshi, M. N. (2012). Performance measurement systems for green supply chains using modified balanced score card and analytical hierarchical process. *Sci Res Essays*, 7(36), 3149–3161.

Leung, C., Lam, K. C. and Cao, D. (2006). Implementing the balanced scorecard using the analytic hierarchy process and the analytic network process. *J Op Res Soc*, 57(6), 682–691.

Liberatore, M. J. and Miller, T. (1998). A framework for integrating activity-based costing and the balanced scorecard into the logistics strategy development and monitoring process. *J Bus Logist*, 19(2), 131–154.

Modak, M., Pathak, K. and Ghosh, K. K. (2017). Performance evaluation of outsourcing decision using a BSC and fuzzy AHP approach: A case of the Indian coal mining organization. *Resour Policy*, 52, 181–191.

Monteiro, S. and Ribeiro, V. (2017). The balanced scorecard as a tool for environmental management: Approaching the business context to the public sector. *Manag Environ Qual Int J*, 28(3), 332–349.

Ravi, V., Shankar, R. and Tiwari, M. K. (2009). Analyzing alternatives in reverse logistics for end-of-life computers: ANP and balanced scorecard approach. *Comp Indus Eng*, 48(2), 327–356.

Rodrigues, A. P., Fernandes, M. L., Rodrigues, M. F. F., Bortoluzzi, S. C., Da Costa, S. G. and De Lima, E. P. (2018). Developing criteria for performance assessment in municipal solid waste management. *J Clean Prod*, 186, 748–757.

Roodposhti, F. R., Jahromi, M. B. and Kamalzadeh, S. (2018). Portfolio selection using analytic hierarchy process and numerical taxonomy analysis: Case study of Iran. *Am J Fin Account*, 5(4), 394–414.

Saaty, T. L. (1990). How to make a decision: The analytic hierarchy process. *Eur J Oper Res*, 48(1), 9–26.

Sedighi, M., Splunter, S. V., Zand, F. and Brazier, F. (2015). Evaluating critical success factors model of knowledge management: An analytic hierarchy process (AHP) approach. *Intl J Know Manage*, 3(11), 17–36.

Singh, S., Olugu, E. U., Musa, S. N. and Mahat, A. B. (2018). Fuzzy-based sustainability evaluation method for manufacturing SMEs using balanced scorecard framework. *J Intell Manuf*, 29(1), 1–18.

Staš, D., Lenort, R., Wicher, P. and Holman, D. (2015). Green transport balanced scorecard model with analytic network process support. *Sustainability*, 7(11), 15243–15261.

Veisi, H., Liaghati, H. and Alipour, A. (2016). Developing an ethics-based approach to indicators of sustainable agriculture using analytic hierarchy process (AHP). *Ecol Indic*, 60, 644–654.

Wu, H.-Y., Tzeng, G.-H. and Chen, Y.-H. (2009). A fuzzy MCDM approach for evaluating banking performance based on balanced scorecard. *Expert Syst Appl*, 36(6), 10135–10147.

Yüksel, I. and Dagdeviren, M. (2010). Using the fuzzy analytic network process (ANP) for balanced scorecard (BSC): A case study for a manufacturing firm. *Expert Syst Appl*, 37(2), 1270–1278.

Chapter 7

Productivity Modeling of Manufacturing Industry Using the Rough Analytic Hierarchy Process

Bolun Liang*, Golam Kabir†,‡ and Guoqing Zhang*

**Department of Mechanical, Automotive & Materials Engineering, University of Windsor, Windsor, ON, Canada, N9B 3P4*

†Industrial Systems Engineering, University of Regina, Regina, SK, Canada, S4S 0A2

‡golam.kabir@uregina.ca

In the manufacturing industry, productivity modeling can be used as an essential tool for predicting future productivity and allocating optimal resources for production. This chapter introduces a rough number set method into a hierarchy manufacturing model to establish a typical productivity model. Rough number set describes a range of uncertain criteria weight. By implementing this method, the criteria in each level of an analytic hierarchy process (AHP) model can be defined with an array with the same space instead of a precise number. After multiplying the rough number sets, the relative global weight of each parameter can be calculated. With this method, a productivity model can be established with imprecise input. This model responds with a predictable change of overall productivity with the change of imprecise input.

1. Introduction

Productivity is the measure of yield efficiency of a manufacturing process, which can be represented by the output (final products) over a certain amount of input (i.e., capital spent, material cost, labor hour). In actual production processes, various parameters exert influence on productivity. For example, a higher utilization of the manufacturing line would yield a higher output rate, thus increasing the productivity. While a higher level of quality of raw materials would reduce waste and scrap, thereby increasing productivity by reducing unnecessary input. Each of these parameters influence final productivity at different levels. Furthermore, as many manufacturing parameters do not directly exert an influence on final productivity, but act through other parameters, productivity could not be determined directly but can be analyzed with a structured model.

Productivity modeling means constructing a model with productivity as the final goal and all other manufacturing parameters as the input to the model. By changing the input, the model can give an estimated final productivity. Meanwhile, with a fixed desired output, such a model can also return the corresponding relative importance of all inputs to achieve the desired productivity. Such prediction can play as a critical role in allocating future resources to maximize the profit.

In the manufacturing industry, the essential elements that are related to productivity are the effectiveness of the product and the efficiency of the processes. These two elements directly decide the output rate of the qualified product and the overall input required. Meanwhile, these two factors are influenced by other manufacturing parameters. As a result, the structure of the entire manufacturing system is often established with a hierarchy or network form. In this case, the production model is actually a multicriteria decision-making problem.

In the early stage, the research of productivity modeling considers output and input as two separate factors (Anyaeche, 2009; Fulenwider, 2004). These models analyze the connection between the productivity with each input individually. These input/output orientated model could reflect a relatively precise impact of every single factor on the final productivity. But they cannot reveal the internal relationship among the factors.

To provide a solution to this problem, the AHP approach was introduced as a method of analysis. An AHP model is a hierarchical structure. The vertex of the model is the overall goal of the model. A group of criteria branches from the overall goal with groups of subcriteria branching from them. The level of criteria is not limited but can be established depending upon the requirement of the problem (Kabir, 2013; Sedaghat, 2012; Ezzat, 2016).

For each criterion in AHP, its direct subcriteria or all alternative choices if it's in the lowest level are supposed to be set into a form of pairwise comparison, with the respective weight or importance between subjects. Geometric mean method and eigenvalue method are two possible means to calculate the relative weight of all the criteria in comparison (Saaty, 1987). By multiplying all the connected criteria weight matrix, the overall influence on the final productivity of each criterion could be calculated.

In practice, the criteria are not always given with a precise value, but often with a value range as there is always error in conducting real processes. In this case, a rough number set (Song *et al.*, 2014; Zhai *et al.*, 2008) could be implemented for solving such a situation.

A rough number set describes a range in which the criteria lies within. It is defined with a set of three numbers standing for the lower limit, acceptable average and higher limit. This method returns an adjusted average lower approximation and higher approximation for each subcriterion. By multiplying the weight range, the global weight of each criterion can be calculated.

In this chapter, a hierarchical model will be built for a typical manufacturing firm with all the criteria represented in the form of rough number sets. The relative importance level of all elementary manufacturing parameters is calculated. And a comparison between AHP with rough number set and traditional AHP method is done and discussed.

2. Methodology

2.1. *Analytical hierarchy process*

In an AHP model, a pairwise comparison matrix is created for each criterion. This matrix is of size $k \times k$ where k is the number of subjects

Table 1: Definitions of intensity importance (Saaty, 1987).

Intensity importance	Definition	Intensity importance	Definition
1	Equal importance	1	Equal importance
2	Slightly more important	1/2	Slightly less important
3	Moderately more important	1/3	Moderately less important
4	Moderate plus	1/4	Moderate plus
5	Strongly more important	1/5	Strongly less important
6	Strong plus	1/6	Strong plus
7	Very strongly more important	1/7	Very strongly less important
8	Very, very strongly more important	1/8	Very, very strongly less important
9	Extremely more important	1/9	Extremely less important

of this criteria. The element of the matrix represents the intensity of importance between any pair of subjects, represented by number 1 to 9 (Table 1). The higher the number is higher is the importance. As all subjects have equal importance to themselves, the diagonal of the matrix is filled with 1.

$$B = \begin{bmatrix} 1 & \cdots & x_{1k} \\ \vdots & \ddots & \vdots \\ x_{k1} & \cdots & 1 \end{bmatrix} \tag{1}$$

As the comparison matrix is established, the maximum eigenvalue $\lambda_{\max}^{e}$ of matrix B is to be calculated for the validation of consistency index (CI):

$$\mathrm{CI} = (\lambda_{\max}^{e} - n)/(n-1), \tag{2}$$

$$\mathrm{CR}^{e} = CI/RI(n). \tag{3}$$

With the random consistency index (RI) table, the consistency ratio CR^{e} could be calculated. If $\mathrm{CR}^{e} < 0.10$, the comparison matrix is acceptable.

Two well-accepted methods have been developed in solving the comparison matrix.

Table 2: Random consistency index table (Saaty and Vargas, 2012).

n	2	3	4	5	6	7	8	9	10
RI	0.00	0.52	0.89	1.11	1.25	1.35	1.40	1.45	1.49

- Geometric mean method:

$$W = \sqrt[k]{\prod_{n=1}^{k} x_{in}/\max(x_{mn})} \tag{4}$$

- Eigen value method:

$$W = \left(\sum_{i=1}^{k} \frac{x_{ij}}{\sum_{j=1}^{k} x_{ij}}\right) /\max(x_{mn}) \tag{5}$$

By multiplying the relative weight matrix of all connected criteria, the relative weight of each elementary parameter with respect to the final productivity could be calculated.

2.2. *Rough number sets*

Rough number (RN) (Zhai *et al.*, 2008) describes a range of possible values fitting the criteria. It is represented by an array of $E = (J_1, J_2, \ldots, J_k)$ where $J_1 < J_2 < \cdots < J_k$. Suppose U is the universe of all objects and X is a random subset of X, then $J_n \in E, 1 < n < k$ and $\forall\ X \in U$ (see Figure 1).

Then the lower approximation and higher approximation (Pawlak, 1991) are represented by

$$\underline{\mathrm{Apr}}(J_q) = \cup\{X \in U | E(X) \leq J_q\}, \tag{6}$$

$$\overline{\mathrm{Apr}}(J_q) = \cup\{X \in U | E(X) \geq J_q\}. \tag{7}$$

The boundary interval of corresponding rough number set is

$$\mathrm{Bnd}(J_q) = \cup\{X \in U | E(X) \neq J_q\}. \tag{8}$$

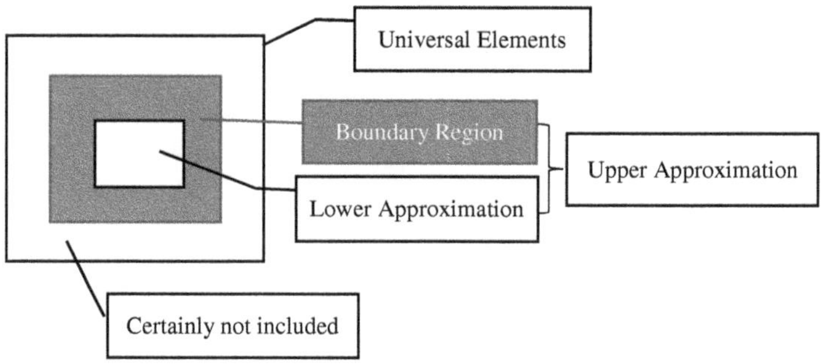

Figure 1: Basic notions of rough set theory (reproduced after Stević *et al.*, 2017).

The rough number of this set can be defined as

$$\mathrm{RN}(J_q) = [\underline{\lim}(J_q), \overline{\lim}(J_q)], \tag{9}$$

where

$$\underline{\lim}(J_q) = \frac{1}{M_L}\sum E(X)|X \in \underline{\mathrm{Apr}}(J_q), \tag{10}$$

$$\overline{\lim}(J_q) = \frac{1}{M_U}\sum E(X)|X \in \overline{\mathrm{Apr}}(J_q). \tag{11}$$

In this case, the intensity importance in pairwise comparison matrix is replaced with $\mathrm{RN}(J_q)$, or $[\underline{\lim}(J_q),\ \overline{\lim}(J_q)]$. To simplify, $RN(J_q) = [\underline{lim}(J_q),\ \overline{lim}(J_q)]$ will be denoted as $RN(a) = [x^L, x^U]$.

Therefore, the criteria matrix with rough number implemented (Roy *et al.*, 2016) could be represented by

$$M = \begin{bmatrix} [1,1] & \cdots & [x_{1k}^L, x_{1k}^U] \\ \vdots & \ddots & \vdots \\ [x_{k1}^L, x_{k1}^U] & \cdots & [1,1] \end{bmatrix} \tag{12}$$

The rough-based weight of each criterion can be calculated with

$$w = \left[\sqrt[k]{\prod_{i=1}^{k} x_{jk}^L},\ \sqrt[k]{\prod_{i=1}^{k} x_{jk}^U}\right]. \tag{13}$$

The normalized weight can be calculated with

$$w' = \frac{w}{\max(\mathrm{w}^{\mathrm{U}})}. \tag{14}$$

By multiplying relative weight matrix of all connected criteria, the relative weight of each elementary parameter with respect to the final productivity could be calculated.

3. Model Framework

A model for a typical manufacturing company can be built with productivity as the final goal and end with all processing input elements.

3.1. *Model description*

The manufacturing productivity can be measured by output over a certain amount of input over a process. The output is demonstrated by the amount of qualified product produced and the input is the time, material and labor cost to produce such a qualified product.

To improve productivity, two main approaches shall be taken: increasing output and decreasing input. It is common in manufacturing industry that machine conditions, management approaches, skills of employees and status are the mostly controllable parameters in routine operations.

Increasing output can be achieved by following approaches:

- Increase the utilization of current processes.
  - Improve warehouse and logistics to achieve higher on-time delivery rate.
  - Implementation of advanced management tool to increase line utilization.
  - Improve line flexibility to satisfy multiple processes with resources requirement unchanged.
- Achieve higher customer satisfaction, thus creating new demands.
  - Satisfy basic requirement on the product.
  - Attract new orders by increasing the variety of product.
  - Achieve fast response to demand changes.

Decreasing input could be achieved by the following methods:

- Decrease the scrap and waste caused by quality problems.

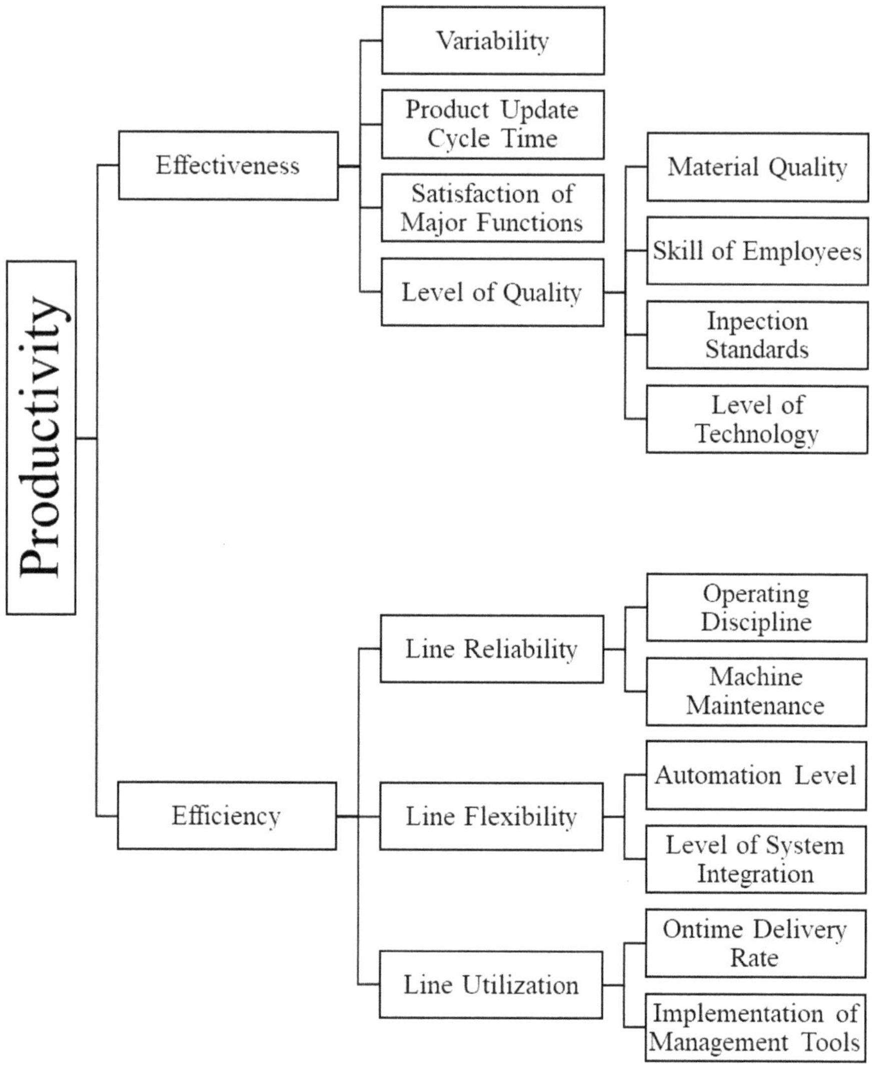

Figure 2: Hierarchy diagram of criteria sets.

— Improve the control of raw material quality level.
— Provide better training for employees to decrease human error.
— Improve quality inspection standard.
— Improve the technology implemented.

- Reduce the machine downtime.
  - Discipline employees to avoid misconduct.
  - Provide better machine maintenance.

Approaches stated above could be categorized into two main aspects: effectiveness of product and efficiency of processes. Therefore, the hierarchy model could be built with stated criteria as shown in Figure 2.

3.2. *Rough weights determination*

In this model, three sets of intensity importance ratio are collected for each pair of criteria. Before the calculation of weight, consistency is supposed to be tested. In this model, CR^e value satisfies the requirement that it is smaller than 1. Therefore, the proposed comparison matrix can be accepted.

3.2.1. *Weight calculation of tier I criteria, effectiveness and efficiency*

According to Table 3, take an example of relative importance of effectiveness over efficiency:

$$\underline{\lim}\left(\frac{1}{5}\right)=\frac{\frac{1}{5}}{1}=\frac{1}{5},$$

$$\overline{\lim}\left(\frac{1}{5}\right)=\frac{\frac{1}{5}+\frac{1}{3}+1}{3}=\frac{7}{15},$$

$$\underline{\lim}\left(\frac{1}{3}\right)=\frac{\frac{1}{5}+\frac{1}{3}}{2}=\frac{4}{15},$$

Table 3: Pair-wise comparison matrix and rough weights of tier I criteria, effectiveness and efficiency.

Tier I criteria	Effectiveness	Efficiency
Effectiveness	(1, 1, 1)	(1/5, 1/3, 1)
Efficiency	(1, 3, 5)	(1, 1, 1)

$$\overline{\lim}\left(\frac{1}{3}\right) = \frac{\frac{1}{3}+1}{2} = \frac{2}{3},$$

$$\underline{\lim}(1) = \frac{\frac{1}{5}+\frac{1}{3}+1}{3} = \frac{7}{15},$$

$$\overline{\lim}(1) = \frac{1}{1} = 1.$$

Therefore,

$$x_{12}^{L} = \frac{\frac{1}{5}+\frac{4}{15}+\frac{7}{15}}{3} = 0.44,$$

$$x_{12}^{U} = \frac{\frac{7}{15}+\frac{2}{3}+1}{3} = 0.84.$$

The pairwise comparison matrix can be converted into:

$$M_1 = \begin{bmatrix} [1.00, 1.00] & [0.44, 0.84] \\ [2.00, 4.00] & [1.00, 1.00] \end{bmatrix},$$

$$w_1 = \begin{Bmatrix} [0.66, 0.92] \\ [1.41, 2.00] \end{Bmatrix},$$

$$w_1' = \begin{Bmatrix} [0.33, 0.46] \\ [0.71, 1.00] \end{Bmatrix}.$$

3.3. *Result*

By multiplying the rough weight of the connected criteria, the relative weight of elementary manufacturing parameters versus the final productivity can be calculated.

In this case, the mean of the rough number boundary interval is taken to sort the rank of all criteria (Vasiljevic *et al.*, 2018).

As Table 4 and Figure 3 show, the implementation of management tools has the highest influence on final productivity, followed by machine maintenance, material quality, on-time delivery rate, satisfaction of major functions, inspection standards, level of system integration, operating discipline, level of technology, update cycle time, automation level and variability. The skill of employees has the lowest influence on productivity.

Table 4: Global rough weights for all final criteria.

Criteria	Lower end	Higher end	Difference	Mean	Rank
Variability	0.0594	0.1288	0.0694	0.0941	12
Update Cycle Time	0.0825	0.161	0.0785	0.12175	10
Satisfaction of Major Functions	0.1947	0.3772	0.1825	0.28595	5
Material Quality	0.208362	0.46	0.251638	0.334181	3
Skill of Employee	0.053361	0.1242	0.070839	0.0887805	13
Inspection Standards	0.162624	0.3588	0.196176	0.260712	6
Level of Technology	0.07623	0.1978	0.12157	0.137015	9
Operating Discipline	0.089815	0.1872	0.097385	0.1385075	8
Machine Maintenance	0.34364	0.72	0.37636	0.53182	2
Automation Level	0.05964	0.14	0.08036	0.09982	11
Level of System Integration	0.10934	0.25	0.14066	0.17967	7
On time Delivery Rate	0.196812	0.46	0.263188	0.328406	4
Implementation of Management Tools	0.423444	1	0.576556	0.711722	1

The data used for this model comes from a plastic injection company following a three-shift work and batch production. This company suffers from long down time resulting from machine failure. So, how to improve process reliability and improve usual line utilization to reduce the loss from machine failure has become a major concern for the company for years. Meanwhile, as the essential process, injection, is totally conducted by machine, the skill of employees has a fairly low impact on productivity. With these facts, it appears that

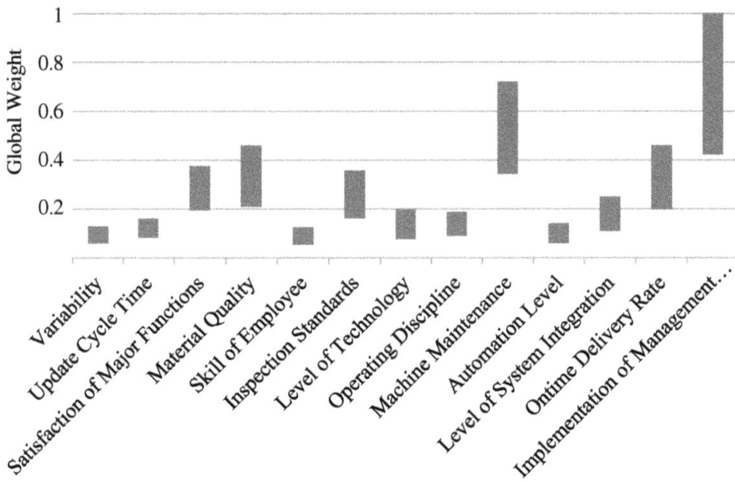

Figure 3: Criteria in rough numbers.

this model could reflect the actual relative importance of elementary parameters on final productivity for this specific case.

4. Case Study

ABC is an American building material company. As this company requires a moderately balanced intervention between automated machine and labor in the manufacturing process, it is considered to be an ideal case to test the productivity model.

From the 2012 annual report of ABC, the productivity was calculated to be 1.784 in fiscal year (FY) 2010 and 1.814 in FY 2011. An increase of 1.70% occurred in this period. To simplify the problem, the mean of each interval is taken in this case. By dividing each criteria weight by their sum, the relative percentage influence of each parameter on the productivity can be calculated (see Table 5).

These percentage of influence can work as a coefficient to describe how the change of each parameter would influence the final productivity.

During FY 2011, this company invested 3.08% more in engineering services and process maintenance and 7.29% in improving operating discipline, as a result of higher operating profits. Research and

Table 5: Percentage influence of parameters.

Criteria	Mean	Percentage influence
Variability	0.0941	0.0284
Update Cycle Time	0.12175	0.0368
Satisfaction of Major Functions	0.28595	0.0863
Material Quality	0.334181	0.1008
Skill of Employee	0.0887805	0.0268
Inspection Standards	0.260712	0.0787
Level of Technology	0.137015	0.0413
Operating Discipline	0.1385075	0.0418
Machine Maintenance	0.53182	0.1605
Automation Level	0.09982	0.0301
Level of System Integration	0.17967	0.0542
On time Delivery Rate	0.328406	0.0991
Implementation of Management Tools	0.711722	0.2149

development on product increased by 22.6% in the corresponding year.

On administrative and management aspect, Company ABC increased expense on employee satisfaction and training by 1.71%. An increase of 1.60% was incurred for material quality management.

With all the other parameters basically unchanged, the predicted productivity change equals 1.94%. In this case, the error between the estimated productivity increase and real value is 14.2%.

This result shows that in this specific case, this model performs effectively in predicting the future productivity change based on the change of other manufacturing parameters. In a different industry, the criteria must be evaluated based on their actual situations.

5. Comparison and Discussion

To compare the differences between rough AHP and conventional AHP methods, the mean of each rough number set interval is taken to conduct the conventional AHP analysis. The result is shown in Table 6 and Figure 4.

To provide a more straightforward demonstration, the mean of rough weight is normalized. As Table 6 and Figure 4 show, rough

Table 6: Comparison of criteria rank between rough AHP and AHP.

	Rough AHP				AHP	
Criteria	Lower end	Higher end	Normalized mean	Rank	Weight	Rank
Variability	0.0594	0.1288	0.0284081	12	0.0237446	11
Update Cycle Time	0.0825	0.161	0.0367554	10	0.0359935	8
Satisfaction of Major Functions	0.1947	0.3772	0.0863262	5	0.0912579	4
Material Quality	0.208362	0.46	0.1008868	3	0.0393786	7
Skill of Employee	0.053361	0.1242	0.0268022	13	0.009535	13
Inspection Standards	0.162624	0.3588	0.0787071	6	0.0331134	9
Level of Technology	0.07623	0.1978	0.0413638	9	0.0160359	12
Operating Discipline	0.089815	0.1872	0.0418144	8	0.0480225	6
Machine Maintenance	0.34364	0.72	0.1605526	2	0.1920899	2
Automation Level	0.05964	0.14	0.0301349	11	0.0304379	10
Level of System Integration	0.10934	0.25	0.0542411	7	0.0608757	5
On time Delivery Rate	0.196812	0.46	0.0991434	4	0.139838	3
Implementation of Management Tools	0.423444	1	0.2148638	1	0.279677	1

AHP and conventional AHP give similar overall trends in sorting the importance of criteria. But due to the precision of data provided, relatively significant differences do exist.

In this case, material quality level, inspection standards and level of technology have the highest differences. Such a result is the direct consequence of using different criteria scales (Mukhametzyamov and Pamucar, 2017).

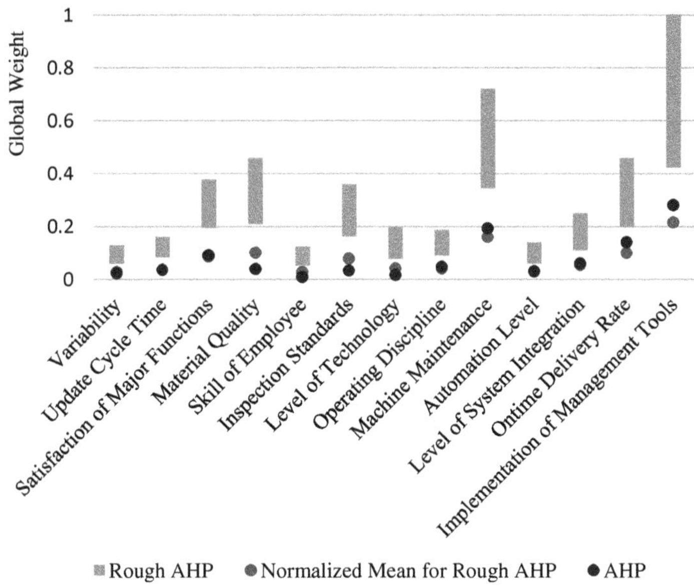

Figure 4: Comparison of criteria rank between rough AHP and AHP.

6. Conclusion

This chapter discussed the implementation of rough AHP in productivity modeling. A model describing the change in productivity with regard to percentage change of manufacturing parameters in the entire process is developed and tested. With the input parameters, the model returns with a relatively precise result.

This model prompts the user to generate a set of rough numbers as the boundary of criteria weight. The rough number theory processes the input set and generates lower approximation and higher approximation for all the criteria in the hierarchy model. By multiplying the weight of each criterion, the global weight of each manufacturing parameter is calculated. After normalization, the global influence would work as the coefficient describing how productivity reflects the change of input parameters. This method provides an effective tool in establishing productivity model with imprecise input parameters.

In practice, manufacturing structures are not always hierarchical. Each manufacturing parameter could exert influence on others and affect productivity at the same time. In this case, a network model is established.

To provide an effective solution for such case, a network model with imprecise approach would be considered to be a better method. A credal network model is a good tool in combining imprecise input with the network model, which could be an essential aspect for further research.

References

Anyaeche, C. and Oluleye, A. (2009). A productivity evaluation model based on input and output orientations. *The South African J Industr Eng,* 20(1).

Eshraghniae, J. A. and Borjalilu, N. (2010). Presenting a model for productivity measurement and improvement in Plastiran. In *8th International Management Conference.*

Ezzat, A. and Hamoud, H. (2016). Analytic hierarchy process as module for productivity evaluation and decision0making of the operation theater. *Avicenna J Med*, 6(1), 3–7.

Fulenwider, M., Helmes, P., Mojtahedzadeh, M. and MacDonald, R. (2004). Operational labor productivity model. In *2004 SD Conference.*

Kabir, A., Latif, H. and Sarker, S. (2013). A multi-criteria decision-making model to increase productivity: AHP and fuzzy AHP approach. *Int. J. Intelligent Systems Technologies and Applications*, 12(3/4).

Mukhametzyanov, I. and Pamučar, D. (2017). Multi-criteria decision-making methods: Sensivity of results in variation of estimaons of alternatives by criteria. In *The 1st International Conference on Management, Engineering and Environment ICMNEE*, pp. 2–24.

Nadiri, M. and Prucha, I. (1999). Dynamic factor demand models and productivity analysis. National Bureau of Economic Research.

Pawlak, Z. (1991). *Rough Sets: Theoretical Aspects of Reasoning about Data.* Kluwer Academic Publishing, Dordrecht.

Saaty, R. W. (1987). *The Analytical Hierarchy Process — What it is and how it is used*, 4922 Ellsworth Avenue, Pittsburgh, PA 15213, USA.

Saaty, T. L. and Vargas, L. G. (2012). Models, methods, concepts & applications of the analytic hierarchy process. *Springer Sci Business Media*, 175(2), 1–346.

Sedaghat, M. (2012). A Productivity Improvement Evaluation Model by Integrating AHP, TOPSIS and VIKOR Methods under Fuzzy Environment (Case Study: State-owned, Partially Private and Private Banks in Iran). *Economic Computation and Economic Cybernetics Studies and Research,* Academy of Economic Studies.

Song, W., Ming, X., Wu, Z. and Zhu, B. A. (2014). Rough TOPSIS approach for failure mode and effects analysis in uncertain environments. *Quality Reliab Eng Int*, 30(4), 473–486.

Stević, Ž. (2017). Criteria for supplier selection: A literature review. *Int J Eng, Business Enterprise Appl*, 19(1), 23–27.
Vasiljević, M., Fazlollahtabar, H., Stević, Z. and Vesković, S. (2018). A rough multicriteria approach for evaluation of the supplier criteria in automotive industry. *Decision Making: Appl Manage Eng*, 1(1), 82–96.
Zhai, L. Y., Khoo, L. P. and Zhong, Z. W. (2008). A rough set enhanced fuzzy approach to quality function deployment. *The Int J Adv Manufact Technol*, 37(5), 613–624.

Appendix A

A.1. *Rough weight determination*

Weight calculation of effective-related criteria (see Table A.1)

$$M_2 = \begin{bmatrix} [1.00, 1.00] & [0.45, 0.79] & [0.33, 0.73] & [0.23, 0.30] \\ [1.50, 2.50] & [1.00, 1.00] & [0.30, 0.43] & [0.27, 0.42] \\ [2.00, 4.00] & [2.50, 3.50] & [1.00, 1.00] & [0.72, 0.94] \\ [3.50, 4.50] & [2.61, 4.11] & [1.11, 1.56] & [1.00, 1.00] \end{bmatrix},$$

$$w_2 = \left\{ \begin{matrix} [0.43, 0.64] \\ [0.59, 0.82] \\ [1.38, 1.90] \\ [1.78, 2.32] \end{matrix} \right\},$$

Table A.1: Pair-wise comparison matrix and rough weights of effective-related criteria.

Effectiveness	Variability	Update cycle time	Satisfaction major functions	Level of quality
Variability	(1, 1, 1)	(1/3, 1/2, 1)	(1/5, 1/3, 1)	(1/5, 1/4, 1/3)
Update Cycle Time	(1, 2, 3)	(1, 1, 1)	(1/4, 1/3, 1/2)	(1/5, 1/3, 1/2)
Satisfaction of Major Functions	(1, 3, 5)	(2, 3, 4)	(1, 1, 1)	(1/2, 1, 1)
Level of Quality	(3, 4, 5)	(2, 3, 5)	(1, 1, 2)	(1, 1, 1)

Table A.2: Pair-wise comparison matrix and rough weights of efficiency-related criteria.

Efficiency	Line reliability	Line flexibility	Line utilization
Line Reliability	(1, 1, 1)	(2, 3, 5)	(1/2, 1/2, 1)
Line Flexibility	(1/5, 1/3, 1/2)	(1, 1, 1)	(1/5, 1/4, 1/3)
Line Utilization	(1, 2, 2)	(3, 4, 5)	(1, 1, 1)

$$w_2' = \begin{Bmatrix} [0.18, 0.28] \\ [0.25, 0.35] \\ [0.59, 0.82] \\ [0.77, 1.00] \end{Bmatrix}.$$

Weight calculation of efficiency-related criteria (see Table A.2)

$$M_3 = \begin{bmatrix} [1.00, 1.00] & [2.61, 4.11] & [0.56, 0.78] \\ [0.27, 0.42] & [1.00, 1.00] & [0.23, 0.30] \\ [1.44, 1.89] & [3.50, 4.50] & [1.00, 1.00] \end{bmatrix},$$

$$w_3 = \begin{Bmatrix} [1.13, 1.47] \\ [0.40, 0.50] \\ [1.71, 2.04] \end{Bmatrix},$$

$$w_3' = \begin{Bmatrix} [0.55, 0.72] \\ [0.20, 0.25] \\ [0.84, 1.00] \end{Bmatrix}.$$

Weight calculation of level of quality-related criteria (see Table A.3)

$$M_2 = \begin{bmatrix} [1.00, 1.00] & [3.50, 4.50] & [1.11, 1.56] & [2.50, 3.50] \\ [0.23, 0.30] & [1.00, 1.00] & [0.30, 0.43] & [0.45, 0.79] \\ [0.72, 0.94] & [2.50, 3.50] & [1.00, 1.00] & [1.50, 2.50] \\ [0.30, 0.43] & [1.50, 2.50] & [0.45.0.79] & [1.00, 1.00] \end{bmatrix},$$

Table A.3: Pair-wise comparison matrix and rough weights of level of quality-related criteria.

Level of quality	Material quality	Skill of employee	Inspection standards	Level of technology
Material Quality	(1, 1, 1)	(3, 4, 5)	(1, 1, 2)	(2, 3, 4)
Skill of Employee	(1/5, 1/4, 1/3)	(1, 1, 1)	(1/4, 1/3, 1/2)	(1/3, 1/2, 1)
Inspection Standards	(1/2, 1, 1)	(2, 3, 4)	(1, 1, 1)	(1, 2, 3)
Level of Technology	(1/4, 1/3, 1/2)	(1, 2, 3)	(1/3, 1/2, 1)	(1, 1, 1)

Table A.4: Pair-wise comparison matrix and rough weights of line utilization-related criteria.

Line utilization	On-time delivery rate	Implementation of management tools
On time Delivery Rate	(1, 1, 1)	(1/2, 1/2, 1)
Implementation of Management Tools	(1, 2, 2)	(1, 1, 1)

$$w_2 = \begin{Bmatrix} [1.76, 2.23] \\ [0.42, 0.57] \\ [1.30, 1.69] \\ [0.67, 0.96] \end{Bmatrix},$$

$$w_2' = \begin{Bmatrix} [0.79, 1.00] \\ [0.19, 0.26] \\ [0.58, 0.76] \\ [0.30, 0.43] \end{Bmatrix}.$$

Weight calculation of line utilization-related criteria (see Table A.4)

$$M_5 = \begin{bmatrix} [1.00, 1.00] & [0.44, 0.84] \\ [2.00, 4.00] & [1.00, 1.00] \end{bmatrix},$$

$$w_5 = \begin{Bmatrix} [0.66, 0.92] \\ [1.41, 2.00] \end{Bmatrix}$$

$$w'_5 = \begin{Bmatrix} [0.33, 0.46] \\ [0.71, 1.00] \end{Bmatrix}.$$

Weight calculation of line reliability-related criteria (see Table A.5)

$$M_6 = \begin{bmatrix} [1.00, 1.00] & [0.23, 0.30] \\ [3.50, 4.50] & [1.00, 1.00] \end{bmatrix},$$

$$w_6 = \begin{Bmatrix} [0.48, 0.55] \\ [1.87, 2.12] \end{Bmatrix}$$

$$w'_6 = \begin{Bmatrix} [0.23, 0.26] \\ [0.88, 1.00] \end{Bmatrix}.$$

Weight calculation of line flexibility-related criteria (see Table A.6)

$$M_7 = \begin{bmatrix} [1.00, 1.00] & [0.45, 0.79] \\ [1.50, 2.50] & [1.00, 1.00] \end{bmatrix},$$

$$w_7 = \begin{Bmatrix} [0.67, 0.89] \\ [1.22, 1.58] \end{Bmatrix},$$

$$w'_7 = \begin{Bmatrix} [0.42, 0.56] \\ [0.77, 1.00] \end{Bmatrix}.$$

Table A.5: Pair-wise comparison matrix and rough weights of line reliability-related criteria.

Line utilization	Operating discipline	Machine maintenance
Operating Discipline	(1, 1, 1)	(1/5, 1/4, 1/3)
Machine Maintenance	(3, 4, 5)	(1, 1, 1)

Table A.6: Pair-wise comparison matrix and rough weights of line flexibility-related criteria.

Line utilization	Automation level	Level of system integration
Automation Level	(1, 1, 1)	(1/3, 1/2, 1)
Level of System Integration	(1, 2, 3)	(1, 1, 1)

Chapter 8

Abrasive Assisted Hybrid EDM of Titanium Alloy for Enhanced Productivity

Preetkanwal Singh Bains* and Sarabjeet Singh Sidhu

Mechanical Engineering Department,
BCET, Gurdaspur, India
**preetbains84@gmail.com*

This investigation aims to predict the optimal machining conditions in abrasive assisted electrical discharge machining (EDM) of Ti-based biomaterial. The present study focuses on tool erosion rate (TER), material erosion rate (MER) and surface roughness (SR) analysis of titanium (Ti–6Al–4V)-based biomaterial in SiC-mixed hybrid ED machining. Copper electrodes with a variable inner hole diameter were selected while machining in SiC_p mixed dielectric medium. The output was globally optimized by desirability function technique to predict the suitable parametric combinations for the selected workpiece machining. This machining route revealed a considerable enhancement in MER and SR in abrasive-mixed dielectric fluid. In this dielectric flushing method, a considerable decline in tool wear affected by peak current while using SiC particulates was observed. Subsequently, SR was predominantly enhanced under the influence of hole size, pulse-off and abrasive mixed dielectric as the control parameters. The scanning electron micrographs revealed vital influence pertaining pulse-on duration, peak current and type of dielectric fluid on the microstructure of machined surfaced of Ti–6Al–4V samples. In a nutshell, the use of novel hybrid machining route and the optimization technique of experimentation proved to be beneficial in preserving product quality and manufacturing equipment, which can be a source for additional environmental sustainability and process costs.

1. Introduction

Titanium and its alloys are considered as the most promising metals finding extensive applications in biomedical industry (Long and Rack, 1998). Ti–6Al–4V, the most common among Ti-alloys, has been preferred since long for these applications. Owing to their distinctive characteristics, i.e. advanced corrosion resistance, fatigue endurance, appreciable strength-to-weight ratio (Abdul-Rani *et al.*, 2018), Ti alloys are significantly employed in modern industries. In addition, nonferromagnetic properties of titanium rods aid in safe and secure patient examinations like magnetic resonance imaging (MRI) and could last there for decades (Nourbakhsh *et al.*, 2013). Besides biomedical domain, myriad other industries like automobile, chemical, electronic, aerospace and food industries utilize titanium in many applications (Gu *et al.*, 2012). However, titanium's high toughness and low thermal conductivity result in hard to machine surface of these materials (Wansheng *et al.*, 2002). To combat such machining problems, nontraditional machining methods, for instance EDM, Laser beam, ECM and abrasive jet machining (AJM), are the reliable and affordable alternatives. Electrical discharge machining (EDM), among these, has been effectively used to machine hard-to-cut and high-strength materials (Kumar *et al.*, 2009). This technique has been well known having a possible deduction in the energy consumption, allowing the manufacturing exceptionally effective in terms of environmental protection and sustainability. The judiciously chosen machining factors help to achieve optimal performance in transferring the molten debris from the gap while machining (Mohal and Kumar, 2017; Bains *et al.*, 2016a). As the low material removal rate (MRR) and poor surface quality had been the major drawback of the simple EDM while large-scale production, aluminum powder in the dielectric fluid (kerosene) was used to augment the material removal capacity. Herein, promising results were achieved in terms of decline in SR (Ra) as compared to traditional ED machining (Talla *et al.*, 2015). An effective and convenient flushing route was recommended to enhance machining execution along with reduced recast layer that made use of an abrasive particle mixed dielectric fluid jet (Arantes *et al.*, 2016). MER of EDM largely depends on the flushing conditions and to enhance this capacity of EDM, tube electrodes of varying inner diameters were employed. In this experiment, the surface quality along with machining efficiency was investigated thereafter. The results showed that increase in internal diameter enhanced the surface quality and material removal efficiency (Zhang *et al.*, 2016). Other than this, a hybrid process of ED machining was investigated (Bains *et al.*, 2018), wherein the researchers employed an external magnetic field coupled with ED machining, which resulted in improving the ejection efficiency of debris. Moving ahead, a setup was proposed (Nguyen *et al.*, 2012) in which both the operations of micro EDM and micro ECM were performed, simultaneously. The authors concluded that this mechanism significantly improved the machining accuracy and surface finish by using de-ionized water.

AJM has been the focus of many researchers (Bains *et al.*, 2016b). It is among the most suitable options for processing of brittle and hard workpiece efficiently in operations like surface finishing, deburring and cleaning of surfaces. Abrasive jet polishing (Tsai *et al.*, 2008) was employed for surface finishing of ED-machined SKD 61 mold steel. In this experiment, the responses registered a significant decline in SR. Furthermore, in a research study conducted, the authors (Lin *et al.*, 2012) had investigated the hybridized EDM technique with AJM considering several parameters. They concluded that abrasive particles used in hybrid process not only improved the finish of machined part but also the material removal capacity.

The powder-mixed EDM has acknowledged the stability of machining process toward surface modification of composites and Ti implants (Bains *et al.*, 2019). The primary aim of this investigation was to analyze the performance of abrasive particle-assisted EDM operation on machining characteristics, for instance, MER, TER and SR. The crucial in-process variables, for instance, peak current, pulse-on/off duration, inner diameter of tube electrode and type of dielectric were selected to analyze their influence on execution of process. The objective of the study is to enhance productivity, i.e., higher MER with lower TWR and also aim to develop biocompatible surface, i.e., higher SR.

2. Materials and Methods

In this experiment, abrasive mixed dielectric was made to flow through the two hollow electrodes of electrolytic copper (Figure 1) having an outer diameter of 15 mm and internal diameters of 2 and 4 mm, respectively. Ti–6Al–4V (ASTM Grade 5) was selected as the workpiece (5 mm × 70 mm × 70 mm). The chemical composition of Ti–6Al–4V is shown in Table 1. EDM oil of commercial grade was chosen as the dielectric fluid. SiC (220 mesh) were mixed in 30 g/L of dielectric and stirred consistently with a pump to ensure uniform mixing.

An orthogonal array of $L_{16}(2)^{15}$ Taguchi design having a five-factor setup was selected for conducting 16 trials (Table 2). The sequence of experiments was performed on EDM (S645, ZNC, OSCARMAX). A specifically designed tank and a pump arrangement for uniform mixing of abrasive particles were selected as the setup. The schematic layout diagram of designed setup has been briefly illustrated in Figure 2. For comparative analysis, experiments were performed with both normal and abrasive particulate-mixed dielectric through the electrode at a pressure of 10 bar. The weights of workpiece and electrodes were recorded with a precision electronic balance (0.001-g precision), before and after machining.

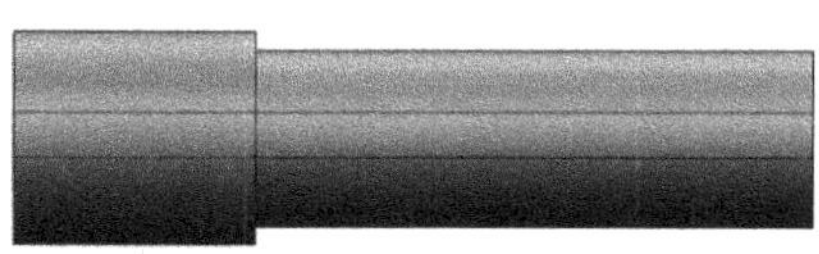

Figure 1: Copper electrode with through hole.

Table 1: Chemical composition of Ti–6Al–4V.

Component	Al	V	Fe	O	C	N	H	Ti
Weight (%)	6.08	4.02	0.22	0.18	0.02	0.01	0.0053	balance

Table 2: Control variables with levels.

Control parameters	Level 1	Level 2
Pulse on (μs)	45	90
Current (A)	6	12
Flushing type	Plain dielectric	Abrasive mixed
Pulse off (μs)	45	90
Electrode hole diameter (mm)	2	4

Figure 2: Pictorial demonstration of experimental setup.

3. Results and Discussion

The planned experimental design and output obtained, thereafter, has been demonstrated in Table 3. Taking into consideration the efficiency of EDM, the optimization of the technological parameters

Table 3: Experimental plan and results.

Trial No.	Hole Dia (mm)	Current (A)	Pulse-on (μs)	Pulse-off (μs)	Dielectric fluid	MER (mg/min)	TER (mg/min)	SR (μm)
1	2	6	45	45	1	1.481	2.463	0.1533
2	2	6	45	90	2	1.750	1.666	0.4175
3	2	6	90	45	2	2.875	1.875	0.5425
4	2	6	90	90	1	1.380	1.921	0.5940
5	2	12	45	45	2	4.600	1.800	1.0333
6	2	12	45	90	1	1.542	2.866	1.0466
7	2	12	90	45	1	2.540	3.200	1.0256
8	2	12	90	90	2	4.058	1.945	1.0060
9	4	6	45	45	2	2.549	1.108	0.2050
10	4	6	45	90	1	3.844	0.645	0.7033
11	4	6	90	45	1	3.698	0.426	0.9820
12	4	6	90	90	2	3.436	0.550	0.2300
13	4	12	45	45	1	3.280	3.714	0.5740
14	4	12	45	90	2	4.575	1.005	0.2233
15	4	12	90	45	2	6.986	0.745	1.1600
16	4	12	90	90	1	1.966	3.579	0.6975

in terms of sustainability were varied toward the maximization of the MER. ANOVA was employed to analyze various outcomes to confirm the effectiveness of the implemented model. The analysis of variances (Table 3) indicated that inner diameter of the electrode, peak current and type of dielectric are the most influencing input parameters. The main effect plot for the MER in Figure 3 shows the effect of discharge current, the inner diameter of the tool, duration of pulse-on/off and type of dielectric media on material removal as the output. In line with previous researches, the rise in discharge energy results in the electric spark getting stronger and generating high temperature in the concerned area, thereby achieving enhanced material removal.

Trial 15 (I: 12A; abrasive mixed dielectric; Φ 4 mm; P-on: 90 μs) witnessed higher MER, ranging from 2.528 to 3.692 mg/min. A 170% rise in the MER values has been recorded in these settings as compared to plain dielectric with 2 mm hole diameter (Trial 7), when other conditions were kept identical. This can be attributed to rise in pulsating current producing higher spark energy along with augmented discharge of abrasive fluid, which resulted in more melting and erosion of molten debris. Hence, an overall 406.2% increase in

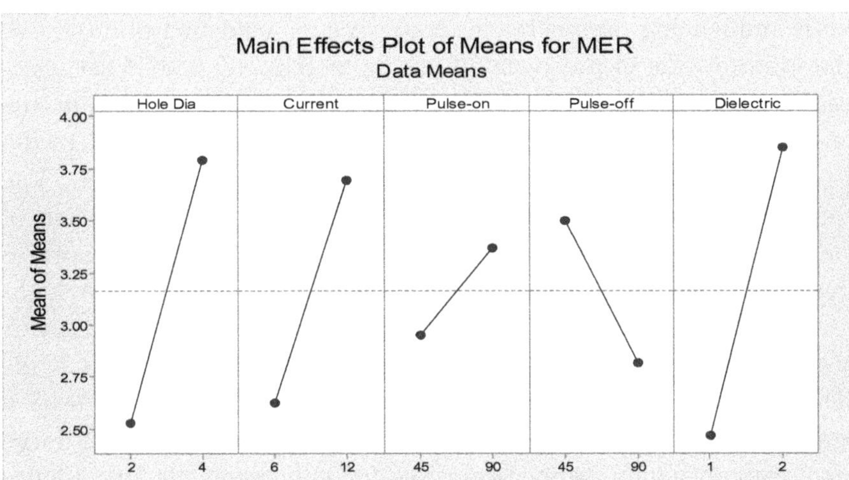

Figure 3: Main effects plot for means for MER.

Table 4: Analysis of variance of means for MER.

Source	DF	Seq SS	Adj SS	Adj MM	F value	p value
Hole Dia (A)	1	6.3857	6.38573	6.38573	9.09	0.030*
Current (B)	1	4.5518	4.55182	4.55182	6.48	0.052*
Pulse-on (C)	1	0.6881	0.68807	0.68807	0.98	0.368
Pulse-off (D)	1	1.8619	1.86186	1.86186	2.65	0.164
Dielectric fluid (E)	1	7.6979	7.69785	7.69785	10.96	0.021*
Hole Dia × Pulse-off	1	0.0003	0.00034	0.00034	0.00	0.983
Hole Dia × Dielectric	1	0.1564	0.15642	0.15642	0.22	0.657
Current × Pulse-on	1	0.0028	0.00281	0.00281	0.00	0.952
Current × Dielectric	1	7.1342	7.13424	7.13424	10.15	0.024*
Pulse-on × Dielectric fluid	1	1.2343	1.23432	1.23432	1.76	0.242
Residual error	5	3.5130	3.51296	0.70259		
Total	15	33.2264				

Note: * Significant at 95%.

material removal capacity has been observed in hybrid ED machining of workpiece with abrasive-mixed dielectric as compared to traditional EDM and AJM, individually.

The wear of tools while machining the workpiece was affected by various machining parameters, such as current, inner diameter of electrode and type of dielectric fluid. Current was the most significantly influencing parameter for tool wear, as evident from Table 4, whereas duration of pulse-on/off and hole diameter were least effective. The tool wear was enhanced sharply with a current in the range of 6–12 A. SiC-mixed abrasive particles in dielectric media also proved significant while enhancing the TER (see Figure 4). The irregular machined surfaces are observed due to the development of craters and pits owing to the sparks and subsequent material erosion during machining. All the chosen control variables, as evident from ANOVA (Table 5), proved to be significant toward the surface quality. Current (12A) had the highest influence on SR while ED machining. Besides, inner diameter of the electrode showed a significant effect on surface quality, resulting in deeper and large sized craters, which were responsible for enhancing SR (see Figure 5). Trial 1 showcased the minimum SR (0.1533 μm) of the sample at the topmost pulse-on setting (90 μs) and current (12A) with 2 mm

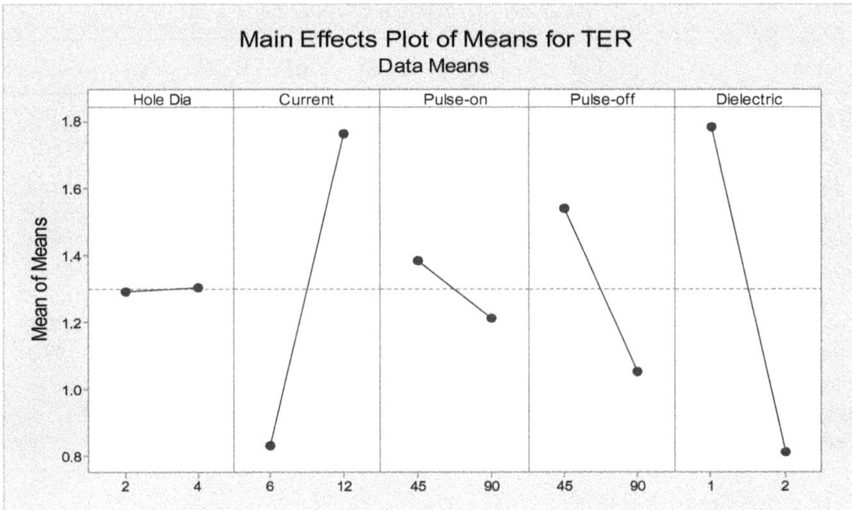

Figure 4: Main effects plot for means for TER.

inner diameter of electrode in normal dielectric media. Deep craters were formed due to the creation of multiple discharges from a single pulse input. Although efficient debris removal was facilitated by crater formation, it also depreciated the quality of machined surface (Bains *et al.*, 2017).

The SR values were significantly increased by hole diameter, pulse-off duration and type of dielectric. The maximum roughness of 1.1600 μm (Trial 15) was recorded with the 4mm inner diameter of the tube electrode, at high-end pulse-off setting (90 μs).

SEM micrographs were captured to examine the morphology of the machined surface. The discharge craters, microcracks and splashed molten debris were spotted on machined faces of a workpiece (Figures 6 and 7) showing the surfaces of the maximum and minimum values of MER and SR, respectively. The generation of craters on the machined surfaces accompanied by the cracks and molten drops was attributed to elevated temperature produced in each discharge.

A higher discharge current created deeper and larger craters, which resulted in enhancement in the roughness of the surface, as

Table 5: Analysis of variance of means for TER.

Source	DF	Seq SS	Adj SS	Adj MM	F value	p Value
Hole dia	1	2.2231	2.22308	2.22308	4.02	0.101
Current	1	4.2025	4.20250	4.20250	7.60	0.040*
Pulse-on	1	0.0658	0.06579	0.06579	0.12	0.744
Pulse-off	1	0.0832	0.08323	0.08323	0.15	0.714
Dielectric fluid	1	4.1209	4.12090	4.12090	7.45	0.041*
Hole Dia × Pulse-off	1	0.0329	0.03294	0.03294	0.06	0.817
Hole Dia × Dielectric	1	0.2007	0.20070	0.20070	0.36	0.573
Current × Pulse-on	1	0.0891	0.08910	0.08910	0.16	0.705
Current × Dielectric	1	3.6176	3.61760	3.61760	6.54	0.051*
Pulse-on × Dielectric	1	0.0006	0.00060	0.00060	0.00	0.975
Residual error	5	2.7639	2.76394	0.55279		
Total	15	17.4004				

Note: *Significant at 95%.

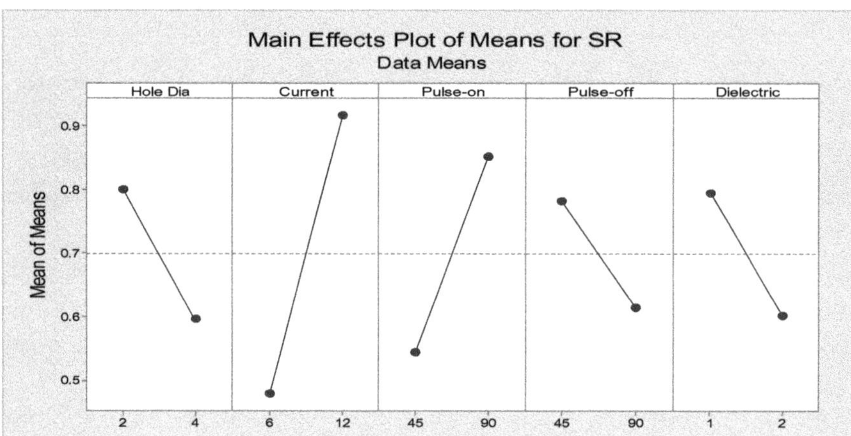

Figure 5: Main effects plot for means for SR.

observed in Figures 7(a) and 7(b) in the current intensity of 6 and 12A, respectively. Similarly, with the decrease in current, surface deterioration was less and surface finish improved. Also, splashes of the molten particles on the machined surface are observed in Figure 7(b), which increased the SR of the sample. As seen in Figure 7(b), some spherically shaped debris is present on the machined surface

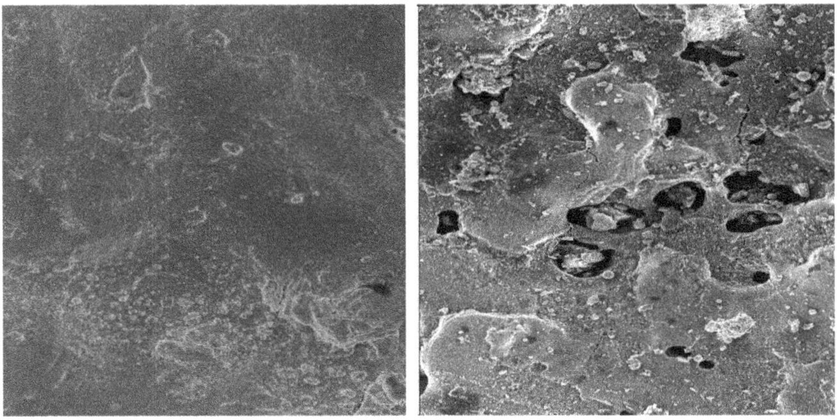

(a) (b)

Figure 6: SEM micrographs (a) minimum MER (Trial 6), (b) maximum MER (Trial 15).

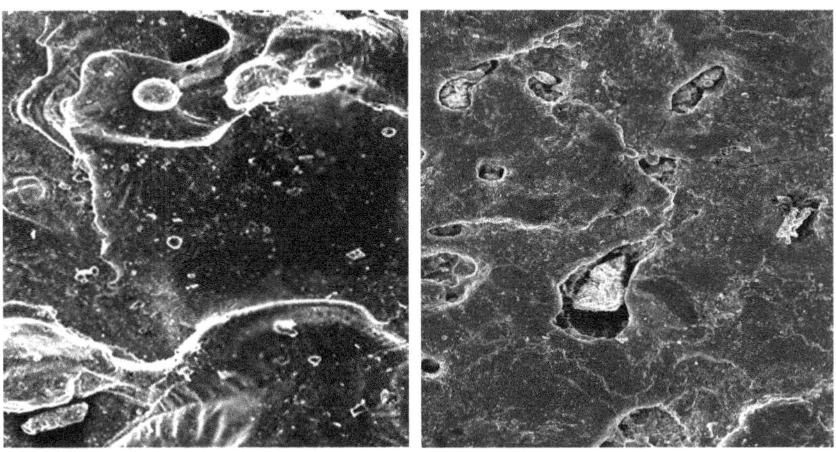

(a) (b)

Figure 7: SEM micrographs (a) minimum SR (Trial 1) and (b) maximum SR (Trial 15).

(Trial 1). This is because flushing with nonabrasive fluid was not effective enough to flush out all the melt debris generated while high current and subsequently some molten material were impregnated onto the surface of the tool.

Table 6: Analysis of variance for means of SR.

Source	DF	Seq SS	Adj SS	Adj MS	F value	p Value
Hole Dia	1	0.1641	0.1641	0.16415	6.53	0.038*
Current	1	0.7725	0.7725	0.77247	30.73	0.001**
Pulse-on	1	0.3777	0.3777	0.37767	15.03	0.006*
Pulse-off	1	0.1113	0.1113	0.11129	4.43	0.073*
Dielectric Fluid	1	0.1474	0.1474	0.14738	5.86	0.046*
Hole Dia × Current	1	0.3741	0.3741	0.37412	14.88	0.006*
Current × Pulse-off	1	0.1329	0.1329	0.13293	5.29	0.055*
Pulse-on × Pulse-off	1	0.2983	0.2983	0.29828	11.87	0.011*
Residual Error	7	0.1759	0.1759	0.02513		
Total	15	2.5542				

Notes: **Most significant at 99%, *Significant at 95%.

4. Multiobjective Optimization for In-process Parameters

4.1. *Desirability function approach*

Harrington (1965) introduced the theory of desirability function approach. In this approach, the response y_i (i.e., MER, TWR and SR) is systematically transformed into d_i as individual desirability function which varies as

$$0 \leq d_i \leq 1.$$

Here, the undesired response of y_i is represented as $d_i = 0$, while $d_i = 1$ signifies that y_i is at its goal/aim. The desirability function has an associated weight factor (r) for each response, which defines its shape. The individually independent desirability of response is defined as per its goal/aim. For the minimized response objective y_i, the desirability function can be determined as

$$d_i = \begin{cases} 0 & y_i < L_i \\ \left(\frac{y_i - L_i}{T_i - L_i}\right)^r & L_i \leq y_i \leq T_i \\ 1 & y_i > T_i. \end{cases}$$

If the target is to minimize the objective response then

$$d_i = \begin{cases} 1 & y_i < T_i \\ \left(\frac{U - y_i}{U_i - T_i}\right)^r & T_i \leq y_i \leq U_i \\ 0 & y_i > U_i, \end{cases}$$

where L_i and U_i demonstrate the upper and lower limits values of response y_i. By maximization of composite desirability (D), the overall optimal control variables can be worked out. The geometric mean of individual desirability function is the overall (composite) desirability function, which is shown by

$$D = (d_1 d_2 d_3 \cdots d_m)^{1/m},$$

where m is the number of responses and $d_1, d_2, d_3, \ldots, d_m$ are the desirability functions.

4.2. *Optimization by desirability function*

The required target for the SR and MER is to become maximized while TER is to be minimized in order to quantify overall optimum results. The TER is considered as the important factor for the machined component, hence, the target of TER is set at 0.426 mg/min. The unit weight factor for each response is same, i.e., r =1. Thus, a criterion utilized to attain the global optimization is represented in Table 7. The overall optimum results for the three responses are shown in Figure 8. In this figure, each column represents process parameters, and each row represents responses. The solution obtained from the software is as follows:

$C_1 = 4.0\,\text{mm}, C_2 = 12\,\text{A}, C_3 = 90\,\mu\text{s}, C_4 = 45\,\mu\text{s}, C_5 = 2$ (i.Abrasive mixed).

Table 7: Criteria for the desirability function approach.

Response	Goal	Lower (L_i)	Target (T_i)	Upper (U_i)	Weight	Importance
MER	Maximize	1.3800	6.986	—	1	1
TER	Minimize	—	0.426	3.714	1	1
SR	Maximize	0.1533	1.160	—	1	1

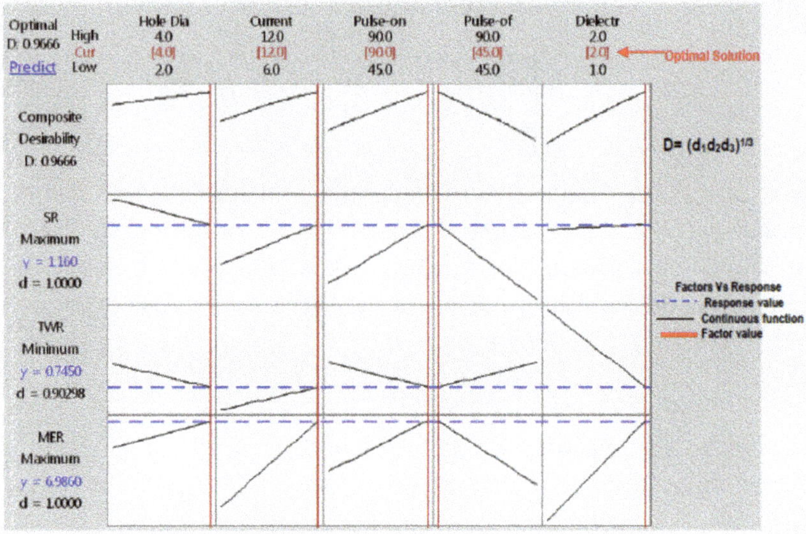

Figure 8: Plot showing the effect of each factor on various responses.

These process parameters setting yields the minimum TER and the maximum possible MER and SR value for sustainable production process.

5. Conclusions

By maximizing the MER and SR, the minimization of TER, the optimized and "sustainable" control parameters can be figured out. In particular, although this design of experiment could eventually result a decrease in MER and SR, the TER minimization can be achieved at the minimum spark energy. In the present work, the effect of machining control variables on the output factors like material erosion rate (MER), tool erosion rate (TER) and the surface roughness (SR) was investigated. The following conclusions were drawn from our research:

1. The material removal rate is significantly affected by the inner diameter of tool, peak current as well as type of dielectric media and enhanced steeply with selected parameters.

2. In case of TER, the parameters peak current and dielectric affected the output most significantly. The tool wear declined with a rise in current from 6 to 12 A, whereas abrasive mixed dielectric fluid enhanced the erosion of tool.
3. The SR of the machined component was affected by all the selected input parameters.
4. Due to the augmented discharge current during abrasive jet-assisted hybrid EDM, deeper and more significant craters were developed that resulted in enhancement of the roughness of the machined surface.
5. The desirability functional approach was successfully applied to obtained globally optimized solution ($D = 96.6\%$) for higher productivity.

References

Abdul-Rani, A. M., Aliyu, A. A. A., Hastuty, S., Ginta, T. L., Rao, T. V. V. L. N. and Ali, S. (2018). Enhancing surface quality of Zr-Cu-Ni-Ti-Be through hydroxyapatite mixed EDM for potential orthopedic application. In *AIP Conference Proceedings,* Vol. 2035, No. 1. AIP Publishing, p. 080010.

Arantes, L. J., da Silva, E. R., dos Santos, R. F., Sales, W. F. and Raslan, A. A. (2016). The electrical discharge machining process aided by abrasive jet. *Int J Adv Manufact Technol*, 1–10.

Bains, P. S., Sidhu, S. S. and Payal, H. S. (2016a). Study of magnetic field assisted ED machining of metal matrix composites. *Mater Manufact Process*, 31(14), 1889–1894.

Bains, P.S., Sidhu, S.S. and Payal, H. S. (2016b). Fabrication and machining of metal matrix composites: A review. *Mater Manufact Process*, 31(5), 553–573.

Bains, P. S., Sidhu, S. S. Payal, H. S. and Kaur, S. (2019). Magnetic field influence on surface modifications in powder mixed EDM. *Silicon*, 11(1), 415–423.

Bains, P. S., Sidhu, S. S. and Payal, H. S. (2018). Investigation of magnetic field-assisted EDM of composites. *Mater Manufact Process*, 33(6), 670–675.

Gu, L., Lei, L., Wansheng, Z. and Rajurkar, K. P. (2012). Electrical discharge machining of Ti6Al4V with a bundled electrode. *Int J Mach Tools Manufact*, 53(1), 100–106.

Harrington, E. C. (1965). The desirability function. *Industr Quality Control*, 21, 494–498.

Kumar, K. L., Sivasubramanian, R. and Kalaiselvan, K. (2009). Selection of optimum parameters in non-conventional machining of metal matrix composite. *Portugaliae Electrochimica Acta*, 27(4), 477–486.

Lin, Y. C., Chen, Y. F., Wang, A. C. and Sei, W. L. (2012). Machining performance on hybrid process of abrasive jet machining and electrical discharge machining. *Trans Nonferrous Metals Soc China*, 22, s775–s780.

Long, M. and Rack, H. J. (1998). Titanium alloys in total joint replacement — a materials science perspective. *Biomaterials*, 19, 1621–1639.

Mohal, S. and Kumar, H. (2017). Parametric optimization of multiwalled carbon nanotube-assisted electric discharge machining of Al-10% SiCp metal matrix composite by response surface methodology. *Mater Manufact Process*, 32(3), 263–273.

Nguyen, M. D., Rahman, M. and San Wong, Y. (2012). Simultaneous micro-EDM and micro-ECM in low-resistivity deionized water. *Int J Mach Tools Manufact*, 54, 55–65.

Nourbakhsh, F., Rajurkar, K. P., Malshe, A. P. and Cao, J. (2013). Wire electro-discharge machining of titanium alloy 2013. *Procedia CIRP*, 5, 13–18.

Talla, G., Kumar, D., Sahoo, S. G. and Biswas, C. K. (2015). Modeling and multi-objective optimization of powder mixed electric discharge machining process of aluminum/alumina metal matrix composite. *Eng Sci Technol Int J*, 18(3), 369–373.

Tsai, F. C., Yan, B. H., Kuan, C. Y. and Huang, F. Y. (2008). A Taguchi and experimental investigation into the optimal processing conditions for the abrasive jet polishing of SKD61 mold steel. *Int J Mach Tools Manufact*, 48(7), 932–945.

Wansheng, Z., Wang, Z., Di, S., Chi, G. and Wei, H. (2002). Ultrasonic and electric discharge machining to deep and small hole on titanium alloy. *J Mater Process Technol*, 120(1), 101–106.

Zhang, Y., Xu, Z., Xing, J. and Zhu, D. Effect of tube-electrode inner diameter on electrochemical discharge machining of nickel-based superalloy. *Chin J Aeronaut*, 29(4), 1103–1110.

Chapter 9

Cell Formation Based on Similarity in Parts Design Using Multiattribute Decision-Making Algorithm

Ali Jahan*,‡ and Sanaz Souri†

*Department of Industrial Engineering,
Semnan Branch, Islamic Azad University,
Semnan, Iran

†Graduate Student of Industrial Engineering,
Semnan Branch, Islamic Azad University,
Semnan, Iran

‡a.jahan@semnaniau.ac.ir

Cell formation and production planning are two important steps in implementing the group technology system. In this research, the use of multiattribute decision-making concepts and similarity coding system is suggested in designing components to create a cellular model. An improved algorithm with rank order clustering (ROC) and row and column masking (RCM) methods was proposed for application in Parizan Sanat Co. as a case study. The results were compared to the productivity index, and the superiority of the improved algorithm was observed compared to the previous methods.

1. Introduction

The cell formation system is one of the techniques increasing the productivity and efficiency in production. This system is defined based on group technology concept as a technique to improve production and productivity by reducing the startup time, delivery time and process, material handling expenses, cost of tools and production equipment (Patel and Patel, 2014). Through group technology, the parts in a family are determined and allocated to the machine cell. The issue of cell formation is the part-machine grouping. The classification methods are based on creating a part-machine crossing matrix, where the row and column are the machine and the part, respectively. The matrix components are represented by 0 and 1 (Gunasingh and Lashkari, 1989).

The concept of group technology and grouping the parts was initially introduced in 1971, then in 1972, the single linkage clustering (SLC) was established based on 0–1 matrix (McAuley , 1972). The SLC based on the 0–1 matrix is applied for manual calculations and small problems. In this algorithm, the formation of machine cells is based on the similarity of equipment, with the advancement of science and adding more inputs such as production volume in ALC algorithm by Seifoddini (1989), while the time and production sequence were proposed by Srinivasan *et al.* (1990). Then in 1972, the bond energy (BE) algorithm was presented by Mecormick (1972). The grouping methods were investigated based on the similarity coefficient in 1973, and based on classification counting technique in 1977 (Burbidge, 1977). Rank order clustering (ROC) algorithms were proposed by King (1980). Also, direct clustering algorithm (DCA) and modified ROC were presented in 1986 (Chandrasekharan and Rajagopalan, 1986). Artificial intelligence is another grouping method presented in 2000 (Akturk and Turkcan, 2000; Kusiak, 1990). Furthermore, innovative methods were proposed for grouping the parts in recent years. So far, most of the proposed methods have been used to solve the problems of cellular formation systems based on the formation of cell machine and the internal layout of cells. In 2009, a method proposed by Ahi *et al.* (2009) used multicriteria decision-making methods for the formation of machine cell and the internal layout of cells. They used the SAW and TOPSIS methods in the proposed approach. In 2012, design, plan and control of the production in cellular manufacturing were proposed (Patel and Patel, 2014). Optimization of cell formation with computer simulation was also proposed by Koupaei *et al.* (2016). The basis of the preceding clustering methods was binary numbers, including a set of machines and components in a family (Ahi *et al.*, 2009). These methods, regardless of production volume, sequence of operations, production costs, inventory and production constraints, categorize the parts and machines as the main objective. The input of information is the base of innovative methods. One of the inputs is the identification of similarities evaluated on two aspects of group technology. In fact, group technology has an engineering approach and a production process. Another categorizing method is the use of coding systems proposed in the two visualization and coding types (Luggen, 1991). In the research on machine grouping issues, graph theory methods and mathematical techniques (Akturk and Turkcan, 2000; Han and Ham, 1986), artificial intelligence techniques such as neural networks, fuzzy logic and genetic algorithm were proposed (Kaparthi and Suresh, 1991; Moon, 1998). In the studied articles, the formation of the machine-part cell was based on the production approach and the manufacturing process. In this research, a grouping of parts with an engineering approach is performed through a case study, and then the cell machines are formed using an improved algorithm based on multiattribute decision-making and existing methods by analyzing the production process. In 2015, in order to solve the cellular formation problem, a quasi-linear mathematical model was proposed in cell production systems, whose results had a significant effect on the block structures of the device matrices (Koupaei *et al.*, 2016; Deep and Singh, 2015a). Then, in 2018, cellular systems were designed to study optimal cells, optimal processing paths and balanced cellular cells (minimum cellular load). The proposed method for solving this problem was based on genetic algorithm, which confirms the result of the calculations and the superiority of the proposed method (Deep and Singh, 2015b; Mohammadi and Forghani, 2017; Rabbani *et al.*, 2017).

2. Part Grouping Methods

2.1. *Coding methods*

The two main methods used in group technology to group the parts are coding and clustering analysis (Kusiak, 1990). The coding method is classified into visual and coding groups. In the visual approach, the components are classified in terms of appearance. In the coding process, the parts are classified based on the shape, dimensions, material, tolerance and the required accuracy. There are three types of code structure methods in the GT system.

2.1.1. *Chain structure*

In this structure, known as a multicode, each digit provides separate information independent of the previous digits (Singh and Rajamani, 2012).

2.1.2. *Hierarchical structure*

This is known as a single code with a tree structure in which each number is determined by the previous characteristics. The single-code structure has considerable complexity compared to the multi-code structure (Singh and Rajamani, 2012).

2.1.3. *Combined structure*

The third type of combined structure is a combination of two previous structures. The current structure benefits from both systems, while it has fewer disadvantages (Kusiak , 1990). As described in Section 3.1, grouping the parts is performed using coding methods through the combined structure.

$$r_{ij} = \begin{cases} r_{ij} = g_{ij} & \text{the number of parts visit machine } i \\ r_{ij} = 0 & \text{otherwise} \end{cases}$$

2.2. *Improved algorithm based on multiattribute decision-making method*

2.2.1. *The first step*

Step 1: Introducing machines and parts as alternatives and criteria (attribute) in the decision-making matrix, respectively.

Step 2: The weight of each criterion is calculated according to equation (1) (Edwards and Barron, 1994). So, that the total weights are equal to one.

$$
\begin{aligned}
wj &= \left(\frac{1}{j}\right)\sum_{i=j}^{j}\frac{1}{i}\\
w_1 &= \left(1+\frac{1}{2}+\frac{1}{3}+\cdots+\frac{1}{j}\right)\Big/ j,\\
w_2 &= \left(0+\frac{1}{2}+\frac{1}{3}+\cdots+\frac{1}{j}\right)\Big/ j,\\
w_3 &= \left(0+0+\frac{1}{3}+\frac{1}{4}+\cdots+\frac{1}{j}\right)\Big/ j,\\
w_j &= \left(0+\cdots+0+\frac{1}{j}\right)\Big/ j.
\end{aligned}
\tag{1}
$$

Step 3: Normalize decision by the following equation:

$$n_{ij} = \left\{\frac{r_{\mathrm{i}}\mathrm{j}}{r_{\mathrm{i}}\mathrm{j}^{\max}}\,\middle|\, j \in b\right\} \quad i = 1, 2, \ldots, m. \tag{2}$$

In the improved algorithm, all the attributes are considered as benefit criteria.

Step 4: Calculating the weighted normalized decision matrix based on the weight of each attribute (equation (3)).

$$v_{\mathrm{i}}\mathrm{j} = wj\, n_{\mathrm{i}}\mathrm{j} \quad i = 1, 2, \ldots, m \quad j = 1, 2, \ldots, n. \tag{3}$$

Step 5: Calculating the score of each alternative using the following equation:

$$s_{\mathrm{i}} = \sum_{j=1}^{n} v_{\mathrm{i}}\mathrm{j} \quad i = 1, 2, \ldots, m. \tag{4}$$

Step 6: Ranking the machines in a descending order.

Step 7: Modifying the initial selection matrix based on ranking of machines.

2.2.2. *The second step*

Step 1: The part and machine are the alternatives and criteria (attributes) in the decision-making matrix, respectively.

Step 2: The weight of each criterion is calculated according to equation (5) (Edwards and Barron, 1994). so that the total weights are equal to 1.

$$
\begin{aligned}
wi &= \left(\frac{1}{i}\right)\sum_{j=i}^{i}\frac{1}{j},\\
w_1 &= \left(1+\frac{1}{2}+\frac{1}{3}+\cdots+\frac{1}{i}\right)\Big/ i,\\
w_2 &= \left(0+\frac{1}{2}+\frac{1}{3}+\cdots+\frac{1}{i}\right)\Big/ i,\\
w_3 &= \left(0+0+\frac{1}{3}+\frac{1}{4}+\cdots+\frac{1}{i}\right)\Big/ i,\\
w_j &= \left(0+\cdots+0+\frac{1}{i}\right)\Big/ i.
\end{aligned}
\tag{5}
$$

Step 3: Normalize decision by the following equation:

$$n_{ij} = \left\{\frac{r_{\mathrm{ij}}}{r_{\mathrm{ij}}{}^{\max}}\,\middle|\, j\in b\right\} \quad i=1,2,\ldots,m. \tag{6}$$

Step 4: Calculating the weighted normalized decision matrix based on the weight of each attribute (equation (7))

$$v_{\mathrm{ij}} = w\mathrm{j}\, n_{\mathrm{ij}} \quad i=1,2,\ldots,m \quad j=1,2,\ldots,n. \tag{7}$$

Step 5: Calculating the score of each alternative using the following equation:

$$s_{\mathrm{i}} = \sum_{j=1}^{n} v_{\mathrm{ij}} \quad i=1,2,\ldots,m. \tag{8}$$

Step 6: Ranking the items in a descending order.

Step 7: Modifying the revised matrix in the first step according to ranking of the parts.

The two steps of the improved time algorithm cease to prevent the change in a row or column.

2.3. *Group effect and group productivity*

Group productivity is a weighted average measuring two factors of idleness and the movement of parts between cells in a 0–1 matrix. Suresh Kumar and Chandrasekharan, (1990) proposed a model to calculate the group productivity. The parameters required for this model are as follows:

m	Number of rows
n	Number of columns
e	Total number of parts
$e1$	Number of 1 in diagonal block
$e0$	Number of 1 out of the diagonal block (exceptional parts)
ev	The number of 0 of the diagonal block (idleness)
Ψ	The rate of intercellular motion to all parts
Ø	The idleness rate to total parts (ev/e)
Γ	Group work
η	Productivity
Q	The weight factor $0< q < 1$

The group productivity of the model (9) is calculated as follows:

$$\eta = q\,\eta_1 + (1-q)\,\eta_2, \tag{9}$$

$$\eta_1 = e_1/(e_1 + e_v), \tag{9a}$$

$$\eta_2 = -e_v/(-e_v + (e - e_1)), \tag{9b}$$

where η_1 is the rate of 1 of dthe iagonal block to the whole parts of diagonal blocks; η_2 is the rate of 0 out of the diagonal block to the whole parts out of diagonal blocks; q is a weight factor assigning a number within 0 and 1, if $q = 0.5$ (the weight of the same blocks), the group productivity is calculated based on model (10).

$$\eta = \left(2 - \left[\frac{e_v}{e} + \frac{e_0}{(mn-e)}\right]\right) \Big/ 2. \tag{10}$$

The optimal group productivity is between 75% and 100%. The number closer to 1 shows the optimal condition, and the number closer to 0 indicates the inappropriate conditions (Edwards and Barron, 1994).

2.3.1. *Group effect*

Ψ represents the rate of intercellular movement to the whole parts, and $\varnothing$ is the idleness rate of the diagonal blocks to all parts in a group effect.

$$\Gamma = \frac{1-\Psi}{1+\varphi} = (e-e_0)/(e+e_v). \tag{11}$$

3. Using the Proposed Algorithm in Parizan Sanat Company

Parizan Co. is a manufacturer of repairing equipment. Considering the maximum use of machinery and manpower capacity, implementation of flexible manufacturing systems and group technology was introduced. Group technology was discussed both in engineering and production. This research is conducted using a combination of both approaches.

3.1. *Designing the coding system*

This section deals with how to create a family of components using coding systems to reduce the design and redesign time in the engineering unit. Parizan Sanat Co. has a diverse production line and moderate production volumes. According to various available products, there are approximately 30 types of products and 5,000 parts. Since there is a system in the engineering unit to identify parts not having the same technical characteristics and appearance, and a part with a high degree of similarity may be redesigned, or the search time to find a similar map of the new part may be high, a specific coding system was designed. In this system, H, W and L were used as the symbols for height, width and length, respectively, and D was used as a symbol of diameter. Table 1 shows the coding structure. The same group technology codes are arranged into a family of parts. Appendix A includes the family of components.

Table 1: Coding structure.

First digit	1	Flat parts	$2h < w$	$2h < d$
	2	Cylindrical and circular parts	$w < 2h$	$h < 2d$
	3	Long parts	$2d < h < 6d$	$2w < l < 6w$
	4	Very long parts	$l < 6w$	$h < 6d$
Second Digit	1	An overall cylinder		
	2	A combination of several cylinders with different diameters		
	3	A flat cylinder		
Third digit	1	An overall hole, with nonround cross-section (rectangular)		
	2	An overall hole with round cross-section		
	3	A nonoverall hole with round cross-section		
	4	A nonoverall hole with non-round cross-section		
Forth digit	1	A combination of cylinder and cube		
	2	Stepped and sloping outer surfaces		
	3	A combination of triangles and cylinders		
	4	Angular outer surfaces		
Fifth digit	1	Smooth and nonstepped cavity surfaces		
	2	Internal surfaces of the angular cavity-imperfect cone or pyramid		
	3	Stepped and edged cavity surfaces		
	4	Cavity surfaces with angular or edges or stairs		
Sixth digit	1	ST37	6	Pipe st37
	2	Trans	7	Steel
	3	Steel pipe	8	Aluminum
	4	Polyamide and Teflon	9	CK45
	5	shafting	0	Cast iron

3.2. *Part categorization methods*

In this subsection, the ROC and RCM algorithms along with the proposed algorithm are applied and the results are compared. First, the machine-part decision-making matrix, of which the components represent the number of parts per machine family per unit family (Table 2). The components are grouped by an improved algorithm shown in Table 3. The matrix of Table 1 is converted to 0–1 binary matrix (Table 4) and, based on the previous methods, the part family and machine are formed (Tables 5 and 6). Table 7 presents the productivity index and the group effect of the solution matrices of all three methods in this chapter.

Table 2: Primary machine-part decision-making matrix.

Parts		44	43	42	41	40	39	38	37	36	35	34	33	32	31	30	29	28	27	26	25	24	23	22	21	20	19	18	17	16	15	14	13	12	11	10	9	8	7	6	5	4	3	2	1
Gt code		4230350	4222120	4134250	4130220	4120460	4100050	3230220	3132150	3122120	3120460	3120150	3120120	3104020	3102020	2424220	2232320	2224120	2222320	2222120	2222110	2202020	2132420	2132220	2130140	2122420	2120340	2120320	2120140	2120120	2120110	2102090	2100140	2100020	2012100	1344410	1330120	1322140	1322110	1320140	1320110	1312010	1304070	1304010	1004010
Machines	1																1		1							3							1			1	1	2		2	2	1	2		1
	2			1				1		1							1		1	1				2		3	1									1			1		1				1
	3	1	1																																				2				1	1	
	4								1																														2						
	5																	1																	3										
	6						2			1		1				1				1		2			1		1		1	1			2		3										
	7																				1										2			1											
	8	1	1	1									2	1	1								2					1				1													
	9				1			1									1		1					2		3																			
	10					1			1		1																																		

Table 3: The improved algorithm solution matrix based on the multi-attribute decision-making method.

	Parts																																											
Machines	44	41	35	40	39	34	33	32	31	30	28	25	24	23	21	18	17	16	43	15	14	12	9	8	6	4	2	42	13	5	38	3	37	36	26	22	19	11	10	7	1	29	27	20
1																							1	2	2	1			1	2		2							1		1	1	1	3
6																													2					1	1		1	3						
2																												1		1	1			1	1	2	1		1	1	1	1	1	3
8	1						2	1	1					2		1			1		1							1																
3	1																		1								1					1								2				
7					2	1						1								2		1																						
9		1																													1					2						1	1	3
4																																	1							2				
10			1	1						1			2		1		1	1															1											
5											1																											3						

Table 4: Primary machine-part matrix (Binary).

. Primary matrix		Machines									
Parts		1	2	3	4	5	6	7	8	9	10
1	1004010	1	1								
2	1304010			1							
3	1304070	1		1							
4	1312010	1									
5	1320110	1	1								
6	1320140	1									
7	1322110		1	1	1						
8	1322140	1									
9	1330120	1									
10	1344410	1	1								
11	2012100					1	1				
12	2100020							1			
13	2100140	1					1				
14	2102090								1		
15	2120110							1			
16	2120120						1				
17	2120140						1				
18	2120320								1		
19	2120340		1				1				
20	2122420	1	1							1	
21	2130140						1				
22	2132220		1							1	
23	2132420								1		
24	2202020						1				
25	2222110							1			
26	2222120		1				1				
27	2222320	1	1							1	
28	2224120					1					
29	2232320	1	1							1	
30	2424220						1				
31	3102020								1		
32	3104020								1		
33	3120120								1		
34	3120150						1				
35	3120460										1
36	3122120		1				1				
37	3132150				1						1
38	3230220		1							1	
39	4100050						1				
40	4120460										1
41	4130220									1	
42	4134250		1						1		
43	4222120			1					1		
44	4230350			1					1		

Table 5: ROC algorithm machine-part solution matrix.

ROC		Parts																																											
		30	21	11	25	15	12	28	40	35	37	33	32	31	23	18	14	39	34	24	17	16	2	7	44	43	41	42	36	26	19	38	22	9	8	6	4	10	5	1	3	13	29	27	20
Machines	1																																	1	1	1	1	1	1	1	1	1	1	1	1
	2																											1	1	1	1							1	1	1			1	1	1
	9																										1	1	1			1	1										1	1	1
	3																						1	1	1	1															1				
	6	1	1	1														1	1	1	1	1									1	1	1									1			
	8											1	1	1	1	1	1								1	1				1															
	4										1													1																					
	10								1	1	1																																		
	5			1				1																																					
	7				1	1	1																																						

Table 6: RCM algorithm machine-part solution matrix.

RCM		Parts																																									
		32	31	14	15	30	44	43	2	1	42	41	40	39	38	37	36	35	34	33	29	28	27	26	25	24	23	22	21	20	19	18	17	16	13	12	11	10	9	8	7	6	5	4	3
Machines	1									1											1		1							1					1			1	1	1		1	1	1	1
	2									1	1				1		1				1		1	1				1		1	1							1					1		1
	3						1	1	1																																1				
	4															1																									1				
	5																					1															1								
	6				1	1								1			1		1					1		1			1		1		1	1	1		1								
	7																								1											1									
	8	1	1	1			1	1			1									1							1					1													
	9											1			1						1		1					1		1															
	10												1			1		1																											

Table 7: Comparison of improved algorithms and existing methods.

Algorithm	e1	e0	ev	Group productivity	Group effect
Improved algorithm solution matrix, Table 3	59	32	94	0.43	0.30
ROC algorithm solution matrix, Table 5	56	10	86	0.33	0.36
RCM algorithm solution matrix, Table 6	52	40	82	0.16	0.20

As Table 7 shows, the efficiency of the improved algorithm is higher than other methods; however, its group effect is lower compared to the ROC. The group effect is directly related to intercellular movements. Machine cell and family of cellular optimum components include the cell with the smallest intercellular motions. As Table 7 shows, the number of critical components (e0) in the improved algorithm solution matrix is higher than the ROC algorithm solution matrix. Therefore, to reduce the ROC critical components, the resources need to be increased. Increasing the resources is based on the volume of production, and in the case of an economical increase in equipment and machinery, or supplied by contractors. However, Parizan Sanat Co. follows these strategies:

- There is no increase in equipment and machinery.
- The components can be produced according to the process of manufacturing and design of a part with a similar machine in the corresponding cell.

After considering the assumptions, the modified matrices in Tables 8–10 should be considered. Table 11 presents the comparison of the solution matrices in the algorithms after considering the problem assumption.

As Table 11 shows, in all the methods, considering the hypothesis of Parizan Sanat Co., the modified matrices are improved compared to the initial solution matrices, and the percentage of productivity and group effect are higher in the improved algorithm compared to the above-mentioned methods.

Table 8: Modified MADM algorithm machine-part solution matrix.

Modified matrix		Parts																																									
		4230350	4130220	3120460	4120460	4100050	3120150	3120120	3104020	3102020	2424220	2224120	2222110	2202020	2132420	2130140	2120320	2120140	2120120	4222120	2120110	2102090	2100020	1330120	1322140	1320140	1312010	1304010	4134250	2100140	1320110	3230220	1304070	3132150	3122120	2222120	2132220	2120340	2012100	1344410	1322110	1004010	2232320	2222320	2122420
		44	41	35	40	39	34	33	32	31	30	28	25	24	23	21	18	17	16	43	15	14	12	9	8	6	4	2	42	13	5	38	3	37	36	26	22	19	11	10	7	1	29	27	20
Machines	1																							1	2	2	1	1		1	2		3							1	2	1	1	1	3
	6																													2					1	1		1	3						
	2																												1		1	1			1	1	2	1		1	1	1	1	1	3
	8	1						2	1	1		1			2		1			1		1							1																
	3	1																		1																									
	7										1		1	2		1		1	1		2		1																						
	9		1																													1					2						1	1	3
	4																																	1							2				
	10			1	1	2	1																											1											
	5																																						3						

Table 9: ROC algorithm modified solution matrix.

Modified ROC	Parts																																											
Machines	30	21	11	25	15	12	28	40	35	37	33	32	31	23	18	14	39	34	24	17	16	2	7	44	43	41	42	36	26	19	38	22	9	8	6	4	10	5	1	3	13	29	27	20
1																																	1	1	1	1	1	1	1	1	1	1	1	1
2																											1	1	1	1	1	1					1	1	1			1	1	1
9																										1	1	1	1		1	1										1	1	1
3																						1	1	1	1															1				
6																	1	1	1	1	1									1											1			
8											1	1	1	1	1	1								1	1																			
4										1													1																					
10			1					1	1	1																																		
5			1				1																																					
7	1	1		1	1	1																																						

Table 10: RCM algorithm modified solution matrix.

Modified RCM	Parts																																											
Machines	32	31	14	15	30	44	43	2	1	42	41	40	39	38	37	36	35	34	33	29	28	27	26	25	24	23	22	21	20	19	18	17	16	13	12	11	10	9	8	7	6	5	4	3
1									1											1		1							1					1			1	1	1		1	1	1	1
2									1	1				1		1				1		1	1				1		1	1							1					1		1
3						1	1	1																																1				
4															1																									1				
5														1					1	1	1	1				1	1		1		1					1								
6													1			1		1					1		1			1		1		1	1	1		1								
7															1		1							1											1									
8	1	1	1			1	1			1																																		
9											1																																	
10				1	1							1																																

Table 11: Comparison of the algorithm's productivity index based on the case study hypotheses.

Algorithm	e1	e0	ev	Productivity	Group effect
Modified algorithm solution matrix, Table 8	73	18	84	0.61	0.46
Modified solution ROC algorithm matrix, Table 8	62	4	80	0.38	0.42
Modified solution RCM algorithm matrix, Table 10	47	19	89	0.30	0.30

4. Conclusion

In this chapter, the group technology focused on engineering and manufacturing activities. Creating a database based on group technology codes would reduce redesigning in new orders. By accessing the parts database, we can control whether a part or a similar part is already designed. This system provides stabilization of the information and the experience after changing the technical staff. This system considerably reduces the repetitive engineering activities and the time for delivery of engineering activities, which leads to an increase in productivity. Moreover, the schemes will be more reliable, and a standardization will occur at all stages. The concepts of multiattribute decision-making method was used to improve the algorithm. Furthermore, to compare the authenticity of the methods, the productivity and group effects were used. In this method, machine-part matrix arrays were determined by the number of parts in the parts group formed by the coding system. The primary machine-part matrix (Table 2) is calculated using previous binary matrix and grouping methods (Table 4), and the results were compared to the proposed algorithm by means of group effect index and group productivity. In previous methods such as ROC and RCM algorithms, the 0–1 matrix and constant weight of the parts were considered while in this research, considering large number of parts (5,000), a lot of calculation is needed. Therefore, to reduce the computations in the matrix, instead of each part, we used the similar parts group and due to the unequal parts in each group, an improved algorithm based on multiattribute decision-making method was proposed. The improved decision-making-based algorithm includes higher productivity and group effect compared to ROC and RCM algorithms. Hence, this is not necessarily the optimal solution; however, it may also lead to better solutions in other algorithms and methods. The cellular production system is an effective way to increase productivity of manpower, material, space and time. Therefore, similar machines are grouped into groups called cells and assigned to the same family process for the production process. The formation of cells based on the family of similar components leads to a reduction in the transportation of materials and semimanufactured goods.

Appendix A Family of Components

GT code	Part family
1320410	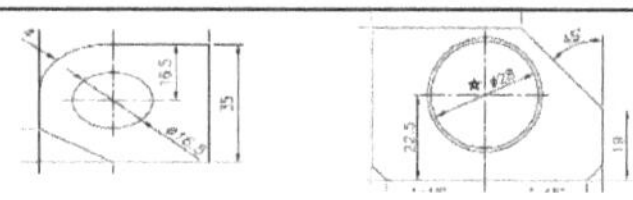
1321190	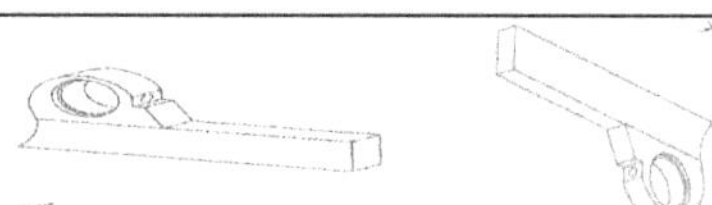
1322110	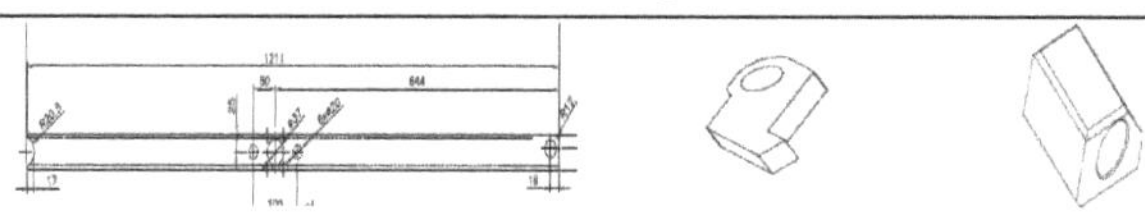
1322140	
2120140	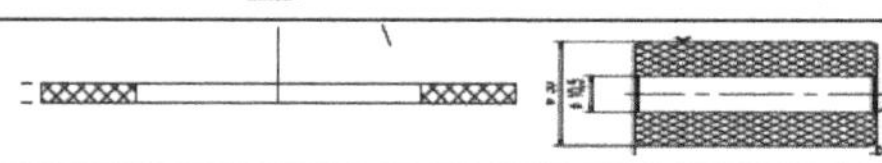
2202050	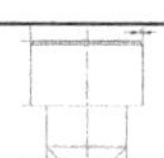
2202220	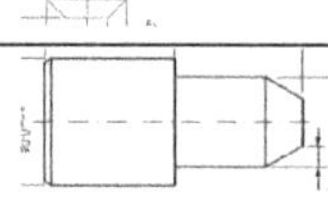
2222220	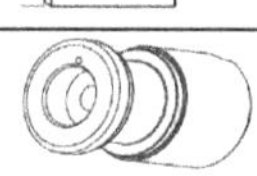

(*Continued*)

(Continued)

GT code	Part family
2222320	 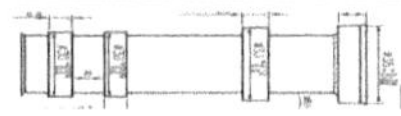 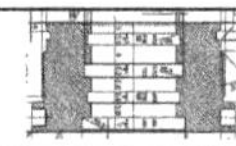
2232320	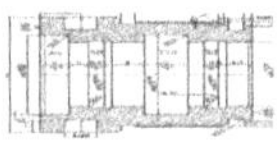 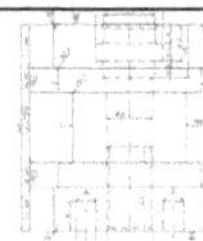
2242210	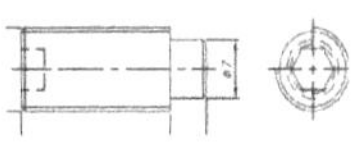 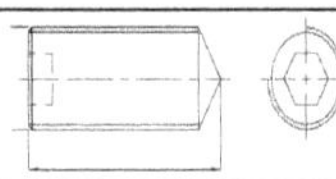
3102050	 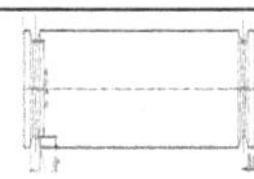
3102150	
3120120	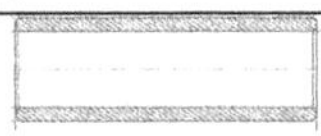 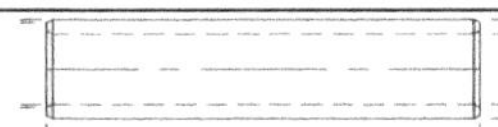
3122120	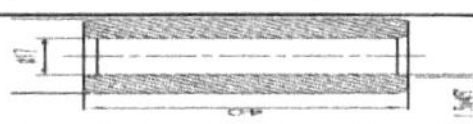 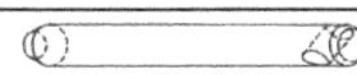
3122250	
3122310	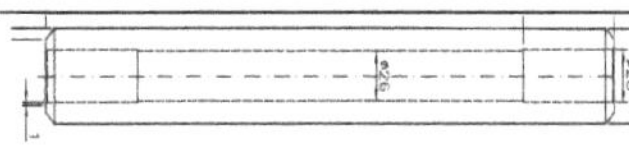
3132250	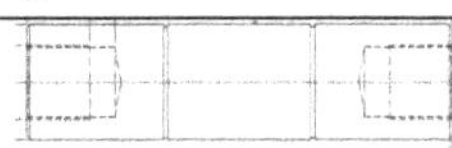 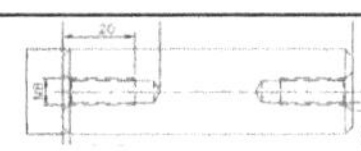
2120210	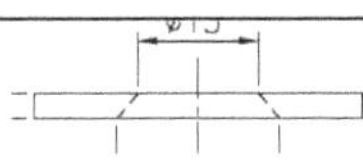
1322190	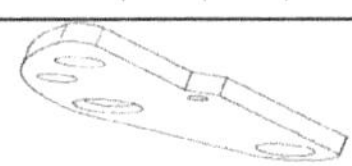 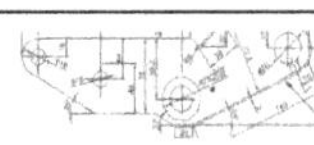
1324320	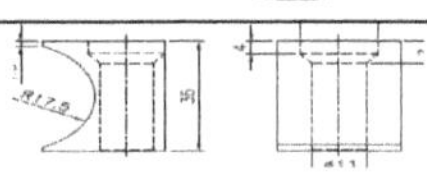
1330220	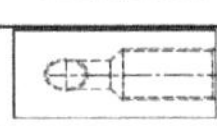

(Continued)

(*Continued*)

GT code	Part family
2102010	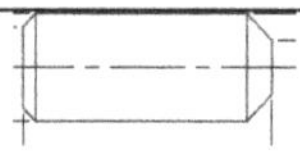
1324110	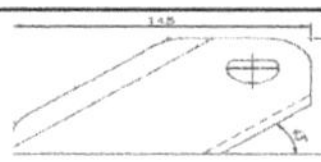
312020	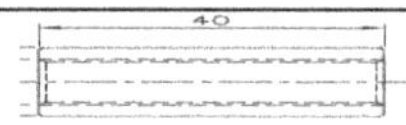
412020	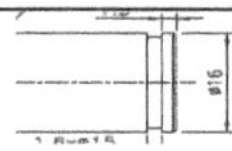
1302010	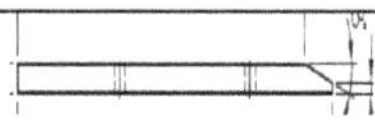
1304010	
1304070	 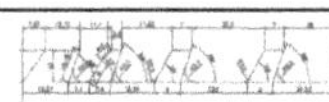
1312010	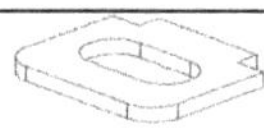
2012100	 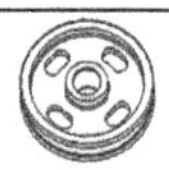 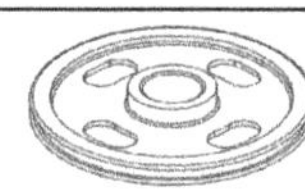
2100020	
2100140	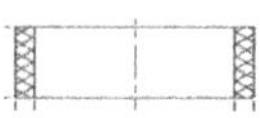 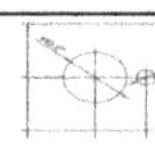
2102090	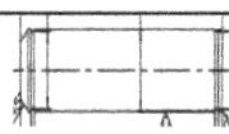
2120110	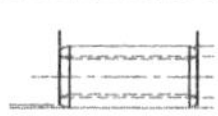 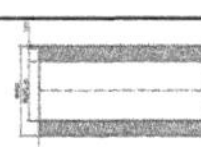
2120120	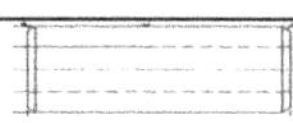
2120320	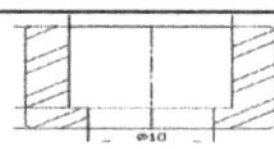
2120340	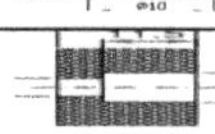
2122420	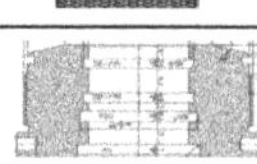 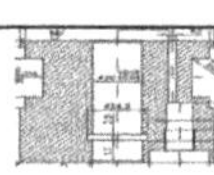
2130140	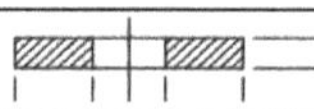

(*Continued*)

(Continued)

GT code	Part family
2132220	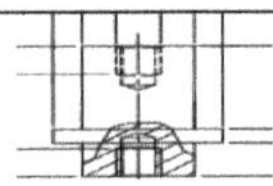
2132420	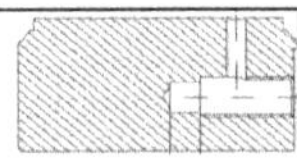
2202020	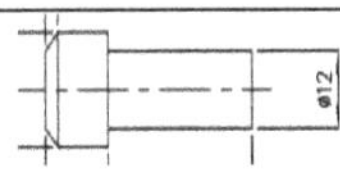
2222110	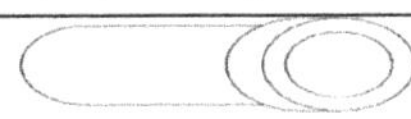
2222120	
2224120	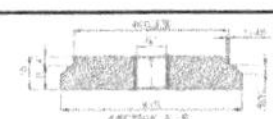
2424220	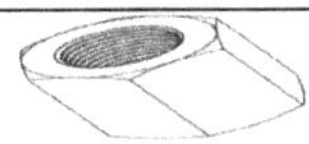
3102020	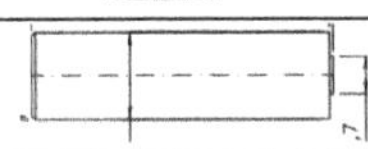
3104020	
3120150	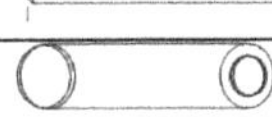
3120460	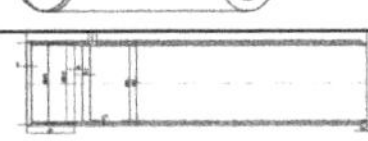
3132150	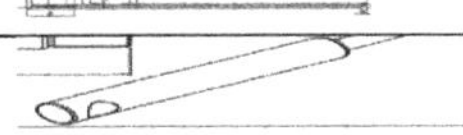
3230220	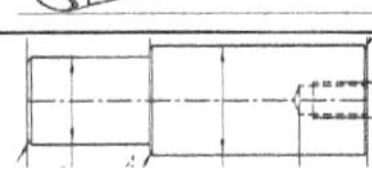
4130220	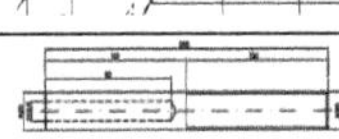
4134250	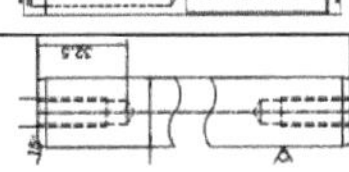
4222120	
4230350	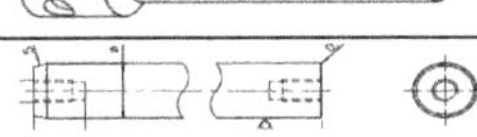

References

Ahi, A., Aryanezhad, M. B., Ashtiani, B. and Makui, A. (2009). A novel approach to determine cell formation, intracellular machine layout and cell layout in the CMS problem based on TOPSIS method. *Comput Oper Res*, 36(5), 1478–1496.

Akturk, M. S. and Turkcan, A. (2000). Cellular manufacturing system design using a holonistic approach. *Int J Product Res*, 38(10), 2327–2347.

Burbidge, J. (1977). A manual method of production flow analysis. *Product Eng*, 56(10), 34–38.

Chandrasekharan, M. and Rajagopalan, R. (1986). MODROC: An extension of rank order clustering for group technology. *Int J Product Res*, 24(5), 1221–1233.

Deep, K. and Singh, P. K. (2015). Design of robust cellular manufacturing system for dynamic part population considering multiple processing routes using genetic algorithm. *J Manufact Syst*, 35, 155–163.

Deep, K. and Singh, P. K. (2015). Machine cell formation: Using genetic algorithm-based heuristic considering alternative route. *Int J Oper Res*, 24(1), 83–101.

Edwards, W. and Barron, F. H. (1994). SMARTS and SMARTER: Improved simple methods for multiattribute utility measurement. *Organiz Behav Human Decision Process*, 60(3), 306–325.

Gunasingh, K. R. and Lashkari, R. (1989). Machine grouping problem in cellular manufacturing systems an integer programming approach. *Int J Product Res*, 27(9), 1465–1473.

Han, C. and Ham, I. (1986). Multiobjective cluster analysis for part family formations. *J Manufact Syst*, 5(4), 223–230.

Kaparthi, S. and Suresh, N. C. (1991). A neural network system for shape-based classification and coding of rotational parts. *Int J Product Res*, 29(9), 1771–1784.

King, J. R. (1980). Machine-component grouping in production flow analysis: An approach using a rank order clustering algorithm. *Int J Product Res*, 18(2), 213–232.

Koupaei, M. N., Mohammadi, M. and Naderi, B. Optimizing flexible manufacturing system: A developed computer simulation model. *Int J Eng — Trans B: Appl*, 29(8), 1112.

Kusiak, A. (1990). *Intelligent Manufacturing Systems*. Prentice Hall Press, NJ, USA.

Luggen, W. W. (1991). *Flexible Manufacturing Cells and Systems*. Prentice Hall.

McAuley, J. (1972). Machine grouping for efficient production. *Product Eng*, 51(2), 53–57.

McCormick Jr, W. T., Schweitzer, P. J. and White, T. W. (1972). Problem decomposition and data reorganization by a clustering technique. *Oper Res*, 20(5), 993–1009.

Mohammadi, M. and Forghani, K. (2017). A hybrid method based on genetic algorithm and dynamic programming for solving a bi-objective cell formation problem considering alternative process routings and machine duplication. *Appl Soft Comput*, 53, 97–110.

Moon, Y. (1998). Part family identification: New pattern recognition technologies. Group Technology and Cellular Manufacturing: A State-of-the-art synthesis of research & practice.

Patel, J. and Patel, S. (2014). Approaches to solve cell formation, machine layout and cell layout problem: A Review. *Trans Mach Learn Artif Intell*, 2(5), 80–96.

Rabbani, M., Farrokhi-Asl, H., Rafiei, H. and Khaleghi, R. (2017). Using metaheuristic algorithms to solve a dynamic cell formation problem with consideration of intra-cell layout design. *Intell Decision Technols*, 11(1), 109–126.

Seifoddini, H. K. (1989). Single linkage versus average linkage clustering in machine cells formation applications. *Comput Industr Eng*, 16(3), 419–426.

Singh, N. and Rajamani, D. (2012). *Cellular Manufacturing Systems: Design, Planning and Control*. Springer Science & Business Media.

Srinlvasan, G., Narendran, T. and Mahadevan, B. (1990). An assignment model for the part-families problem in group technology. *The Int J Product Res*, 28(1), 145–152.

Suresh Kumar, C. and Chandrasekharan, M. (1990). Grouping efficacy: A quantitative criterion for goodness of block diagonal forms of binary matrices in group technology. *Int J Product Res*, 28(2), 233–243.

Chapter 10

Electrical Discharge Micromachining with Sustainable Dielectrics

G. Kibria*,‡, T. D. Sherpa†, I. Shivakoti† and B. B. Pradhan†

*Mechanical Engineering Department, Aliah University, Kolkata, India
†Mechanical Engineering Department,
Sikkim Manipal Institute of Technology,
Sikkim Manipal University, East Sikkim, India
‡prince_me16@rediffmail.com

The chapter presents the advances in microelectrical discharge machining process with various dielectrics used during micromachining so as to improve the machining performance. Microelectrical discharge machining is an established method and has a huge demand in the manufacturing sector in addition to its appeal among the researchers for exploring its potential so as to produce microfeatures with higher precision, accuracy and surface integrity. The dielectric fluid employed in the micro-EDM process plays a major function for achieving the desired features. Therefore, it is important to understand its role in micro-EDM machining. The chapter attempts to review the gradual evolution of micro-EDM machining with different dielectrics and their effects on the variety of performance measures such as machining rate, rate of tool wear and geometric accuracy and the sequential development that has taken place over the years on the use of dielectrics during micromachining of various conducting materials.

1. Introduction

In recent times, with rapid growth in micro-machining technology and miniaturization, the demand of micro-products has increased tremendously in numerous fields like aerospace, automotive, biomedical and electronics. Micromachining can be defined as the process in which the products are fabricated in the range of 1–999 μm or it is the process in which material removal (unit removal) is achieved at the micro- or nanolevel (Jain, 2010). Micromachining is the primary technology for the fabrication of miniaturized parts and components. Compared to optical or X-ray lithography, generation of microstructures using micromachining requires simple but precision equipment and machinery. For effective micromachining of a wide range of materials, the development of high-precision machine tools with advanced control unit and power generators are essentially required. Limited availability of microtool has always been a major concern in micromachining. Micro-EDM is the most prominent technique for micromachining among various nonconventional methods by virtue of its inherent ability to remove materials at micro- to nanometer range. However, it has been a challenging task for manufacturing engineers to produce micro parts with higher accuracy and precision by micromachining of difficult-to-machine materials.

Micro-EDM comes under the tool-based micromachining using nonconventional sources of energy, i.e., electro-thermal energy is utilized for effective removal of the material (Jain, 2010; Prihandana *et al.*, 2009; Lim *et al.*, 2003; Pham *et al.*, 2004; Katz and Tibbles, 2005; Tansel *et al.*, 2000). Micro-EDM is an important and an established method for micromachining of components. Since the material removal is based on thermal erosion, micro-EDM has the capabilities to produce complex shapes irrespective of material hardness (Prihandana *et al.*, 2009). An attempt has been made in this chapter to review the advances in the micro-EDM process with different pure and powder mixed dielectrics that has taken place so far to improve performance during micromachining.

2. Overview of Micro-EDM

Micro-EDM is one of the successfully developed thermoerosive processes. In this process, a material is removed by applying repetitive pulse discharges between two electrodes. Generally, both micro-EDM and conventional EDM work on the same principle. However, some of the differences studied in the past research between EDM and micro-EDM are that specific energy for material removal in microEDM is in the range of microjoules (μJ), which is quite low compared to the EDM. In addition, the interelectrode gap in micro-EDM is about 1–5 μm whereas in EDM, it is in the range of tens of μm, which has a significant effect on machining (Jain, 2010). In micro-EDM, the tool electrode and workpiece do not come into contact during discharges and therefore there is no force involved during material removal hence any conductive materials (ductile and brittle) can be machined efficiently with no constraint of the material's hardness (Prihandana *et al.*, 2009; Lim *et al.*, 2003). However, there are various process parameters that influence the performance characteristics of the micro-EDM process.

3. Challenges in Micro-EDM

Micro-EDM is one of the best modern micromachining technologies that has enabled access to an advanced miniaturized product at a reasonably low cost. However, still there are many problems that need to be addressed and research is continuously in progress in order to make μ-EDM a reliable process. In micro-EDM, the manufacturing engineers are faced with great challenges to produce micro parts with a high degree of accuracy. On-machine fabrication of microelectrodes is an important issue in micro-EDM to avoid microelectrode handling problems and clamping error issues (Lim *et al.*, 2003). Therefore, efforts are primarily made to minimize the erosion of workpiece in single discharge and the electrode wear has also been reported as an important criterion to be considered in the micro-EDM process (Pham *et al.*, 2004). However, electrode heating has also been a problem due to the limiting size of tool electrode in micro-EDM (Katz and Tibbles, 2005).

3.1. *Tool handling*

Electrode wear has been reported as one of the major challenges when it comes to micro-EDM (Jain, 2010). As reported by few authors, the handling of electrodes have been problematic due to the reduction of wire diameter in the micro-EDM process (Pham *et al.*, 2004). In this process, prevention of the distortion of tool electrode can be done by using dielectrics with high heat capacity as revealed by a few studies (Katz and Tibbles, 2005). In micromachining, premature tool failure is an important issue because fabrication of such microtools is costly. Moreover, vibration and chip flow characteristics are not achieveable without high-precision equipment or accessories. The cutting tools fail early due to premature breaking of cutting edges (Tansel *et al.*, 2000).

3.2. *Pulse generator*

There are various types of pulse generators used in micro-EDM, namely RC pulse generator, transistor generator, RC relaxation generator and iso-pulse generator with their own capabilities and limitations. Mostly, RC pulse generator is found to be used in micro-EDM. The RC pulse discharge circuit was suitable for producing high-frequency discharges (Chung *et al.*, 2007). It was observed by some researchers that erosion value is directly proportional to energy of the spark. A single capacitor discharge would be the simplest form of circuit to fulfil the need of micro-EDM. Few authors have reported that single discharge can be controlled in iso-pulse generator, which also aids in determining the discharge energy. However, discharge duration was found to be influenced by ignition delay in equi-frequency pulse generator (Cao *et al.*, 2005). Applying transistor type iso-pulse generator with servo feed control, MRR as high as 24 times is achieved compared to RC pulse generator in micro-EDM (Han *et al.*, 2004). It was revealed that power transistor can be operated with large current with high response. Power transistor itself was found to have a certain amount of delay time (Kunieda *et al.*, 2005). Several attempts were made by researchers to reveal the capabilities of transistor type in micro-EDM, but a minimum discharge time of only 30 ns were achieved (Chow *et al.*, 2000).

In relaxation-type pulse generator, some researchers reported stray capacitance in micro-EDM. This stray capacitance survives between tool electrode and workpiece or electrode holder and workpiece (Dhanik and Joshi, 2005). Experimentally, it is found that at low pulse energy, the size of microcrater generated was consistent than higher discharge energies while using RC pulse discharge circuit and the estimated erosion efficiency was found to be higher at lower discharge energies (Wong *et al.*, 2003). The RC pulse generator is able to generate short pulse-on-time. When experiments were conducted using a transistor type iso-pulse generator, higher machining speed was obtained compared to the RC pulse generator. However, pulse-on-time of about 80–30 ns was obtained (Han *et al.*, 2007).

Later on, transistor-type pulse generator with improved capability of power transistors was developed, which replaced relaxation-type pulse generator with the aim of improving production efficiency (Chung *et al.*, 2009). While using water as a dielectric, to prevent electrolytic corrosion a high-frequency bipolar pulse generator was developed for micro-EDM (Chung *et al.*, 2011). Using the RC generator, improved quality of machined surface is obtained by employing a hydrocarbon-based dielectric (Ekmekci *et al.*, 2009). The view of microcraters at various discharge energies using RC discharge circuit is depicted in Figure 1.

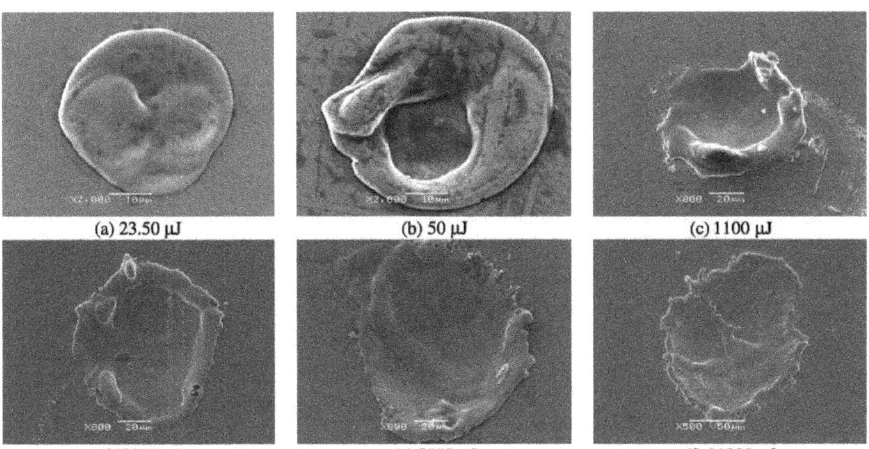

Figure 1: Micro-craters at 2 μm spark gap and different discharge energies using RC discharge circuit (Wong *et al.*, 2003).

3.3. *Flushing of dielectrics*

It has been reported that flushing condition of the dielectric may cause variation in the gap between the electrode and workpiece in the micro-EDM process (Pham *et al.*, 2004). It has also been reported by few authors that dielectric with low viscosity was found to have effective flushing (Katz and Tibbles, 2005). However, one of the inherent problems in micro-EDM was found to be the ineffective removal of ejected debris. Due to this, regular adhesion of tool electrode with workpiece, i.e., short-circuit, occurs. These results in longer machining time due to unstable machining. To clear the interelectrode gap, the tool electrode has to be withdrawn back and again, the spark discharge is maintained (Kunieda *et al.*, 2005). Increase in concentration of the powder particles in the dielectric decreases dielectric strength, which leads to an increase in the spark gap, thus providing proper flushing (Jahan *et al.*, 2010).

3.4. *Tool development*

On the machine, manufacturing of tool electrode has been a challenge in micro-EDM as fabrication of micro-tool is done on machine itself. Figure 2 shows the micro-tool fabrication method on-the-machine itself. Among several micro-tool fabrication techniques, it was found that fabrication of microtool electrode using rotating method is mostly efficient compared to other technologies (Lim *et al.*, 2003).

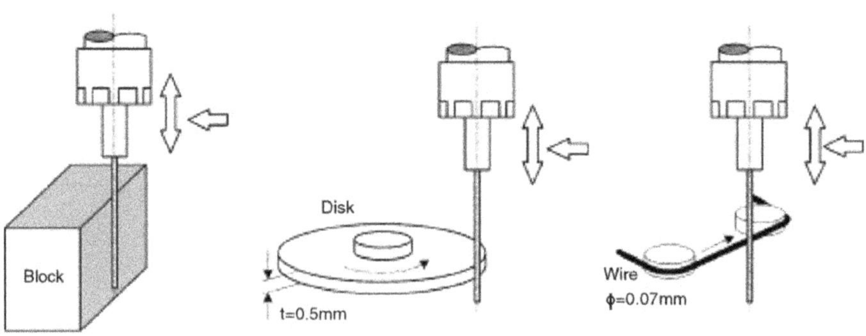

Figure 2: Tool fabrication methods "on-the-machine (Pham *et al.*, 2004).

However, it has been problematic while performing on-the-machine manufacturing since a number of additional subsystems were required to be incorporated to hold the microelectrodes (Pham *et al.*, 2004).

It was observed that microtool fabrication in air as well as in deionized water was found to be difficult due to higher cooling capacity. Keeping in view the problem of cost and time consumption in fabricating the microelectrode, some authors have proposed a creative way of fabricating microelectrode in a single discharge using the principle of pressure difference in plasma (Cao *et al.*, 2005). Microtool electrode was found to be off centred from axis of rotation of the spindle when a microtool was made using another machine. However, no run out of electrode was observed in wire electrodischarge grinding (Kunieda *et al.*, 2005). In Figure 3, the schematic representation of micro-tool electrode fabrication technique in wire electro discharge grinding process is observed.

3.5. *Geometry and surface quality*

Extensive research has been carried out on improvement of EDMed surface. However, very limited research has been conducted for micro-EDM surface. Observations were made by the researchers that due to hardly any change in grain size, there was no recast layer on the surface machined by micro-EDM (Jain, 2010). However, the surface

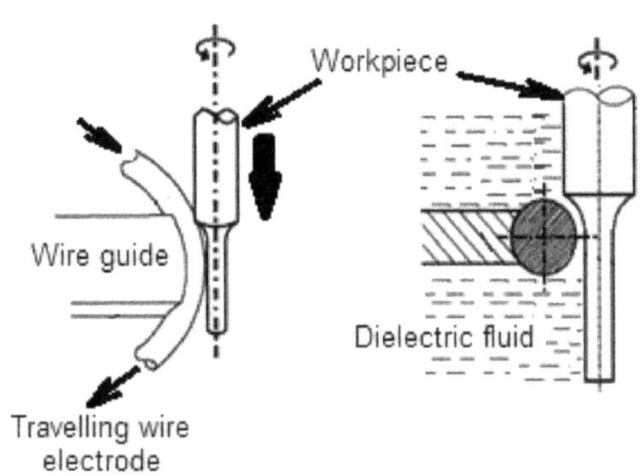

Figure 3: Schematic of wire electro discharge grinding method.

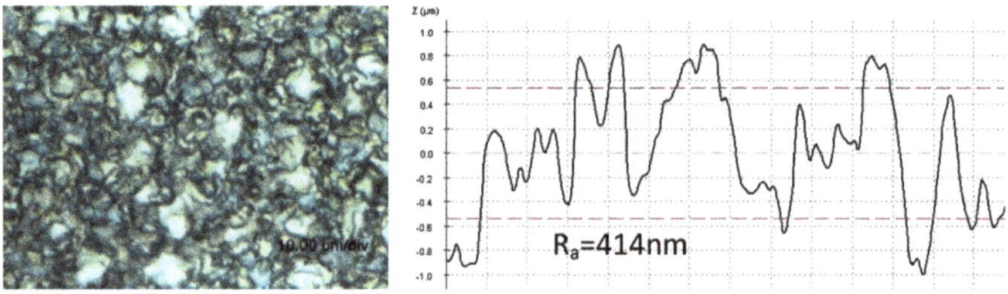

Figure 4: Microscopic images of machined surface by micro-EDM and corresponding surface profile (Nguyen *et al.*, 2012).

machined by micro-EDM was found to have high surface roughness due to asperities generated by vaporization and melting, which also affect machining shape and accuracy (Nguyen *et al.*, 2012). The microscopic view of EDMed surface with surface profile is observed in Figure 4. Tubular electrode with high melting point was reported to have better surface finish than the circular electrode (D'Urso and Merla, 2014).

4. Dielectric Fluid in Micro-EDM

Dielectric fluid is an insulating medium which takes in account the effects of alternating electric field. Dielectric fluid plays an important role in micro-electrical discharge machining process. Even for micro-tool fabrication, cooling capacity of dielectrics was reported as an influencing factor (Cao *et al.*, 2005). In micro-EDM, the most significant aspect to certify the effectiveness of the feed is the dielectric fluid. Mostly, improvement of machining rate, geometry and accuracy of surface features, corrosion and wear resistance depend on the type of dielectric fluid employed during micro-EDM (Chung *et al.*, 2009). Using proper dielectric fluid with good flushing condition would surely improve MRR and TWR. In the literature, several researchers have carried out in-depth studies to quantify the effect of dielectric fluid on various performance measures. However, the number of such research is still limited and inadequate. Therefore, at this stage, in-depth investigation and experimentation is still required for finding out the environment-friendly dielectric fluid, which can be

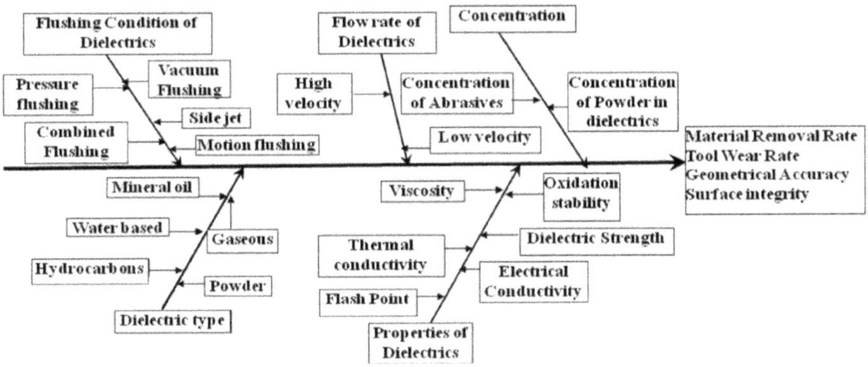

Figure 5: Influences of various parameters associated with dielectric on machining characteristics of micro-EDM process.

applied for micromachining of a wide range of engineering materials. Figure 5 shows the influences of various parameters associated with dielectric fluid in fish-bone diagram that affect the performance characteristics of micro-EDM during the machining process. Various researches on the effects of dielectrics on EDM and micro-EDM have been undertaken for the past few years. The graphical representation of a number of publications on the effect of dielectrics from 2000 to 2018 year-wise is shown in Figure 6.

4.1. *Functions of dielectric fluid*

One of the most important functions of dielectric fluid is that a discharge gap must be established between electrodes and therefore premature discharges occurring between the electrodes can be avoided. In addition, proper flushing of dielectric fluid in the small IEG helps the ejected debris to flush away and thus efficient discharges are established in the IEG. Moreover, the machining stability is obtained when dielectric fluid is responsible to cool the tool electrode surface due to heat convention phenomenon in the boundary layer (Kunieda *et al.*, 2005). Since a very high temperature is generated in the plasma channel of discharges, the dielectric fluid helps to carry out additional heat from tool tip and discharge craters. It was shown that the variation of dielectric fluid properties and flushing conditions

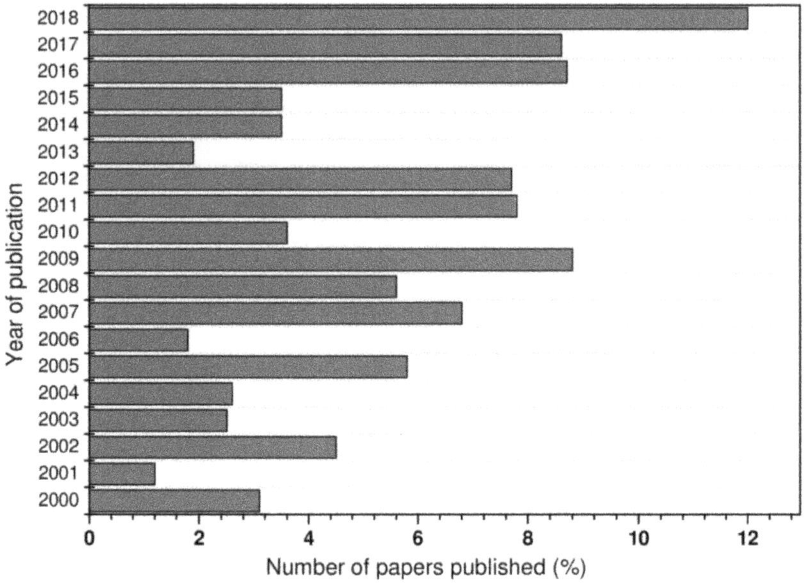

Figure 6: Percentage of publications on the effect of dielectric fluid on micro-EDM for the year 2000–2018.

result in a variation in the rate of material removal and amount of tool wear (Kunieda *et al.*, 2005). It helps in the restrictions of spark, i.e., the localization of thermal energy in order to have higher density of energy and this consequently leads to a higher performance (Kunieda *et al.*, 2005).

4.2. *Properties of dielectric fluids*

There are various properties of dielectric fluids that affect the machining performance. Some of the important properties of dielectric that are considered often are dielectric strength, viscosity, oxidation stability, thermal conductivity and thermal capacity.

4.2.1. *Dielectric strength*

A good dielectric fluid should be such that it can maintain the resistivity before discharges occur and also it must have the ability to recover the initial resistivity after discharge in IEG. A higher value of dielectric strength takes less time to recover its dielectric properties for further discharges. Therefore, it is sufficient to keep low values of pulse-off-time for discharges. Powder additives decrease the dielectric strength of dielectric fluid and increase the spark gap (Jahan *et al.*, 2010).

4.2.2. *Viscosity*

Viscosity is the property that describes a fluid resistance to flow. The dielectrics with low viscosity were effective in flushing as well as removal of debris in the micro-EDM process (Katz and Tibbles, 2005). Due to better flow capabilities of dielectric fluids having lower values of viscosity, flushing of debris occurs efficiently and thus the machining efficiency and accuracy are improved.

4.2.3. *Thermal conductivity*

To take out excess heat from IEG, the dielectric fluid plays an important role. For that, the dielectric fluid should possess high heat capacity and high thermal conductivity properties. Thus, the thermal damage of machined surface due to excess heat can be minimized. Higher conductivity of a dielectric fluid extends the gap distance, which helps to easily collapse the insulation of the dielectric fluid (Chow *et al.*, 2000). Pour points, flash point and volatility are also the other properties that may affect the performance. However, they are often not listed in the specifications for dielectric fluids.

4.3. *Chemical aspects of dielectrics*

The presence of metallic elements in workpiece that react with oxidized oil at high discharge energy prevents a stable discharge between the electrodes producing organometallic compounds on the job surface. These compounds are formed on both tool electrode and workpiece surfaces. Thus, it was proposed by researchers that use of these dielectrics develops stable plasma channels in the IEG since some ejected debris are converted into organometallic compounds. Thus, short-circuiting or arcing in IEG is avoided. Considering fire hazards and the release of harmful vapors such as CO and CH_4, use of deionized water is found to be much more preferable to hydrocarbon-based dielectrics (Kunieda *et al.*, 2005). In addition, several hazardous fumes are formed while using mineral oil or organic fluid. Some of these fumes are benzene, aromatic hydrocarbons, mineral aerosols and polycyclic aromatic hydrocarbons. Since these fumes are hazardous for the environment around the EDM setup, use of mineral oil or an organic fluid is a risk for the environment (Tonshoff *et al.*, 1996). Boron carbide powder has some extraordinary properties that includes wear and abrasion resistance, high chemical resistance and high hardness (third behind diamond). These properties aid in better performance of micro-EDM when mixed with dielectrics (Kibria *et al.*, 2010). The hydrocarbon oil kerosene produces air pollution by degrading itself when used in the long run. However, the use of pure water and SiC powder has been proven to produce surfaces without any environmental pollution. Moreover, aluminum powder with water as a dielectric results in environmental pollution as aluminum powder has smaller density due to which stirring effect was not good enough (Chow *et al.*, 2008).

5. Development in Research of Micro-EDM with Different Dielectrics

With the advancement of research in micro-EDM, various dielectrics have been used by researchers during experiments to get most from the machining process so that the performance characteristics considered during the process gets improved. In general, a combination of kerosene and deionized water is used as a dielectric fluid in the EDM process (Masuzawa *et al.*, 1989). However, there are many other dielectrics that were employed in the micro-EDM process so that the performance of the process would improve.

5.1. *Hydrocarbon-based dielectrics*

5.1.1. *Kerosene*

Kerosene was one of the first popular dielectric oil and was used in most die sinking process. Kerosene was found to be a suitable dielectric as well for the fabrication of micro-electrode as compared to deionized water (Cao *et al.*, 2005). Use of EC-CVD diamond electrode with dielectric kerosene oil (vitol2) was highly efficient in terms of material removal rate as compared to other electrodes (Sharma *et al.*, 2005). Use of kerosene has an advantage of having low viscosity, which leads to proper flushing (Masuzawa *et al.*, 1989). It was observed that impulsive force of discharge was higher using kerosene as a dielectric. While using kerosene as a dielectric during micromachining, discharge energy was found to decrease as carbon adherence to the electrode was high, resulting in electrode wear as compared to water (Lin *et al.*, 2007). Figure 7 shows the qualitative effect of dielectric fluid (kerosene and de-ionized water) at the same parametric setting.

5.1.2. *Kerosene with additives*

Considering some of the limitations of kerosene into account, kerosene with various additives has been used as dielectrics in micro-EDM expecting an improvement in the performance. Crater dimensions were observed to have reduced by the addition of silicon particles having average size of 10 μm in kerosene–Castrol (Kagaya *et al.*, 1986). While producing blind micro hole of 500 μm depth and aspect ratio of 1.67 only 7-min machining time was taken

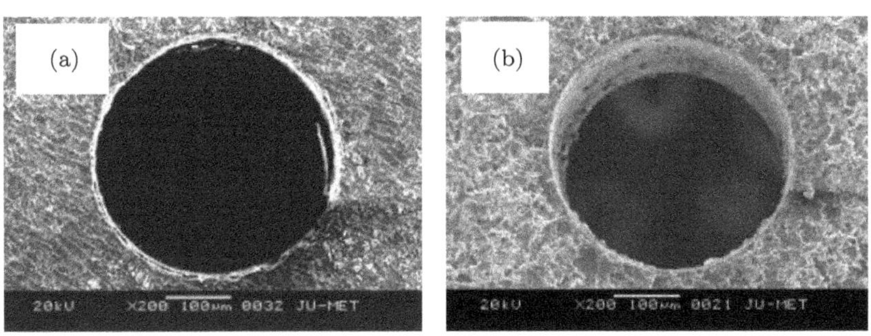

Figure 7: SEM images of micro-hole produced using (a) kerosene and (b) deionized water when machined at 1.5 A (current)/5 μs (pulse-on-time)/0.5 kgf/cm^2 (flushing pressure)/95% (duty ratio) (Kibria *et al.*, 2010).

with nanographite powder added to kerosene as compared to 34 min with a normal dielectric (Prihandana *et al.*, 2012).

5.1.3. *Mineral oil*

While using hydrocarbon-based oil, debris particles formed during machining piled up with work material due to fluid vortex at the interelectrode gap due to which work material produces its complementary shape on the tool electrode (Ekmekci and Sayar, 2013). When EDM oil was used as dielectric while machining nonconducting zirconia ceramics, carbon deposition on surface took place, which may be attributed to thermal splitting of hydrocarbon oil molecules caused by the plasma discharge (Schubert *et al.*, 2013). When an experiment was conducted using different nanopowder-mixed EDM 3 oil dielectric on a tungsten carbide material, semi-conductive graphite nanopowder demonstrated best performance followed by aluminum powder and alumina (Jahan *et al.*, 2010). Carbon-enriched layers were formed due to formation of craters when pulse duration was increased while using hydrocarbon oil as a dielectric (Ekmekci *et al.*, 2009).

5.2. *Water-based dielectrics*

5.2.1. *Deionized water*

It was reported that low resistivity of deionized water causes dielectric breakdown leading to large machining gap. Hence tool wear is reported to have reduced with decreasing resistivity (Chung *et al.*, 2007). Using deionized water in finishing process after micro-hole machining with certain resistivity, voltage, tool rotation and finishing time reduced roughness from 0.225 to 0.066 μm (Chung *et al.*, 2009). While using deionized water as dielectric, it was reported that machining process becomes more stable since deionized water does not produce carbon and also the conductivity of deionized water is low (Masuzawa *et al.*, 1989). Moreover, electrode feed rate in deionized water was observed to be higher compared to mineral oil (Yu *et al.*, 2002).

5.2.2. *Pure water*

It was observed that the use of water as a dielectric gives good discharge since no carbon deposition was found using water during machining (Chow *et al.*, 2000). Electrolytic corrosion was observed as a main problem while using water as a dielectric in micro-EDM. However, with the development of high-frequency bipolar pulse generator micro-holes with high surface quality was observed (Chung *et al.*, 2011). Use of pure water as a dielectric in micro-slit EDM machining of titanium alloy has proved to be stable with high removal rate and low electrode wear throughout the process regardless of pulse duration (Lin *et al.*, 2007). It was observed that increasing resistivity of water to 10^7 ohm-cm minimizes the effect of electrochemical action (Yu *et al.*, 2002). In addition, during the long run, pure water has no such diverse effect by the temperature generated at discharge zone (Chow *et al.*, 2008).

5.3. *Micro-EDM with liquid and gaseous dielectrics*

Micro-tool fabrication was found to be difficult while using air as dielectric due to the higher cooling capacity of air (G. Cao *et al.*, 2005). In liquid dielectric, it was observed that discharge energy used for melting work piece was constant regardless of pulse duration. However, discharge energy was found to decrease with increase in pulse duration in air (M. Yoshida *et al.*, 2011).

5.4. *Powder mixed dielectrics*

5.4.1. *Ultrasonic vibration with powder-mixed dielectric*

The ultrasonic vibration during micro-EDM process is an innovative technique to improve the machining efficiency and other geometrical features. In addition, several researchers applied micro-tool vibration/dielectric vibration and even workpiece vibration when employing a powder-mixed dielectric fluid. Researchers found that applying ultrasonic vibration in dielectric is cheaper than applying vibration to tool or work piece as vibration on work piece and tool can only prevent the deposition of debris but not the agglomeration of powder (Prihandana *et al.*, 2009). Improved micro-EDM performance was observed along with material removal rate, machining depth and surface topography (Liew *et al.*, 2014).

5.4.2. *Dielectrics with various powder additives*

Use of MoS_2 powder in the dielectric was found to improve the μ-EDM process by decreasing the dielectric strength of the dielectric fluid. Moreover, it is also easily removable and less expensive as compared to nanopowder (Prihandana *et al.*, 2009). Introduction of graphite powder (10 g/L) in kerosene with workpiece vibration reduced the machining time (Prihandana *et al.*, 2012). Some researchers have done comparative study on boron carbide (B_4C) powder mixed with kerosene and deionized water. It was observed that B_4C powder mixed with deionized water has positive influence concerning surface quality and accuracy while machining Ti–6Al–4V (Kibria *et al.*, 2010). Stepped circular electrode performed better during microhole machining in terms of surface finish using an abrasive slurry (Yu *et al.*, 2002). However, it was observed that a higher concentration of abrasives increases MRR owing to the fact that it bridges the gap between tool and workpiece (Yoshida *et al.*, 2011).

Some of the important characteristics like particle size and concentration, electrical resistivity, particle density, thermal conductivity, etc. severely affect the performance characteristics of micro-EDM process (Jahan *et al.*, 2010). It has been found that a reduction of 14–24% in the value of surface roughness is obtained using nanopowder-suspended dielectrics than pure dielectric fluid (Tan *et al.*, 2008). During micro-EDM, drilling and milling of cemented tungsten carbide material and the use of suspended graphite nanopowder in dielectric produced a smooth and defect-free surface with high material removal rate (Jahan *et al.*, 2011).

It was found that the rotation of tool electrode using SiC powder mixed dielectric oil resulted in flushing effects, which further affected the accumulation of powder in the discharge gap (Tan and

Yeo, 2011). Over a period of time, various researchers carried out studies to improve the machining performance of micro-EDM process using different dielectrics with or without additives. In this review, it is felt necessary to consolidate them so as to understand the usage of various dielectrics with or without additives. Figure 8 shows different dielectrics used so far in micro-EDM process and powder mixed dielectric has a greater proportion amongst all the other dielectrics that has been used in micro-EDM. Furthermore, there are various powder additives having different properties and among all these powders, silicon carbide and graphite account for the highest usage in the field of micro-EDM and carbon, copper, and molybdenum disulfide being the lowest as per the data given in Figure 9.

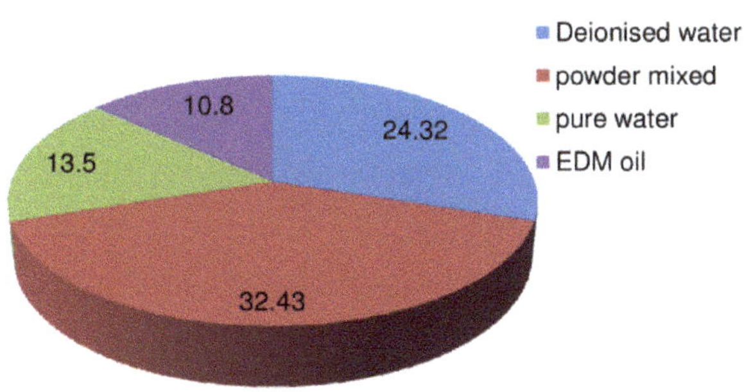

Figure 8: Percentage of different dielectrics used during year 2000–2018.

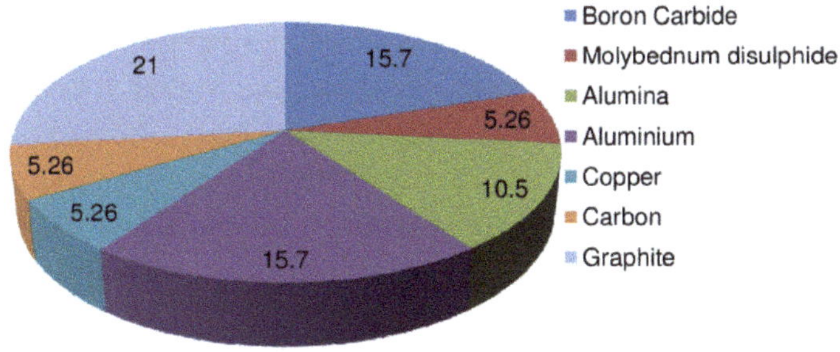

Figure 9: Percentage of various powder mixed dielectrics used in micro-EDM.

5.5. *Dielectrics with changing polarity*

Due to lower stiffness of electrode, polarity of workpiece was made negative in micro-EDM process to prevent the formation of insulating layer (Lim *et al.*, 2003). However, combination of right electrode with polarity ensures good performance. Straight polarity with use of tungsten electrode provided high MRR and low electrode wear (Kagaya *et al.*, 1986). Using electrically conductive chemical vapor-deposited diamond electrode at negative polarity with kerosene (Vitol 2) as a dielectric, no tool wear was observed as compared to straight polarity (Sharma *et al.*, 2005). High removal rate and low electrode wear were obtained with negative polarity (Lin *et al.*, 2007). Negative polarity machining using pure water as a dielectric provided higher MRR, low electrode wear and less burr formation on the surface as compared to positive polarity regardless of pulse duration (Chow *et al.*, 2008).

6. Effects of Various Dielectrics in Micro-EDM Process Performances

6.1. *Effects of dielectrics in material removal mechanism*

Various researchers studied the effects of ultrasonic vibration on powder-mixed dielectric during the micro-EDM process and reported that using MoS_2 powder in dielectric with copper as the workpiece and brass as a tool increases MRR and in addition ultrasonic vibration prevents adhesion of carbon particles at bottom of the tank and also improves MRR. MRR was found to be higher for the higher concentration of powder (Prihandana *et al.*, 2009). Use of deionized water as a dielectric fluid increases machining gap than that obtained with the kerosene, which further increases the machining rate (Chung *et al.*, 2007). It was observed that SiC powder can machine higher depth compared to Al powder mixed in kerosene oil (Chow *et al.*, 2000). One of the dielectric fluids that replaced kerosene in most die sinking process was pure water whose temperature was not affected by long working time. Hence improvement in MRR was also observed

(Chung *et al.*, 2009). It was also observed that deionized water with B_4C powder showed an excellent increase in MRR compared to the B_4C abrasive mixed kerosene (Kibria *et al.*, 2010). Removal ratio of the process using liquid was found to be comparatively lower than air as the removed material remains on the bottom of crater (Yoshida *et al.*, 2011). Use of silicon carbide and water as a dielectric during micro-slit EDM machining provided higher MRR compared to pure water alone (Chow *et al.*, 2008). Using boron carbide mixed additive, the stability of pulse discharge occurs due to more uniform distribution of energy across the IEG. At higher pulse duration, the formation of carbon particles that adhere to tool tip and titanium oxide that adhere to work surface restrict the material to remove easily and thus MRR decreases.

6.2. *Effects of dielectric fluid in tool wear rate*

When deionized water with low resistivity was employed as a dielectric, tool wear was very small and the machining time was also observed to decrease. With an increase of resistivity of the deionized water, tool wear observed to be high as well as machining gap was comparatively less (Chung *et al.*, 2007). Using tungsten electrode and water as a dielectric fluid, both electrode wear and working time were reduced (Kagaya *et al.*, 1986). Using kerosene dielectric, the rate of electrode wear is less than using Al- and SiC-suspended kerosene (Chow *et al.*, 2000). Tool wear was found to be higher using both deionized water and B_4C mixed deionized water than kerosene while machining titanium super alloy as no adhesion of carbon particles occurred while using deionized water. Using boron carbide mixed dielectric, the tool wear is less due to adhesion of carbon particles produced by decomposing the carbide additive (Kibria *et al.*, 2010). During micro-EDM milling, use of Al powder mixed EDM 3 oil as a dielectric provided lower electrode wear rate among alumina and graphite powder (Jahan *et al.*, 2010). While using pure water and SiC added to pure water as a dielectric during micro-slit EDM machining of titanium alloy, low electrode wear was provided by SiC-mixed dielectric regardless of pulse duration as compared to pure water

(Chow *et al.*, 2008). With increase in pulse-on-time, the duration of the presence of carbide particles in pulse discharge is high and it results in more adhesion of carbon particles on the tool electrode surface. This adhesion of carbon particles is completely avoided utilizing oxygen-based deionized water.

6.3. *Effects of dielectrics in geometrical accuracy*

While using micro-MoS_2 powder mixed dielectric with ultrasonic vibration on dielectric fluid during micro-EDM process, it was reported by the authors that tool material was not so significant factor for MRR but affects the geometrical accuracy. Some researchers have compared the surface profile of grooves machined using kerosene and deionized water and it was found that the grooves machined using deionized water have a better profile and smaller pits compared to the latter one. Specific resistivity of water was found to have a profound effect on the surface quality and shape of the tool electrode and holes generated. Hence, resistivity of water was controlled to 10^5 to 10^6 Ω-cm (Kagaya *et al.*, 1986).

While machining titanium alloy with copper as an electrode using three different types of dielectrics namely kerosene, SiC-mixed kerosene and Al-powder-mixed kerosene, Al powder with kerosene appeared to be best among the three dielectrics. This is due to the fact that at very low value of powder concentration, the surface quality obtained is excellent in kerosene dielectric compared to other dielectrics (Chow *et al.*, 200). Figure 10 shows the comparison of the surface machined by using kerosene, SiC-mixed kerosene and Al-powder-mixed kerosene dielectric. The fine surface quality was obtained using high-frequency bipolar pulse generator in tap water as compared to deionized water (Chung *et al.*, 2011). When B_4C additive was used with kerosene and deionized water, overcut was found to be larger with both powder-mixed dielectrics compared to pure kerosene and deionized water (Kibria *et al.*, 2010). Semi-conductive graphite nanopowder at a concentration of 0.2–0.6 g/L with EDM 3 oil provided lowest roughness value during micro-EDM milling compared to aluminum and alumina powder materials (Jahan *et al.*, 2010).

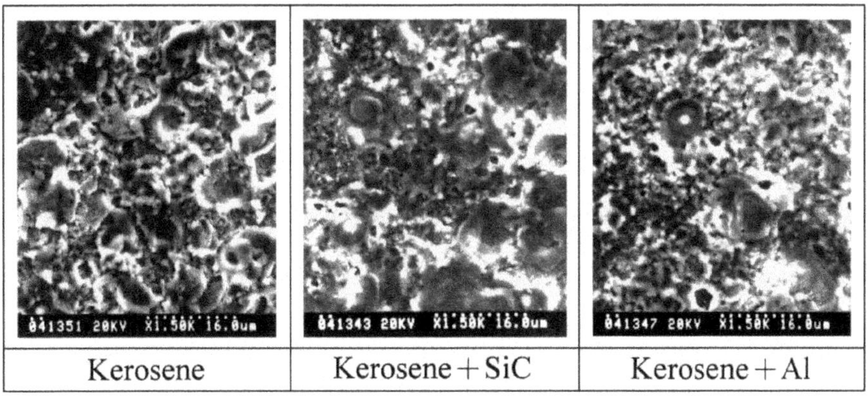

Figure 10: SEM images of EDMed surface using different dielectric fluids (Chow *et al.*, 2000).

Few micro-slit burrs were formed using SiC and water as SiC disperses energy and improves roughness as compared to pure water (Chow *et al.*, 2008). Recast layer thickness increases for both the dielectric with increase in pulse on time (Tan and Yeo, 2011). When micro-drilling operation of Ti–6Al–4V alloy using B_4C powder mixed deionized water, the microhole taper is less compared to using pure deionized water. From various studies, it was found that additives mixed with normal dielectric, i.e., powder mixed dielectric have a positive influence on geometrical accuracy. Employing powder mixed dielectric, the size and depth of crater is less and also the thickness of white layer formed is less compared to pure dielectrics.

7. Summary

Due to deposition of carbon nuclides on the tool and highly volatile nature of kerosene leads to unstable machining, which is undesirable. In micro-EDM, various researchers have diagnosed tool wear as the major problem to be addressed for precise 3D machining. So, it would not be an enhanced notion to use kerosene as a dielectric in micro-EDM. Many researchers have found deionized water and distilled water as a proper alternative for kerosene. The effects of most of the used dielectrics on various process performances have been discussed possibly. However, further continuous experimental investigation is required to utilize the potential of micro-EDM process that can make a remarkable advancement in the field of machining process.

References

Cao, G., Zhao, W., Wang, Z. and Guo, Y. (2005). Instantaneous fabrication of tungsten microelectrode based on single electrical discharge. *J Mater Process Technol*, 168(1), 83–88.

Chow, H. M., Yan, B. H., Huang, F. Y. and Hung, J. C. (2000). Study of added powder in kerosene for the micro-slit machining of titanium alloy using electro-discharge machining. *J Mater Process Technol*, 101(1–3), 95–103.

Chow, H. M., Yang, L. D., Lin, C. T. and Chen, Y. F. (2008a). The use of SiC powder in water as dielectric for micro-slit EDM machining. *J Mater Process Technol*, 195(1–3), 160–170.

Chow, H. M., Yang, L. D., Lin, C. T. and Chen, Y. F. (2008b). The use of SiC powder in water as dielectric for micro-slit EDM machining. *J Mater Process Technol*, 195(1–3), 160–170.

Chung, D. K., Kim, B. H. and Chu, C. N. (2007). Micro electrical discharge milling using deionized water as a dielectric fluid. *J Micromech Microeng*, 17(5), 867.

Chung, D. K., Shin, H. K., Kim, B. H., Park, M. S. and Chu, C. N. (2009). Surface finishing of micro-EDM holes using deionized water. *J Micromech Microeng*, 19, 045025.

Chung, D. K., Shin, H. K., Park, M. K. and Chu, C. N. (2011). Machining characteristics of micro EDM in water using high frequency bipolar pulse. *Int J Precis Eng Manuf*, 12(2), 195–201.

D'Urso, G. and Merla, C. (2014). Workpiece and electrode influence on micro-EDM drilling performance. *Precis. Eng.* 38(4), 903–914.

Dhanik, S. and Joshi, S. S. (2005). Modeling of a single resistance capacitance pulse discharge in micro-electro discharge machining. *J Manuf Sci Eng*, 127(4), 759–767.

Ekmekci, B., Sayar, A., Öpöz, T. T. and Erden, A. (2009). Geometry and surface damage in micro electrical discharge machining of micro-holes. *J Micromech Microeng*, 19, 105030.

Ekmekci, B. and Sayar, A. (2013). Debris and consequences in micro electric discharge machining of micro-holes. *Int J Mach Tools Manuf*, 65, 58–67.

Han, F., Wachi, S. and Kuneida, M. (2004). Improvement of machining characteristics of micro-EDM using transistor type isopulse generator and servo feed control. *Precis Eng*, 28(4), 378–385.

Han, F., Chen, L., Yu, D. and Zhou, X. (2007). Basic study on pulse generator for micro-EDM. *Int J Adv Manuf Technol*, 33(5–6), 474–479.

Jahan, M. P., Rahman, M. and Yong, Y. S. (2010). Modelling and experimental investigation on the effect of nanopowder-mixed dielectric in micro-electrodischarge machining of tungsten carbide. *Proc Inst Mech Eng B J Eng Manuf*, 224(11), 1725–1739.

Jahan, M. P., Rahman, M. and Wong, Y. S. (2011). Study on the nano-powder-mixed sinking and milling micro-EDM of WC-Co. *Int J Adv Manuf Technol*, 53(1–4), 167–180.

Jain, V. K. (2010). *Introduction to Micromachining.* Narosa Publishing House Pvt. Ltd., New Delhi.

Kagaya, K., Oishi, Y. and Yada, K. (1986). Micro-electrodischarge machining using water as a working fluid-I: Micro-hole drilling. *Precis Eng*, 8(3), 157–162.

Katz, Z. and Tibbles, C. J. (2005). Analysis of micro-scale EDM process. *Int J Adv Manuf Technol*, 25(9–10), 923–928.

Kibria, G., Sarkar, B. R., Pradhan, B. B. and Bhattacharyya, B. (2010). Comparative study of different dielectrics for micro-EDM performance during microhole machining of Ti-6Al-4V alloy. *Int J Adv Manuf Technol*, 48(5–8), 557–570.

Kunieda, M., Lauwers, B., Rajurkar, K. P. and Schumacher, B. M. (2005). Advancing EDM through fundamental insight into the process. *Ann CIRP*, 54(2), 64–87.

Liew, J. A., Yan, J. and Kuriyagawa, T. (2014). Fabrication of deep micro-holes in reaction-bonded SiC by ultrasonic cavitation assisted micro-EDM. *Int J Mach Tools Manuf*, 76, 13–20.

Lim, H. S., Wong, Y. S., Rahman, M. and Edwin Lee, M. K. (2003). A study on the machining of high-aspect ratio micro-structures using micro-EDM. *J Mater Process Technol*, 140(1–3), 318–325.

Lin, C. T., Chow, H. M., Yang, L. D. and Chen, Y. F. (2007). Feasibility study of micro-slit EDM machining using pure water. *Int J Adv Manuf Technol*, 34(1–2), 104–110.

Masuzawa, T., Tsukanoto, J. and Fujino, M. (1989). Drilling of deep micro holes by EDM. *Ann. CIRP*, 38(1), 195–198.

Nguyen, M. D., Rahman, M. and Wong, Y. S. (2012). Simultaneous micro-EDM and micro-ECM in low-resistivity deionised water. *Int J Mach Tools Manuf*, 54–55, 55–65.

Pham, D. T., Dimov, S. S., Bigot, S., Ivanov, A. and Popov, K. (2004). Micro-EDM-recent developments and research issues. *J Mater Process Technol*, 149(1–3), 50–57.

Prihandana, G. S., Mahardika, M., Hamdi, M., Wong, Y. S. and Mitsui, K. (2009). Effect of micro-powder suspension and ultrasonic vibration of dielectric fluid in micro-EDM processes-Taguchi approach. *Int J Mach Tools Manuf*, 49(12–13), 1035–1041.

Prihandana, G. S., Sriani, T. and Mahardika, M. (2012). Improvement of machining time in micro-EDM with workpiece vibration and graphite powder mixed in dielectric fluid. *Ind J Eng Mater Sci*, 19(6), 375–378.

Schubert, A., Zeidler, H., Hahn, M., Oschatzchen, M. H. and Schneider, J. (2013). Micro-EDM milling of electrically nonconducting zirconia ceramics. *Proc CIRP*, 6, 297–302.

Sharma, A., Iwai, M., Suzuki, K. and Uematsu, T. (2005). Potential of electrically conductive chemical vapor deposited diamond as an electrode for micro-electrical discharge machining in oil and water. *New Diam Front C Technol*, 15(4), 181–194.

Tan, P. C., Yeo, S. H. and Tan, Y. V. (2008). Effects of nanopowder additives in microelectrical discharge machining. *Int J Precis Eng Manuf*, 9(3), 22–26.

Tan, P. C. and Yeo, S. H. (2011). Investigation of recast layers generated by a powder- mixed dielectric micro electrical discharge. *Proc Inst Mech Eng B J Eng Manuf*, 225(7), 1051–1062.

Tansel, I. N., Arkan, T. T., Bao, W. Y., Mahendrakar, N., Shisler, B., Smith, D. and McCool, M. (2000). Tool wear estimation in micro-machining.: Part I: Tool usage-cutting force relationship. *Int J Mach Tools Manuf*, 40(4), 599–608.

Tonshoff, H. K., Egger, R. and Klocke, F. (1996). Environmental and safety aspects of electrophysical and electrochemical processes. *Ann CIRP*, 45(2), 553–568.

Wong, Y. S., Rahman, M., Lim, H. S., Han, H. and Ravi, N. (2003). Investigation of micro-EDM material removal characteristics using single RC-pulse discharges. *J Mater Process Technol*, 140(1–3), 303–307.

Yoshida, M., Hanaoka, D., Flynn, B. and McGeough, J. A. (2011). Observation of craters formed by single pulse discharge by stacking cross sectional shapes: Comparison of craters in liquid and air. *Proc Inst Mech Eng B J Eng Manuf*, 225(8), 1311–1318.

Yu, Z. Y., Rajurkar, K. P. and Shen, H. (2002). High aspect ratio and complex shaped blind micro holes by micro EDM. *Ann CIRP*, 51(1), 359–362.

Chapter 11

Dynamic Decision-Making Method for Design Concept Evaluation Based on Sustainability Criteria

Daniel O. Aikhuele* and Desmond E. Ighravwe

Department of Mechanical & Biomedical Engineering, Bells University of Technology, Ota, Nigeria

**danbishop_22@yahoo.co.uk*

This chapter presents the implementation of a new multicriteria group decision-making (MCGDM) model; this is based on intuitionistic fuzzy technique for order of preference by similarity to ideal solution (IFTOPSIS) method, dynamic intuitionistic fuzzy Einstein geometric averaging (DIFWGϵ) operator and intuitionistic fuzzy entropy (IFE) method for the selection of a sustainable power transformer design concepts under a dynamic environment. The model, which addresses the dynamic issues in the development of new products, also contributes to the sustainable design study by accounting for the uncertainty in the decision-making process. To the best of my knowledge, this is the first study to select sustainable design concepts under a dynamic environment or consider the dynamic issues in the development of a new product while meeting sustainability goal.

1. Introduction

There is a growing consensus in the engineering community on how to achieve a sustainable design when stakeholders must consider a series of decisions (Alexander, 1982; Hazelrigg, 1998; Hansen and Andreasen, 2004; Belinda and Amaresh, 2007; Xia *et al.*, 2001; Pandey and Mourelatos, 2015). Decision-making can be referred to as the procedure to find the best alternative among a set of feasible alternatives. Von Neumann and Morgenstern (1947) were among the first group of researchers who attempted to solve decision-making problems using mathematical models. In their work, an axiomatic decision-making approach was proposed, after which several other decision-making methods have emerged. And most of the works that emerged have been applied to sustainable design evaluation and decision-making.

Some of the authors have used the non-numerical approach to solve decision-making problems, for instance Ulrich and Eppinger (2000) designed concept screening, while Pugh (1996) presented design concept selection method. Other decision-making methods include AHP (Saaty, 2008; Laemlaksakul and Bangsarantrip, 2008; Hambali *et al.*, 2009) — analytical hierarchy process, design decision matrixes (King and Sivaloganathan, 1999), quality function deployment (Liu, 2011; Justel *et al.*, 2007; Marini *et al.*, 2016), fuzzy set concepts (Liu, 2011; Wang, 2002; Tauhid and Okudan, 2007; Akay *et al.*, 20111; Yan *et al.*, 2006; Jenab *et al.*, 2013; Vinodh *et al.*, 2013) and gray relation analysis (Zhai *et al.*, 2009). Core sustainability researchers such as Stratton *et al.* (2014) incorporate economic and environmental sustainability attributes into the decision-making phases of design using a hierarchical concept selection method. Buchert *et al.* (2014) suggested the use of combined methodologies for different phases in the selection of sustainability design.

Among the various methods that have been proposed and applied in sustainable design evaluation literature, the fuzzy set concept has remained the most widely used approach for sustainable design evaluation and for design decision-making. This is due to its ability to handle uncertainty and vagueness. The vagueness in the evaluation of sustainable designs, in general, has made the fuzzy set concepts a topic of great interest to many researchers. However, in an effort to further address the vagueness and uncertainty in decision-making process during sustainable design selection, the current study is proposing the incorporation of an intuitionistic fuzzy set (IFS) in a multicriteria decision-making method. The IFS theory is an extension of the traditional fuzzy set (FS) theory (Atanassov, 1986). It is characterized by a membership function and a non-membership function, which are used to deal with ambiguity caused by vagueness and uncertain information. According to Datta *et al.* (2013), the IFS theory is more capable than the traditional FS theory in handling vagueness and uncertain information in practice.

The benefits of applying the IFS have been described by Xu and Liao (2015) and Xu *et al.* (2014).

(1) It has the ability to model unknown information using hesitation degree; in a situation where decision-makers (DMs) are unsure about the preferences of an assessment, the IFS theory method is most suitable to get the opinions of the experts as compared to the fuzzy sets.

(2) It represents three grades of membership function, which include membership degree, non-membership degree and hesitancy degree. Hence, the IFS can be said to consider opinions from three sides to arrive at the preferred one.

(3) All the fuzzy numbers in the IFS theory can all be used to represent vagueness of "agreement"; however, they cannot be used to depict the "disagreement" of the DMs.

As a mathematical tool, the IFS has proven to be able to deal with inaccuracy, fuzziness and uncertain information in practice, and these are shown in its many applications in several different fields and for solving many different problems including MCDM problems related to engineering design evaluation (Hansen and Andreasen, 2004; Li, 2005; Lin *et al.*, 2007; Jahromi, 2012; Bai, 2013; Xu, 2014; Liu and Ren, 2014; Chen and Chiou, 2015).

Decision-making for a sustainable product design concept can be said to be a very critical and an active research area. Hence, in this chapter, the study will be concerned with the introduction of a new approach and method for evaluating and selecting sustainable designs concepts and for making design decisions, with a special interest in (1) modeling and addressing the fuzziness and hesitation of the design stakeholder's subjective assessments in the selected concept design, which seems to have hardly been fully addressed or accounted for in literature and (2) the dynamic issues in the sustainable design concept selection process. New product development processes are challenging and involve risks that are mainly due to unknown or unforeseen obstacles, which may be in the form of technology or sometimes due to business risks (Karniel and Reich, 2011). The process activities for the development of new products that are highly dependent on evolving product knowledge are determined during the product development process (Ottosson, 2004). The development process can be said to be inherently dynamic, thus it requires the adaptation to product knowledge changes as well as other changes that include the influences of resources, scope and targets on performance (Ford and Sterman, 1998). The dynamic nature of these four features can greatly affect the decisions about the product being developed. These features have been modeled extensively in the literature by system dynamics researchers for the later part of product development (Ottosson, 2004; Ford and Sterman, 1998). However, to the best of knowledge, these dynamic features and their influences have not been accounted for when selecting sustainable design concepts.

In addressing the fuzziness and hesitation of the design stakeholder's subjective assessments and the dynamic issues in the selection process of a sustainable design concept, an integrated model based on an intuitionistic fuzzy technique for order of preference by similarity to ideal solution (IFTOPSIS) method, dynamic intuitionistic fuzzy Einstein geometric averaging (DIFWGϵ) operator and the intuitionistic fuzzy entropy (IFE) method is proposed.

The rest of the chapter is organized as follows. In Section 2, the intuitionistic fuzzy set, the dynamic intuitionistic fuzzy Einstein geometric averaging (DIFWGϵ) operator and the IFTOPSIS model are introduced. This is followed by the algorithm of the proposed integrated model in Section 3. In Section 4, a real-life numerical case study of a sustainable design concept is presented to demonstrate the feasibility of the model. Finally, concluding remarks are presented in Section 5.

2. Preliminaries

In this section, the concept of the IFS as described in Aikhuele and Turan (2016) and Despic and Simonovic (2000) is presented. Furthermore, the dynamic intuitionistic fuzzy Einstein geometric averaging (DIFWGϵ) operator (42) and the FTOPSIS model (43) is introduced.

2.1. *Intuitionistic fuzzy set*

Definition 1. If *IFS* A in $X = \{x\}$ is a non-empty fuzzy set with a closed unit interval $I = [0, 1]$. Then the *IFS* can be described as

$$A = \{\langle x, \mu_A(x), v_A(x)\rangle | x \in X\}, \tag{1}$$

where the mapping $\mu_A : X \to [0, 1]$ is the membership function of the fuzzy set A; and $v_A : X \to [0, 1]$ is the non-membership function of the fuzzy set, such that $0 \leq \mu_A(x) + v_A(x) \leq 1, \forall x \in X$ for each element $x \in X$ in A.

From the above, the IFSA, on the non-empty set $X = \{x\}$ can be further defined as

$$A = \{\langle x, \mu_A(x), 1 - v_A(x)\rangle | x \in X\}. \tag{2}$$

The intuitionistic fuzzy hesitation (or the nondeterminacy or uncertainty) degree of whether x belong to A or not is given as

$$\pi_A(x) = 1 - \mu_A(x) - v_A(x) \tag{3}$$

This degree will arise only when there is a relative lack of knowledge, personal error or uncertainty of any form, particularly when $1 - \mu_A(x) - v_A(x) = 0$. For every element $x \in X$ in A, the IFS A belongs to the fuzzy set, where $0 \leq \pi_A \leq 1$, hence, the intuitionistic fuzzy number(s) (IFN(s)), which is used to represent IFS is given as $\alpha = (\mu_A, v_A, \pi_A)$ or as $\alpha = (\mu_A, v_A)$.

2.2. *Intuitionistic fuzzy aggregation operators*

Aggregation operators are used for combining information gathered from diverse sources for onward decision-making. Some of the popular aggregation operators that have found application in engineering design management include the ordered weighted averaging (OWA) operator (Yager, 1988), the weighted OWA (WOWA) operator and its interpolation function (Torra, 2000), and the ordered weighted geometric averaging (OWGA) operator (Xu and Da, 2002).

Others include the generalized ordered weighted proportional averaging (GOWPA) operator (Zhou and Chen, 2014), the generalized triangular intuitionistic fuzzy geometric averaging operator (Aikhuele and Odofin, 2017), dynamic intuitionistic fuzzy weighted averaging (DIFWA) operator and uncertain dynamic intuitionistic fuzzy weighted averaging (UDIFWA) operator (Xu and Yager, 2008), the dynamic intuitionistic fuzzy power geometric weighted average (DIFPGWA) operator (Yin *et al.*, 2018) and the dynamic intuitionistic fuzzy Einstein geometric averaging (DIFWGϵ) operator (Gumus and Bal, 2017). Details of the DIFWGϵ operator (Gumus and Bal, 2017) used for the sustainability design is given in the definition below.

Definition 2 [see Gumus and Bal (2017)]If $\boldsymbol{\alpha(t_i)} = (\boldsymbol{\mu_\alpha(t_i)}, \boldsymbol{v_\alpha(t_i)})$ for all $\boldsymbol{(i = 1, 2, 3, \ldots, p)}$ are set of Intuitionistic Fuzzy Numbers (IFN) for P different periods $\boldsymbol{(i = 1, 2, 3, \ldots, p)}$,

where $\boldsymbol{\delta} = ((\boldsymbol{\delta(t_1)}, \boldsymbol{\delta(t_2)}, \boldsymbol{\delta(t_3)}, \ldots, \boldsymbol{\delta(t_p)})^{\boldsymbol{T}}$ is the weight vector of periods such that $\sum_{\boldsymbol{i=1}}^{\boldsymbol{n}} \boldsymbol{\delta_(t_i)} = \boldsymbol{1}$, then DIFWG$\epsilon$: $\boldsymbol{\Omega^n \rightarrow \Omega}$, if

$$\begin{aligned}
&DIFWG \in_{\delta(t)} (\delta(t_1), \delta(t_2), \ldots, \delta(t_p)) \\
&\quad = \otimes_{i=1}^{p} \in .(\delta_{ij}(t_i))^{\delta(t_i)} \\
&\quad = (\delta_{1j}(t_1))^{\delta(t_1)} \otimes \in (\delta_{2j}(t_2))^{\delta(t_2)} \otimes \in \ldots (\delta_{pj}(t_p))^{\delta(t_p)} \\
&\quad = \left(\frac{2\prod_i^p \mu_{ij}(t_i)^{\delta(t_i)}}{\prod_i^p (2 - \mu_{ij}(t_i))^{\delta(t_i)} + \prod_i^p \mu_{ij}(t_i)^{\delta(t_i)}}, \right. \\
&\qquad \left. \frac{\prod_i^p (1 + v_{ij}(t_i))^{\delta(t_i)} - \prod_i^p (1 - v_{ij}(t_i))^{\delta(t_i)}}{\prod_i^p (1 + v_{ij}(t_i))^{\delta(t_i)} + \prod_i^p (1 - v_{ij}(t_i))^{\delta(t_i)}} \right) \qquad (4)
\end{aligned}$$

2.3. *The intuitionistic fuzzy entropy*

If an IFS A in $X = \{x\}$ is represented with an entropy measure such that

$$\mu_{\bar{A}}(x_i) = \frac{\mu_A(x_i) + 1 - v_A(x_i)}{2}$$

Then based on the definition of fuzzy information entropy by Ye (2010), the IFE is given as follows:

$$E(A) = \frac{1}{n}\sum_{i=1}^{n}\left\{\left\{Sin\frac{\pi^*[1 + \mu_A(x_i) - v_A(x_i)]}{4} + Sin\frac{\pi^*[1 - \mu_A(x_i) - v_A(x_i)]}{4} - 1\right\} * \frac{1}{\sqrt{2} - 1}\right\} \quad (5)$$

However, when the attribute weights are completely unknown, the weight vectors of the attributes can be determined using

$$W_j = \frac{1 - H_j}{n - \sum_{j=1}^{n} H_j}, \quad (6)$$

where $W_j \in [0, 1], \sum_{j=1}^{n} W_j = 1, H_j = \frac{1}{m}E(A_j)$ and $0 \le H_j \le 1$ for $j = 1, 2, \ldots, n$.

2.4. *Intuitionistic fuzzy technique for order of preference by similarity to ideal solution*

Hwang and Yoon (1981) proposed technique for order preference by similarity to the ideal solution (TOPSIS) method. It has remained one of the most widely used MCDM methods with many papers published on its applications (Bulgurcu, 2012; Jadidi *et al.*, 2008; Pakpour *et al.*, 2013; Soufi *et al.*, 2015; Yang and Wu, 2008; Ghazanfari *et al.*, 2014; Zhu *et al.*, 2012; Chou *et al.*, 2012). This concept is that the chosen alternative should have the shortest distance from the positive ideal solution and the farthest distance from the negative ideal solution. The method has been extended by various researchers in the fuzzy environment (Chou *et al.*, 2012; Chen, 2000; Tan *et al.*, 2010) and recently in the intuitionistic environment (IFTOPSIS) (Aikhuele and Turan, 2017; Hung and Chen, 2009). The algorithm of the intuitionistic fuzzy TOPSIS model is given in Table 1.

Table 1: The main implementation steps.

Computational steps of the IFTOPSIS method	
Step 1	The construction of decision matrix
Step 2	The normalization of the decision matrix
Step 3	The weighted normalization of the decision matrix
Step 4	Determination of the positive ideal solution (PIS) and the negative ideal solutions (NISs)
Step 5	Calculation of the distances of each alternative to the positive and negative ideal solutions
Step 6	Calculation of the closeness coefficient (CC) of all the alternatives and finally
Step 7	Ranking of the alternatives

3. Proposed Model

Consider an MCGDM problem with a set of alternatives ($T = T_1, T_2, \ldots, T_z$), which can be evaluated with respect to a set of criteria ($A = A_1, A_2, \ldots, A_z$). If the weight vectors of the criteria is unknown and needs to be evaluated mathematically under a dynamic process due to changes in the criteria as well as in process, scope and resources involved in the evaluation, then the algorithm of the proposed model based on the IFTOPSIS method, DIFWGϵ operator and IFE is given in the following steps:

Step 1: Recruit a group of experts with the relevant experience and expertise in the research subject to evaluate the alternatives with respect to the criteria. The result should be used to construct the individual intuitionistic fuzzy decision matrixes of the experts ($S_{mxn}(x_{ij})$).

The information and evaluation result are collected using a scale of linguistic variables, which are then converted to the intuitionistic fuzzy number (IFN) used in the model. A sample of the intuitionistic fuzzy decision matrix is given below, while the scale of linguistic variables is given in Table 2.

Remark. To ensure accuracy in their evaluation, choose experts from the academic and from the industry depending on the research

Table 2: Scare of linguistic variables.

Linguistic term	Intuitionistic fuzzy numbers (IFN)
Very low (VL)	(0.65, 0.10)
Low (L)	(0.70, 0.20)
High (HG)	(0.80, 0.10)
Very High (VH)	(0.85, 0.15)
Really Very High (RH)	(0.95, 0.20)

subject.

$$S_{mxn}(x_{ij}) = \begin{bmatrix} (\mu_{11}, v_{11}, \pi_{11}) & (\mu_{12}, v_{12}, \pi_{12}) & \cdots & (\mu_{n1}, v_{1n}, \pi_{1n}) \\ (\mu_{21}, v_{21}, \pi_{21}) & (\mu_{22}, v_{22}, \pi_{22}) & \cdots & (\mu_{2n}, v_{2n}, \pi_{2n}) \\ \vdots \quad \vdots \quad \vdots & \vdots \quad \vdots \quad \vdots & \vdots & \\ (\mu_{m1}, v_{m1}, \pi_{m1}) & (\mu_{m1}, v_{m2}, \pi_{m2}) & \cdots & (\mu_{m1}, v_{m2}, \pi_{m2}) \end{bmatrix}. \tag{7}$$

Step 2: Based on the preference information and results from the evaluation by the individual experts, the information is aggregated to construct the collective decision matrix $(R_{pk}(x_{ij}))$, using the DIFWGϵ operator — equation (4).

Step 3: Using the IFE equations (i.e., equations (5) and (6)), compute the weight vectors of each of the evaluating criteria.

Step 4: With the weight vectors in step 4, the weighted normalization is constructed from the collective decision matrix $(R_{pk}(x_{ij}))$ in step 3.

Step 5: Determination of the positive ideal solution (PIS) and the negative ideal solutions (NISs), are done using

$$T^- = \left\{ \left\langle \begin{matrix} A_j, ((\min \mu_{ij}(A_{ij}) | j \in N), (\max \mu_{ij}(A_{ij}) | j \in G)), \\ ((\max v_{ij}(A_{ij}) | j \in N), (\min v_{ij}(A_{ij}) | j \in G)) \end{matrix} \right\rangle | i \in z \right\}, \tag{8}$$

where N be a collection of benefit criteria and G the collection of cost criteria when the criteria are categorized into benefit and cost.

Step 6: Calculate the distances of each of the alternative to the positive and negative ideal solutions using the intuitionistic Euclidean distance.

$$d_{\mathrm{IFS}}(T^{+}) = \sqrt{\sum_{j=1}^{n}\left[\left(\mu_{ij}(A_j) - \mu_{(T^{+})}(A_j)\right)^2 + \left(v_{ij}(A_j) - v_{(T^{+})}(A_j)\right)^2\right]}, \tag{9}$$

$$d_{\mathrm{IFS}}(T^{-}) = \sqrt{\sum_{j=1}^{n}\left[\left(\mu_{ij}(A_j) - \mu_{(T^{-})}(A_j)\right)^2 + \left(v_{ij}(A_j) - v_{(T^{-})}(A_j)\right)^2\right]}. \tag{10}$$

Step 7: Calculation of the closeness coefficient (CC) of all the alternatives

$$CC_i = \frac{d_{\mathrm{IFS}}(T^{-})}{d_{\mathrm{IFS}}(T^{-}) + d_{\mathrm{IFS}}(T^{+})}. \tag{11}$$

Step 8: Use the alternatives' CC values to rank the alternatives. The alternatives with the lowest and highest coefficients are the least and most preferred alternatives.

4. Numerical Illustration of the Model Performance

The authority of Lagos central district, Nigeria, new power investment company is planning to invest in power technology by manufacturing new power transformers locally. To achieve this goal, the company is required to select a sustainable product design concept from a list of designs presented by the company's design team. This is mainly to achieve the sustainability goal of the district and of the company by addressing the changing environmental and power issues in the district and their impacts and to develop a new product that can lead to the economic growth of the company and its expansion.

The numerical illustration considered five power transformers design concepts (T_1, T_2, T_3, T_4, and T_5). These designs were evaluated using four sustainability criteria: environmental impact (pollutant emission) — A_1, social–political impact (social and political acceptance) — A_2, economical impact (cost of the project) —

A_3 and technology (reliability, safety and technical maturity, and feasibility) — A_4.

Since the development process of the power transformers is inherently dynamic and requires the adaptation of knowledge changes as well as other changes such as the influences of resources, scope and targets on performance (Ford and Sterman, 1998) during the development process, the influences of these changes that are dynamic are accounted for in the selection of the power transformers design concepts. To achieve this, the influences of dynamic feature in the power transformers design concept selection are evaluated for a period of 2 years within the last 10 years (i.e., 2009–2010, 2011–2012, 2013–2014, 2015–2016, and 2017–2018) by five experts (E_1, E_2, E_3, E_4, and E_5) in the academic and industry. The weighting vectors for the periods of evaluation are given as $\delta(t) = 0.250, 0.300, 0.350, 0.450, 0.500$, respectively. Using the algorithm presented in Section 3, the experts' designs preference ratings are shown in the intuitionistic fuzzy decision matrix, in Table 3.

Steps 1 and 2: The linguistic variables in Table 3 are replaced with the IFN in Table 2, and then they are aggregated to construct the collective decision matrix, using the DIFWGϵ operator. The result of the constructed collective decision matrix (aggregation), which

Table 3: Intuitionistic fuzzy decision matrix.

	E_1	E_2	E_3	E_4	E_5		E_1	E_2	E_3	E_4	E_5
			A_1						A_2		
T_1	VL	L	L	HG	L		VH	L	RH	HG	L
T_2	VL	HG	HG	VH	HG		RH	HG	VL	VH	HG
T_3	HG	VH	VH	L	VH		HG	HG	L	RH	VH
T_4	VH	HG	RH	VL	RH		VH	L	HG	VL	RH
T_5	RH	HG	HG	VH	VH		RH	HG	VH	L	RH
			A_3						A_4		
T_1	RH	HG	VL	VH	HG		VL	HG	HG	VH	HG
T_2	HG	VH	VH	L	VH		HG	VH	VH	L	VH
T_3	VL	VH	HG	VH	HG		VH	L	HG	VL	RH
T_4	RH	HG	VH	VL	VH		VL	L	L	HG	L
T_5	L	VH	RH	L	RH		RH	HG	VL	VH	HG

Table 4: The collective decision matrix and the sustainability criteria weight vectors.

	A_1	A_2	A_4	A_4	A_5
T_1	(0.4972, 0.3294)	(0.6251, 0.3060)	(0.6455, 0.2299)	(0.6140, 0.2423)	(0.6381, 0.3716)
T_2	(0.6140, 0.2423)	(0.6455, 0.2299)	(0.6741, 0.2822)	(0.6741, 0.2822)	(0.6595, 0.4081)
T_3	(0.6741, 0.2822)	(0.7157, 0.2848)	(0.6273, 0.2566)	(0.6277, 0.2731)	(0.5721, 0.2492)
T_4	(0.7036, 0.2779)	(0.6277, 0.2731)	(0.6493, 0.2491)	(0.4972, 0.3284)	(0.5438, 0.3143)
T_5	(0.7198, 0.2538)	(0.7174, 0.3152)	(0.7169, 0.3221)	(0.6455, 0.2299)	(0.5490, 0.2652)
A_i	0.245969	0.256249	0.267223	0.212037	0.043631

addresses and accounts for the evaluated periods for the power transformers designs concepts evaluation, is given in Table 4. Furthermore, the weight vectors of the sustainability criteria are calculated using equations (5) and (6) from the collective decision matrix. The results of the criteria weights vectors are given in the later part of Table 4.

Using the result of the aggregation and that of the criteria weight vectors, the weighted normalization matrix is constructed. Furthermore, the positive and the negative ideal solutions for the alternatives are determined and used for calculations of the distances of each of the alternatives using equations (9) and (10). Finally, the closeness coefficients of the alternatives are calculated using equation (11). The results of the evaluation and ranking are presented in Table 5.

From the result of the power transformers designs concepts evaluation as shown in Table 5, it is not hard to see that the concept design (T_5) is the most desirable based on the sustainability criteria under the dynamic process of its development and production. The study not only revealed the suitability of the proposed model for ranking sustainable power transformers designs concepts, but reveals its suitability for selecting power transformers designs concepts under a

Table 5: The alternatives' closeness coefficients and ranks.

	$d_{\text{IFS}}(T^+)$	$d_{\text{IFS}}(T^-)$	CC_i	Ranking
T_1	0.11927	0.05267	0.3063	5
T_2	0.08489	0.08769	0.5081	3
T_3	0.06669	0.10265	0.6062	2
T_4	0.09308	0.07353	0.4413	4
T_5	0.04015	0.14262	0.7803	1

Table 6: Alternatives' closeness coefficients and ranks under different models.

	Proposed model		Gümüş and Bali's model (2017)		Liu *et al.*'s model (2012)		Zhang *et al.* (2012) and Aikhuele's (2018) models	
	CC_i	Ranking	$S(r_i)$	Ranking	Q_i	Ranking	$S(r_i)$	Ranking
T_1	0.3063	5	0.3246	5	1.0000	5	0.6440	5
T_2	0.5081	3	0.4015	2	0.8325	4	0.6844	3
T_3	0.6062	2	0.3965	3	0.7798	3	0.6863	2
T_4	0.4413	4	0.3493	4	0.7652	2	0.6590	4
T_5	0.7803	1	0.4264	1	0.0000	1	0.7103	1

dynamic environment, as well as address uncertainty in the decision-making process.

To check the feasibility and rationality of the model, which is based on the IFTOPSIS, DIFWGϵ operator and IFE method, the result of the power transformers designs concepts evaluation under the dynamic environment presented in Table 4 is compared with similar computational models in the literature (e.g., the dynamic aggregation operators based on intuitionistic fuzzy tools and Einstein operations by Gümüş and Bali (2017), the extended VIKOR method under fuzzy environment by Liu *et al.* (2012) and the score function based on intuitionistic fuzzy sets (Zhang *et al.*, 2012; Aikhuele, 2018)). The comparison that focuses on the ranking result shows total agreement with the proposed integrated model; the results are presented in Table 6.

5. Conclusions

A new multicriteria group decision-making (MCGDM) model based on IFTOPSIS method, DIFWGϵ operator and IFE has been presented in this chapter. The model was used to select a sustainable power transformer design concept under a dynamic environment. It considers the dynamic issue in the development of new products; it also accounts for uncertainty in the decision-making process for a new product. To the best of my knowledge, this is the first attempt to select sustainable design concepts under these conditions while meeting the sustainability goal.

To check the feasibility of the model, the results from the evaluation in Table 5 have been compared with similar models in the literature: Gümüş and Bali (2017), Liu *et al.* (2012), Zhang *et al.* (2012) and Aikhuele (2018). The results obtained from these authors models are consistent with the model that was presented in this chapter. In the future, the presented model will be applied in other domains.

References

Aikhuele, D. O. (2018). Reliability evaluation using MAGDM based on triangular intuitionistic attitudinal ranking and aggregating model. *J Mod Tech Eng*, 3(2), 165–178.

Aikhuele, D. O. and Odofin, S. A. (2017). Generalized triangular intuitionistic fuzzy geometric averaging operator for decision-making in engineering and management. *Information*, 8(3), 1–17.

Aikhuele, D. O. and Turan, F. M. (2017). A modified exponential score function for troubleshooting an improved locally made Offshore Patrol Boat engine. *J Mar Eng Technol* [Internet] (February). Available from: http://dx.doi.org/10.1080/20464177.2017.1286841.

Aikhuele, D. O. and Turan, F. M. (2017). Extended TOPSIS model for solving multi-attribute decision making problems in engineering. *Decis Sci Lett*, 6, 365–376.

Aikhuele, D. O. and Turan, T. B. M. (2016). Intuitionistic fuzzy-based model for failure detection. *Springerplus*, 5(1), 1–15.

Akay, D., Kulak, O. and Henson, B. (2011). Conceptual design evaluation using interval type-2 fuzzy information axiom. *Comput Ind*, 62(2), 138–146.

Alexander, E. R. (1982). Design in the decision-making process, *Policy Sci*, 14(3), 279–292.

Atanassov, K. T. (1986). Intuitionistic fuzzy sets. *Fuzzy Sets Syst*, 20(1), 87–96.

Bai, Z. (2013). An interval-valued intuitionistic fuzzy TOPSIS method based on an improved score function. *Sci World J*, 1–9 http://dx.doi.org/10.1155/2013/879089.

Belinda, L. M. and Amaresh, C. (2007). A model of decision-making knowledge in conceptual design. *Guidelines for a Decision Support Method Adapted to NPD Processes*, pp. 661–662.

Buchert, T., Kaluza, A., Halstenberg, F. A., Lindow, K., Hayka, H. and Stark, R. (2014). Enabling product development engineers to select and combine methods for sustainable design. *Procedia CIRP*, 15, 413–418.

Bulgurcu, B. (2012). Application of TOPSIS technique for financial performance evaluation of technology firms in Istanbul stock exchange market. *Procedia — Soc Behav Sci*, 62, 1033–1040.

Chen, C.-T. (2000). Extensions of the TOPSIS for group decision-making under fuzzy environment. *Fuzzy Sets Syst*, 114, 1–9.

Chen, S. M. and Chiou, C. H. (2015). A new method for multiattribute decision making based on interval-valued intuitionistic fuzzy sets, PSO techniques and evidential reasoning methodology. *Proc — Int Conf Mach Learn Cybern*, 1(6), 403–409.

Chou, S. Y., Yu, V. F., Dewa Bharata, A. and Dat, L. Q. (2012). A fuzzy TOPSIS approach for medical provider selection and evaluation. In *International Conference on Fuzzy Theory its Appl Fuzzy*, pp. 322–326.

Datta, S., Samantra, C., Mahapatra, S. S., Mondal, G., Chakraborty, P. S. and Majumdar, G. (2013). Selection of internet assessment vendor using TOPSIS method in fuzzy environment. *Int J Bus Perf and Sup Chain Model*, 5(1), 1–27.

Despic, O. and Simonovic, S. P. (2000). Aggregation operators for soft decision making in water resources. *Fuzzy Sets Syst*, 115(1), 11–33.

Ford, D. N. and Sterman, J. D. (1998). Dynamic modeling of product development processes. *Syst Dyn Rev*, 14(1), 31–68.

Ghazanfari, M., Rouhani, S. and Jafari, M. (2014). A fuzzy TOPSIS model to evaluate the business intelligence competencies of port community systems. *Polish Marit Res*, 21(2), 86–96.

Gumus, S. and Bal, O. (2017). Dynamic aggregation operators based on intuitionistic fuzzy tools and einstein operations. *Fuzzy Inf Eng*, 9(1), 45–65.

Hambali, A., Sapuan, S. M., Ismail, N. and Nukman, Y. (2009). Application of analytical hierarchy process in the design concept selection of automotive composite bumper beam during the conceptual design stage. *Sci Res Essay*, 4(4), 198–211.

Hansen, C. T. and Andreasen, M. M. (2004). A mapping of design decision-making. In *DS 32: Proceedings of DESIGN 2004, the 8th International Design Conference*, Dubrovnik, Croatia, pp. 1409–1418.

Hazelrigg, G. A. (1998). A framework for decision-based engineering design. *J Mech Design*, 120(4), 653–658.

Hung, C.-C. and Chen, L.-H. (2009). A fuzzy TOPSIS decision making model with entropy weight under intuitionistic fuzzy environment. In *Proceedings of the International MultiConference of Engineers and Computer Scientists*, pp. 18–21.

Hwang, C. L. and Yoon, K. (1981). *Multiple Attribute Decision Making Methods and Applications*. Springer, Berlin.

Jadidi, O., Hong, T. and Firouzi, F. (2008). TOPSIS and fuzzy multi-objective model integration for supplier selection problem. *J Achiev Mater Manufacturing Eng*, 31(2), 762–769 (2008).

Jahromi, M. K. (2012). Multiattribute decision making models and methods using intuitionistic fuzzy sets. *Int Math Forum*, 7(57), 2847–2851.

Jenab, K., Sarfaraz, A. and Ameli, M. T. (2013). A Fuzzy conceptual design selection model considering conflict resolution. *J Eng Des*, 24(4), 293–304.

Justel, D., Vidal, R., Arriaga, E., Franco, V. and Val-jauregi, E. (2007). Evaluation method for selecting innovative product concepts with greater potential marketing success. In *International Conference of Engineering Design*, 1–12.

Karniel, A. and Reich, Y. (2011). *Managing the Dynamics of New Product Development Processes: A New Product Lifecycle Management Paradigm*, 1st edn. Springer-Verlag, London, pp. 1–257.

King, A. M. and Sivaloganathan, S. (1999). Development of a methodology for concept selection in flexible design strategies. *J Eng Des*, 10(4), 329–349.

Laemlaksakul, V. and Bangsarantrip, S. (2008). Analytic hierarchy process for design selection of laminated bamboo chair. In *Proceedings of the International MultiConference of Engineers and Computer Scientists*, Vol. II IMECS. Hong Kong, 19–21 March 2008, pp. 19–21.

Li, D.-F. (2005). Multiattribute decision making models and methods using intuitionistic fuzzy sets. *J Comput Syst Sci*, 70(1), 73–85.

Lin, L., Yuan, X. H. and Xia, Z. Q. (2007). Multicriteria fuzzy decision-making methods based on intuitionistic fuzzy sets. *J Comput Syst Sci*, 73(1), 84–88.

Liu, H. T. (2011). Product design and selection using fuzzy QFD and fuzzy MCDM approaches. *Appl Math Model*, 35(1), 482–496.

Liu, H.-C., Liu, L., Liu, N. and Mao, L.-X. (2012). Risk evaluation in failure mode and effects analysis with extended VIKOR method under fuzzy environment. *Expert Syst Appl*, 39(17), 12926–12934.

Liu, M. and Ren, H. (2014). A new intuitionistic fuzzy entropy and application in multi-attribute decision making. *Information*, 5(4), 587–601.

Marini, C. D., Fatchurrohman, N., Azhari, A. and Suraya, S. (2016). Product development using QFD, MCDM and the combination of these two methods. *IOP Conf Ser Mater Sci Eng*, 114, 012089.

Ottosson, S. (2004). Dynamic product development — DPD. *Technovation*, 24(3), 207–217.

Pakpour, S., Olishevska, S. V., Prasher, S. O., Milani, A. S. and Chénier, M. R. (2013). DNA extraction method selection for agricultural soil using TOPSIS multiple criteria decision-making model. *Am J Mol Biol*, 3, 215–228 http://dx.doi.org/10.4236/ajmb.2013.34028.

Pandey, V. and Mourelatos, Z. P. (2015). A new method for making design decisions: Decision topologies. *J Mech Design*, 137(3), 031401.

Pugh, S. (1996). *Creating Innovative Products Using Total Design: The Living Legacy of Stuart Pugh.* Addison-Wesley, Reading, MA.

Saaty, T. L. (2008). Decision making with the analytic hierarchy process. *Intl J Ser Sci,* 1(1), 83–98.

Soufi, M. D., Ghobadian, B., Najafi, G., Sabzimaleki, M. R. and Yusaf, T. (2015). TOPSIS multi-criteria decision modeling approach for bio lubricant selection for two-stroke petrol engines. *Energies*, 8(12), 13960–13970.

Stratton, D., Behdad, S., Lewis, K. and Krishnamurty, S. (2014). A multi-level approach to concept selection in sustainable design. In *19th Design for Manufacturing and the Life Cycle Conference*, pp. 1–14.

Tan, Y., Shen, L., Langston, C. and Liu, Y. (2010). Construction project selection using fuzzy TOPSIS approach. *J Model Manag*, 5(3), 302–315.

Tauhid, S. and Okudan, G. (2007). Fuzzy information axiom approach for design concept evaluation. In *International Conference on Engineering Design, ICED'07*, Cite Des Sciences Et De L'industrie, Paris, France, 28–31 August 2007, pp. 1–12.

Torra, V. (2000). The WOWA operator and the interpolation function W*: Chen and Otto's interpolation method revisited. *Fuzzy Sets Syst*, 113(3), 389–396.

Ulrich, K. T. and Eppinger, S. D. (2000). *Product Design and Development* [Internet]. McGraw-Hill, New York. [cited 2015 Dec 5]. Available from: http://scitation.aip.org/content/aip/proceeding/aipcp/10.1063/1.4707639.

Vinodh, S., Mulanjur, G. and Thiagarajan, A. (2013). Sustainable concept selection using modified fuzzy TOPSIS: A case study. *Int J Sustain Eng*, 6(2), 109–116.

Von Neumann, J. and Morgenstern, O. (1947). *Theory of Games and Economic Behavior*, 2nd edn. Princeton University Press, New Jersey, United States, 641 p.

Wang, J. (2002). Improved engineering design concept selection using fuzzy sets. *Int J Comput Integr Manuf*, 15(1), 18–27.

Xia, B., Chan, A. P. and Yeung, J. F. (2001). Developing a fuzzy multicriteria decision-making model for selecting design-build operational variations. *J Constr Eng Mgt,* 137(12), 1176–1184.

Xu, Z. (2014). Intuitionistic preference modeling and interactive decision making. *Studies in Fuzziness and Soft Computing*, Springer-Verlag Berlin Heidelberg, pp. 195–223.

Xu, Z. and Liao, H. (2015). A survey of approaches to decision making with intuitionistic fuzzy preference relations. *Knowledge-Based Syst*, 80, 131–142.

Xu, Z. and Yager, R. R. (2008). Dynamic intuitionistic fuzzy multi-attribute decision making. *Int J Approx Reason*, 48(1), 246–262.

Xu, Z. S. and Da, Q. L. (2002). The ordered weighted geometric averaging operators. *Int J Intell Syst*, 17(7), 709–716.

Xu, Z., Member, S. and Liao, H. (2014). Intuitionistic fuzzy analytic hierarchy process. *IEEE Trans Fuzzy Syst*, 22(4), 749–761.

Yager, R. R. (1988). On ordered weighted averaging aggregation operators in multi-criteria decision making. *IEEE Trans Syst Man Cybern*, 18(1), 183–190.

Yan, W., Chen, C. H. and Shieh, M. D. (2006). Product concept generation and selection using sorting technique and fuzzy c-means algorithm. *Comput Ind Eng*, 50(3), 273–285.

Yang, C. and Wu, Q. (2008). Decision model for product design based on fuzzy topsis method. *Int Symp Comput Intell Des*, 342–345, Wuhan, China. http://dx.doi.org/10.1109/ISCID.2008.220.

Ye, J. (2010). Two effective measures of intuitionistic fuzzy entropy. *Comput (Vienna/New York)*, 87(1–2), 55–62.

Yin, K., Wang, P. and Jin, X. (2018). Dynamic intuitionistic fuzzy multi-attribute group decision-making or and prediction model. *Symmetry (Basel)*, 10(11), 536.

Zhai, L. Y., Khoo, L. P. and Zhong, Z. W. (2009). Design concept evaluation in product development using rough sets and grey relation analysis. *Expert Syst Appl*, 36(3/2), 7072–7709.

Zhang, Z., Yang, J., Ye, Y., Hu, Y. and Zhang, Q. (2012). A type of score function on intuitionistic fuzzy sets with double parameters and its application to pattern recognition and medical diagnosis. In *Procedia Engineering*, pp. 4336–4342.

Zhou, L. and Chen, H. (2014). Generalized ordered weighted proportional averaging operator and its application to group decision making. *Informatica*, 25(2), 327–360.

Zhu, X., Wang, F., Liang, C., Li, J. and Sun, X. (2012). Quality credit evaluation based on TOPSIS: Evidence from air-conditioning market in China. *Procedia Comput Sci*, 9(10), 1256–1262.

Chapter 12

Challenges in Reducing Seaweed Supply Chain Risks Arising Within and Outside Remote Islands in Indonesia: An Integrated MCDM Approach

Wellem Anselmus Teniwut

Fisheries Agribusiness Study Program,
Tual State Fisheries Polytechnic,
Langgur-Sathean Ringroad Km 6, Southeast Maluku District,
Maluku Province 97611, Indonesia
wateniwut@polikant.ac.id

In Indonesia, which has the largest archipelago in the world, most coastal and small island regions in the country rely upon their daily economic activities in fisheries and marine resource utilization, including seaweed cultivation. Some of the regions are still entirely situated in remote areas such as Kei Islands in Maluku Province, which have relatively smallest islands in the region lacking in infrastructure. The location of these islands brings a whole new challenge for maintaining not only an ideal interisland supply chain but also transportation within the islands. Kei Islands is one of the leading producers of seaweed from eastern Indonesia and therefore reducing the risks of seaweed supply chain within and outside these islands has become a necessity to sustain productivity of seaweed. In this chapter, an integrated multicriteria decision-making (MCDM) with fuzzy logic is used to provide a better decision based on a large number of related complex factors. House of risk (HOR) is performed in the MCDM tool, which was developed based on quality function deployment (QFD) and FMEA concept, and fuzzy AHP to provide a better understanding and give the best alternative to minimize the risks encountered in the seaweed supply chain on the intra-island and inter-island regions in Kei Islands, Indonesia.

1. Introduction

Ideal connectivity from downstream to upstream in the supply chain has become one of important factors for businesses to stay competitive and survive in a highly competitive environment. The bigger the business means the more complex the supply chain hierarchy and therefore the dependence on each party in the supply chain also grows higher. Each party has to become highly reliable in order to maintain the fluidity of their production, distribution and minimize the cost and time of storage, although in practice an ideal supply chain is far from reality, especially in the agriculture sector that includes fisheries. Unpredictable weather and inconsistency of supply in response to customer demand, in addition to fishers, raw material suppliers, transportation, export–import, retailers, suppliers and companies are factors that make an ideal supply chain in fisheries sector profoundly challenging.

Resources utilization in fisheries in most cases is the primary and only source of revenue for many coastal communities around the world. For instance, in the Kei Islands, Maluku Province, Indonesia, the remote islands located far east in the archipelago of Indonesia, most of the people living in this coastal region have a low formal education level and rely much on the sea for their economic welfare. The commodity that gives them a high economic multiplier impact is seaweed. The region is known to be one of the largest seaweed producers from the eastern region of Indonesia where most of the coastal villages in Kei Islands resort to seaweed cultivation. Consistency of production and revenue from seaweed cultivation is very volatile because transportation and connectivity inside the islands and connecting to buyers from other regions is hampered by lack of infrastructure in the islands. Also Kei Islands is not one of the routes in the "sea toll" program launched by the government of Indonesia in order to minimize delivery time and cost.

Based on the understanding that flow of supply chain is crucial for business sustainability, it is important to minimize and prevent the risks that occur within and outside Kei Islands with regard to seaweed supply chain. The study aims to reduce supply chain risks in the seaweed business within the Kei Islands and, between Kei Islands and other regions in the country. The model is based on the integrated multicriteria decision-making (MCDM) with the house of risk (HOR), developed by Pujawan and Geraldine (2009), based on quality function deployment (QFD) and FMEA to find the best practical solution for handling the seaweed supply chain risks that are currently observed and those that might happen in the future. By doing so, the goal of sustainability of seaweed-related business activities in Kei Islands can be ensured by minimizing the risks of disjoint connectivity inside the islands and poor transportation connectivity between Kei Islands and Sulawesi and Java Islands. The chapter also demonstrates the power and usefulness of MCDM portfolios for solving a complex problem, especially in the fisheries sector.

2. Previous Related Work

In the last decade, the attention on supply chain risk has rapidly grown. The current study focuses on supply chain risks in the fisheries sector and the application of MCDM on mitigating supply chain risks in the fisheries sector.

2.1. *Managing supply chain risk*

The potential risk in a supply chain usually triggers an imbalance and uncertainty inside the supply chain. Alora and Barua (2019) mention two causes of supply chain uncertainty, which are the risks due to the environment and those within the supply chain. Also, Jiang *et al.* (2018) stressed the importance of managing supply chain risk by improving efficiency, ability, management and mechanism of the supply chain of a port in China. Chand *et al.* (2017) indicated the hard task of management of supply chain risks because of different forms of uncertainty in each sector, while Ferreira *et al.* (2018) added the impact of failure in supply chain risk management in the context of a bankruptcy of the Brazilian shipbuilding industry, which happened due to an inability to initiate an immediate risk mitigation action. Oliveira *et al.* (2019) reiterated the importance of a modified model and simulation for mitigating supply chain risk. They also stated that the real-time application of the previous models still had some limitations and weaknesses.

2.2. *MCDM in managing supply chain risk*

Efforts to manage supply chain risk are being widely undertaken based on various methodologies, while at the same time there is an increasing use of MCDM for mitigating supply chain risk (Lei and Zhaocai, 2007; Song *et al.*, 2017; Soni and Vaishnavi, 2019; Saxena and Seth, 2014; Zheng, 2010; Vishwakarma *et al.*, 2016). As the complexity of supply chain disruption seems to have increased because of the broader business coverage and network, many studies also modified the use of MCDM tool by integrating several MCDM tools to manage and mitigate various risks in the supply chain (Kavilal *et al.*, 2016; Rostamzadeh *et al.*, 2018; Khemiri *et al.*, 2018; Abdel-Basset *et al.*, 2019; Yazdani *et al.*, 2019).

2.3. *Managing supply chain risk in the fisheries sector*

Empirical studies in the supply chain, which included mitigating the risk of the supply chain in the fisheries sector up to recently, are very few compared to other industries. This lack attention by researchers on the supply chain, especially with a focus on the fisheries sector, or the complexity of supply chain in fisheries that produces a different challenge, as too many moving factors have to be taken into account, has to be noted. There are only a few studies that purposely tried to study and mitigate the supply chain risks in the fisheries sector. For instance, studies by Wang (2010a, 2010b, 2010c) sought to minimize the risks in the fisheries supply chain contracts. Yan *et al.* (2013) stated the importance of limiting the seafood supply chain risk by proposing barcode to monitor the quality of the aquatic product delivery from the producer to the customer. A similar study was also conducted by Al-Busaidi *et al.* (2016) where they stated the importance of maintaining seafood quality in order reduce one seafood supply risk that probably occurred during distribution. Furthermore, there were also specific studies related to supply chain risk in fisheries in Indonesia using the HOR approach by Karningsih *et al.* (2018) and Nguyen *et al.* (2018).

Although the latest study on the use of HOR still has a limitation as stated by Ma and Wong (2018), the lack of accuracy and precision of the original HOR should be diminished by integrating it with a fuzzy language. Thus, this study uses HOR by integrating it with MCDM-based fuzzy logic, where the MCDM tool used in this study was the analytical hierarchy process (AHP). The integration of HOR and AHP-based fuzzy logic would cover the weakness of each approach. HOR provided empirical criteria based on SCOR and most criticism of AHP on sources of criteria relates to the fact that some are too theoretical, while at the same time AHP-based fuzzy logic enables a better calculation of the weight of criteria, which increases accuracy and precision.

3. Seaweed Supply Chain Risk Mitigating Model

3.1. *Empirical problem: A case study in the remote islands of Indonesia*

Indonesia has one of the largest archipelagoes in the world. Some of the islands like Kei Islands in Maluku Province are located in the southeast region of Maluku Province and one of the islands in this region is located in the border of Indonesia and Australia (Figure 1). Both these are remote islands. The location of Kei Islands is problematic in terms of the path of the sea toll proposed by the President of Indonesia, Joko Widodo. As a result, Kei Islands is not one of the destinations of shipments for distribution. The latter problem made the distribution of supplies from Kei Islands to big cities in the country in particular very expensive and time-consuming. The economy in Kei Islands, is highly dependent on supply and distribution via sea

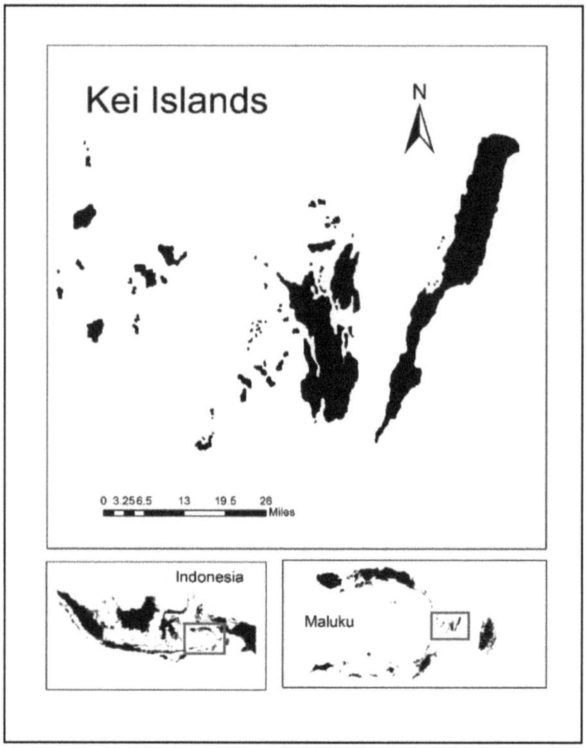

Figure 1: Case study location: Kei Islands, Maluku Province, Indonesia.

transportation for the fisheries sector, in particular the distribution seaweed products from the islands.

Out of many small islands in the Kei Islands, only two islands have a better road infrastructure and telecommunication connectivity. Even in the largest island in Kei Islands, sufficient road infrastructure and telecommunication cover only below 10% of the island. In addition, due to geographical location, more than half of the villages in the islands cannot be accessed through sea transportation for a certain period of time. For connectivity with other regions, there is one public harbor for commercial ships and one airport. Despite the availability of harbor, there is no dedicated section of the harbor for logistic shipments. The cost of distributing any product from and to Kei Islands is very high. The delivery time from Kei Islands to other larger islands like Sulawesi and Java takes between 4 and 6 days where the rent of one ship container is around IDR 11–20 million. This logistics condition proved to be challenging for the seaweed business community in the region, which had a hard time maintaining consistency of their revenue because of the disruption of supply chain inside and outside the islands.

3.2. *Modeling seaweed supply chain risk mitigation*

The model for managing and mitigating seaweed supply chain risk in this study consists of two scenarios. The first was to mitigate the risk of seaweed supply chain inside the islands (intra-island) and the second was to mitigate the risk of seaweed supply chain between the islands (inter-islands), in particular with relation to the buyer and logistic distributor to other big islands in Indonesia. The process of MCDM analysis was also divided into two stages. In the first stage, the HOR was used to provide the risk agents that are to be used as the criteria for each intra- and inter-islands seaweed supply chain risk. The second stage, which based on the criteria provided by HOR (HOR1), calculations are done with AHP-based fuzzy logic (Figure 2) to select the best alternative for mitigating seaweed supply chain risk in both intra- and inter-island scenarios. All processing was done using Microsoft Excel.

3.2.1. *House of risk*

HOR as a model was developed and first introduced by Pujawan and Gelardin (2009). It is an integration of the failure modes and effects analysis model (FMEA) and QFD. This method consists of two parts: first, HOR1 is used to prioritize the significant cause of supply chain risk and the second, HOR2 is used to select the most cost-effective mitigation measures. In this study, the role of HOR2 was replaced by fuzzy AHP. HOR1 was used to prioritize the causes of seaweed supply chain risk for both intra- and inter-island scenarios while for risk mitigation, fuzzy-based AHP was employed to obtain better results by taking advantage of fuzzy logic and AHP.

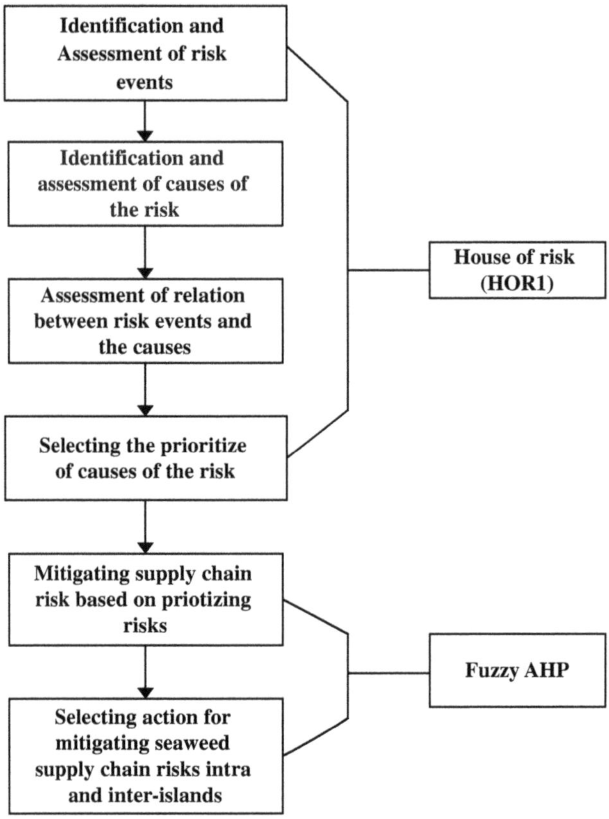

Figure 2: Framework risk mitigating process-based integrated MCDM.

The first step in HOR1 is to identify the risk events (E_i). The next step is to identify risk agents that were the possible causes of risk events identified. The complete steps for the HOR1 procedure are as follows:

(1) Identify risks that may occur in each business process. The identification can be done by mapping supply chain processes such as planning, sourcing, delivery, making and return (model supply chain operations reference) and then identifying "what risks can occur" in each of these processes. E_i (risk events) shows the risks that can occur.

(2) Rate on a scale of 1–10 the severity due to the risks that occur. S_i (severity) shows "how severe each risk is" (in Table 1).
(3) Identify risk agents and assess the possibility of their occurrence. A_j (risk agents) shows risk agents and their occurrence. A scale of 1–10 is also applied where 1 means almost never occurred and a value of 10 means almost certain to happen (in Table 2).
(4) Develop a correlation matrix between each risk agent with each risk. R_{ij} (relationship) {0, 1, 3, 9} with a value of 0 indicating that there is no correlation (no correlation) and values 1, 3, and 9

Table 1: Risk events and severity in the seaweed supply chain.

Process	Sub-process	Risk event	Code	Si
Plan	Demand forecasting	Error forecasting	E1	9
	Production planning	A sudden change in production	E2	9
	Inventory control	Too many inventories	E3	3
		Difference between stock and recorded	E4	5
Source	Procurement	Delay sending purchasing order	E5	9
		Late Delivery	E6	9
		Incorrect purchasing order	E7	2
		Inaccurate owner estimate	E8	4
Make	Production	Delay production	E9	9
		Insufficient material	E10	7
		Damaged product	E11	7
		Shortage material	E12	4
Delivery	Selection shipping	Shortage capacity	E13	7
		No shipping schedule	E14	7
	Product delivery to customers	Shortage of product received by customers	E15	5
		Late delivery to customers	E16	9
		Wrong delivery	E17	7
	Warehouse of the finished product	Shortage of product available	E18	9
	Packing	Damaged packing	E19	5
Return	Reject from customer	A large amount of reject product	E20	5
	Return from supplier	Delay in handling reject product to the supplier	E21	6

Table 2: Risk agents and their occurrence in seaweed supply chain.

Categories	Risk agent	Occurrence	Code
Seaweed related business	Lack of knowledge	9	A1
	Lack of information	9	A2
	Lack of sufficient tool	3	A3
	Inaccurate price	8	A4
	Uncertainty of demand	8	A5
	Short supply capacity	8	A6
	Inconsistency production schedule	4	A7
	Lack of business connection	9	A8
	Dependence single supplier	1	A9
	Dependence single distributor	9	A10
	Untrained worker	7	A11
	Lack of IT' savvy	9	A12
Natural and economic' policy	Natural disaster	3	A13
	Seasoning	5	A14
	Rise oil and gas' price	3	A15
	Inflation	3	A16
	Different Government policies	5	A17
Supplier	Damage materials from supplier	4	A18
	Inaccurate supply materials	4	A19
	Inaccurate exchange information	7	A20
Customers	Significant damage goods	5	A21
	A significant change in demand	6	A22
Others	Shipments do not arrive on time	9	A23
	No regular schedule logistic shipments	9	A24

showing a low (low), moderate (moderate) and high (high) correlation.

(5) Calculate the aggregate risk potential (ARP_j)

$$ARP_j = O_j \sum_i S_i R_{ij}. \tag{1}$$

The rate of severity and occurrence for both risk events and risk agents were obtained by in-depth interviews and through questionnaires from 200 seaweed farmers, distributors, a supplier in addition to two local government officials and two researchers in the Kei Islands. Based on the risk events and risk agents of seaweed supply chain in Kei Islands, Indonesia, the process continued with calculating the ARP value and plotting a Pareto diagram in order to select risks agents that contribute to risks in operational seaweed supply chain Kei Islands by measuring the correlation between the risk agents and risk events, and a value of 0, 1, 3, or 9 was assigned.

Based on the ARP calculations (in Table 3), the Pareto diagram illustrates the magnitude of the causal effects of risks in the seaweed supply chain in the Kei Islands. As seen in Figure 3, 10 causes of risks contribute to 80% of the seaweed supply chain risk in the Kei Islands. Furthermore, the result indicated that five of the 10 risks contribute 54% of the causes of all the existing risks. So in the next calculation process with fuzzy AHP, these five risks are used as a criteria to

Table 3: ARP results.

Risk agent	ARP	Cum (%)
A2	4212	15.06
A11	3339	26.99
A1	3015	37.77
A24	2646	47.23
A5	1896	54.00
A4	1800	60.44
A23	1764	66.74
A6	1320	71.46
A12	1296	76.09
A13	1062	79.89
A20	777	82.67
A18	756	85.37
A7	720	87.94
A3	684	90.39
A19	504	92.19
A10	486	93.93
A8	432	95.47
A14	432	97.02
A21	225	97.82
A22	216	98.59
A15	117	99.01
A17	105	99.39
A9	87	99.70
A16	85	100.00

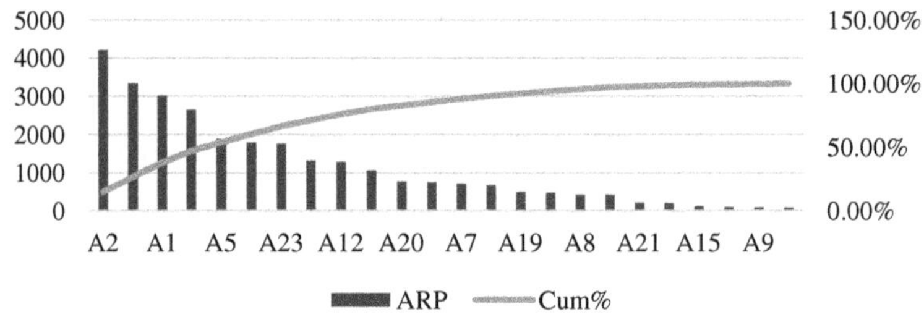

Figure 3: Pareto diagram of risk agents.

mitigate the risks in the supply chain based on two scenarios: intra-island and inter-island seaweed supply chain in the Kei Islands.

3.2.2. *Fuzzy AHP*

In the second stage, the weight of each criterion was obtained to get the final rank of each criterion. For prioritizing each criterion, this study sought three categories of experts: academicians, government and practitioners. Van Laarhoven and Pedrycz (1983) proposed FAHP, which is the combination of the analytic hierarchy process (AHP) and fuzzy theory to use linguistic ratings in the calculations by giving it a certain range. The pairwise comparisons are established using a nine-point scale that converts experts' preferences into available alternatives such as equally, moderately, strongly, very strongly or extremely preferred.

The fuzzy AHP analysis in the study is based on the approach of Sun (2010), where there are two steps in the analysis.

Step 1: Pairwise comparison matrix of all criteria by asking which criterion is more important, as shown in matrix $\tilde{A}$:

$$\tilde{A} = \underset{(1)}{\begin{bmatrix} 1 & \tilde{a}_{12} & \cdots & \tilde{a}_{1n} \\ \tilde{a}_{21} & 1 & \cdots & \tilde{a}_{n2} \\ \vdots & \vdots & \ddots & \vdots \\ \tilde{a}_{n1} & \tilde{a}_{n2} & \cdots & 1 \end{bmatrix}} = \begin{bmatrix} 1 & \tilde{a}_{12} & \cdots & \tilde{a}_{1n} \\ 1/\tilde{a}_{12} & 1 & \cdots & \tilde{a}_{n2} \\ \vdots & \vdots & \ddots & \vdots \\ 1/\tilde{a}_{1n} & 1/\tilde{a}_{2n} & \cdots & 1 \end{bmatrix} \tag{2}$$

where

$$\bar{a}_{ij} = \begin{cases} \tilde{9}^{-1}, \tilde{8}^{-1}, \tilde{7}^{-1}, \tilde{6}^{-1}, \tilde{5}^{-1}, \tilde{4}^{-1}, \tilde{3}^{-1}, \tilde{2}^{-1}, \ \tilde{1}^{-1}, \tilde{1}, \tilde{2}, \tilde{3}, \\ \quad \tilde{4}, \tilde{5}, \tilde{6}, \tilde{7}, \tilde{8}, \tilde{9} & i \neq j \\ 1 & i = j. \end{cases}$$

Step 2: To define fuzzy geometric mean and fuzzy weights of each criterion, we use the geometric mean (Hsieh *et al.*, 2004):

$$\tilde{r}_i = (\tilde{a}_{i1} \otimes \ldots \otimes \tilde{a}_{ij} \otimes \ldots \otimes \tilde{a}_{in})^{1/n}, \tag{3}$$

$$\tilde{w}_i = \tilde{r}_i \left[\tilde{r}_i \otimes \ldots \otimes \tilde{r}_i \otimes \ldots \otimes \tilde{r}_n\right]^{-1}, \tag{4}$$

where $\bar{a}_{in}$ is fuzzy comparison value of criterion i to criteria n, thus, $\tilde{r}_i$ is geometric mean of fuzzy comparison criterion i to each criterion, $\tilde{w}_i$ is the fuzzy weight of the ith criterion, indicated by TFN, $\tilde{w}_i = (Lw_1, Mw_1, Uw_1)$. Where Lw_1 is lower values, Mw_1 is middle value and Uw_1 represent upper values of fuzzy weight of the ith criterion.

The consistency of the matrix we used is the standard consistency ratio (CR) using the following equation:

$$\mathrm{CR} = \frac{\mathrm{CI}}{\mathrm{RI}}, \tag{5}$$

where RI is a random index and CI is consistency index. In addition to determining CI, we used the following equation:

$$\mathrm{CI} = \frac{\lambda_{\max} - n}{n - 1}, \tag{6}$$

where $\lambda_{\max}$ the maximum is the value of eigenvector and n is the number of criteria. The value of CR is acceptable when it is below 0.1 (Saaty, 1980).

Based on the results from the HOR, the fuzzy AHP hierarchy framework was formed by prioritizing the risks obtained as criteria. Also, factors that are related to seaweed supply chain risks in Kei Islands, in particular along with alternatives for mitigating seaweed supply chain risk in the region, were taken into consideration (Figure 4). Furthermore, fuzzy AHP was carried out with two scenarios, namely intra-island seaweed supply chain risk and inter-island

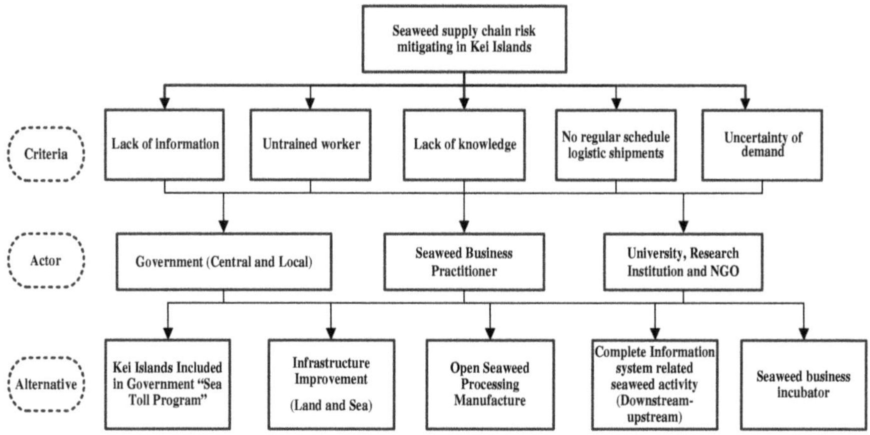

Figure 4: The hierarchical structure of mitigation seaweed supply chain risk in Kei Islands.

seaweed supply chain risk with the aim of obtaining two alternative solutions, each of which corresponds to supply chain connectivity within the Kei Islands and its connection with other islands in Indonesia.

4. Mitigating Seaweed Supply Chain Risk in the Kei Islands

4.1. *Intra-island seaweed supply chain risks*

Based on empirical geographical conditions and quality of infrastructure in the Kei Islands, there exists a high possibility for disruption of seaweed supply chain in this region due to seasonality, which affects the time of farming and distribution and supply of seaweed. The disruption could be avoided if farmers have sufficient knowledge of seaweed cultivation. Currently, seaweed farmers in this region are mostly placed at a very low in socio-economic condition, which means they lack formal and informal education related to seaweed cultivation. In addition, the scheduling for seaweed farming throughout the year so often has not been done consistently, because of various factors including the supply of seeds, which in some cases are not suited

Table 4: FAHP weights for criteria and actor intra-islands seaweed supply chain risk.

Criteria[a]	Weight	Rank	Actor[b]	Weight	Rank
Lack of information	0.156	4	Government (local and central)	0.540	1
Untrained worker	0.218	3			
Lack of knowledge	0.382	1	Seaweed business practitioner	0.267	3
No regular schedule logistic shipments	0.093	5			
Uncertain of demand	0.250	2	Universities, research institution and NGO	0.275	2

Notes: [a]λ: 5.4423; CI: 0.1106; CR: 0.0987 (<10%). [b]λ: 3.0894; CI: 0.0447; CR: 0.0771 (<10%).

to the conditions of the location of seaweed cultivation. At the production stage, the quality of harvested seaweed does not possess a standard overall quality throughout a year. For distribution, seaweed distributors are unable to fulfill the seaweed supply according to customer demand because of the varying quality of seaweed received with fluctuating production quantity throughout the year.

The seaweed farmers play a significant role in maintaining the flow of seaweed supply chain in this region where there are no seaweed users (e.g., seaweed processing company). The seaweed supply chain is limited to the distribution of raw materials to other islands. The availability of constant seaweed supply in this region is very dependent on the performance of seaweed farmers. The low level of education of the coastal community, which is mostly made up of seaweed farmers, is one factor contributing to the risk of seaweed supply chain in this region (Table 4). Expert opinion and fuzzy AHP calculations pointed that lack of knowledge of seaweed cultivation is a risk within the islands. This seaweed production is related to the ability of farmers to carry out seaweed cultivation with the right techniques and technology, including the management of seaweed business.

Regarding the current condition, the role of the central and regional governments assumes importance in improving the capacity of farmers in this region. The problems that exist in this region are not only limited to infrastructure but also increasing the knowledge of seaweed farmers. Based on alternative results from fuzzy AHP calculations as shown in Table 6, it is clear that the opening of seaweed business incubators can help resolve the weaknesses of seaweed farmers in improving their ability in professional and sustainable seaweed business management. In addition, the results also show that besides establishing a seaweed business incubator, activating the seaweed processing company, which was opened some years back, owned by the regional government of Southeast Maluku can also help overcome the risk of seaweed supply chain within the islands because it can trigger farmers to maintain the quantity and quality of seaweeds.

4.2. *Inter-islands seaweed supply chain risks*

With regard to connectivity with other islands in Indonesia, shipping is the preferred option for logistics, for which passenger ships with very limited capacity are used. This is because there is a fixed schedule of these shipments every month at a relatively low cost compared to the cost of containers on freight ships at the port of Kei Islands (Tual City). In accordance with the experts' assessment of the most significant risk with regard to the source of seaweed supply chain, as shown in Table 5, the absence of ships that regularly transport seaweeds on a monthly basis at relatively affordable costs was found. This situation leads to making shipping goods out of the Kei Islands very expensive and most of the time, the shipping pattern does not follow the order schedule from the seaweed customers.

Despite the abundance of fisheries and marine resources in the islands, the role of the private sector is still very limited, which makes the role of the government key to economic growth and acceleration in this region. Thus, the central government can help drive the economy in this region by including Kei Islands as one of the toll sea transit routes from the western part of Indonesia, which is the center of the country's economy as shown in Table 6. On the other

Table 5: FAHP weights for criteria and actor inter-islands seaweed supply chain risk.

Criteria[a]	Weight	Rank	Actor[b]	Weight	Rank
Lack of information	0.216	2	Government (local and central)	0.537	1
Untrained worker	0.123	5			
Lack of knowledge	0.100	5	Seaweed business practitioner	0.244	3
No regular schedule logistic shipments	0.438	1			
Uncertain of demand	0.215	3	Universities, research institution and NGO	0.293	2

Notes: [a]λ: 5.3890; CI: 0.0973; CR: 0.0868 (<10%). [b]λ: 3.1087; CI: 0.0543; CR: 0.0937 (<10%).

Table 6: FAHP weights alternatives intra- and inter-islands seaweed supply chain risk.

Intra-islands			Inter-islands		
Alternatives[a]	Weight	Rank	Alternatives[b]	Weight	Rank
Kei Islands Included in Government "Sea Toll Program"	0.103	5	Kei Islands Included in Government "Sea Toll Program"	0.328	1
Infrastructure Improvement (Land and Sea)	0.164	3	Infrastructure Improvement (Land and Sea)	0.189	3
Open Seaweed Processing Manufacture	0.308	2	Open Seaweed Processing Manufacture	0.107	5
Complete Information system related seaweed activity (Downstream-upstream)	0.154	4	Complete Information system related seaweed activity (Downstream-upstream)	0.323	2
Seaweed business incubator	0.357	1	Seaweed business incubator	0.151	4

Notes: [a]λ: 5.4146; CI: 0.1037; CR: 0.0926 (<10%). [b]λ: 5.3383; CI: 0.0846; CR: 0.0755 (<10%).

hand, in the meantime, the local government can also help to boost the regional economy and facilitate the shipping of goods from Kei Islands to the nearest port that has been included in the sea toll by providing shipping vessels owned by the regional government. This would render the shipping costs to be more affordable with a fixed schedule, compared to shipping operated by a private entity.

By establishing connections with the Kei Islands, which tend to be isolated from big islands in Indonesia, the risk of inter-island seaweed supply chain can be mitigated. This is important because under certain conditions, abundant seaweed supply results in very low selling prices for farmers and also leads to accumulation of large inventory by distributors. By connecting Kei Islands with other islands in Indonesia, this condition can be improved.

5. Conclusions

The risks in seaweed supply chain in regions with unique characteristics such as the Kei Islands must be addressed using an approach that is appropriate to the conditions of the region. In more than half of the regions of the world, the problem of the isolation of regions due to geographical conditions does not occur. In the Kei Islands, apart from its location int he archipelago, limited infrastructure and low socio-economic conditions prevail. As a result, unique supply chain risks have to be dealt with. In a situation where many complex factors must be considered in decision-making with regard to the existing problems in seaweed supply chain, the role of MCDM becomes very important and crucial in supporting the sustainability of fisheries in this region.

The problem of seaweed supply risk in the Kei Islands must be comprehensively resolved, where focus is not only on the connectivity in the region, but also its connection with other regions in Indonesia. By utilizing powerful MCDM integration between HOR and fuzzy-based AHP, two solutions for intra-islands seaweed supply chain are proposed. The first is opening a seaweed business incubator to improve the ability of seaweed farmers to become more professional in seaweed business management. In addition activating the company that specialized in seaweed processing, built a few years ago, will also help reduce the risk of seaweed supply chain in the Kei Islands region. To reduce the risk of the inter-island seaweed supply chain, Kei Islands should be included as one of the routes for the sea toll. As an alternative, if the central government does not support the solution, the local government in Kei Islands must provide ship transportation to the nearest port in the sea toll lane. This will ensure a regular and low-cost connectivity between Kei Islands and other regions in Indonesia and the world.

References

Abdel-Basset, M., Gunasekaran, M., Mohamed, M. and Chilamkurti, N. (2019). A framework for risk assessment, management and evaluation: Economic tool for quantifying risks in supply chain. *Futur Gener Comput Syst,* 90, 489–502.

Al-Busaidi, M. A., Jukes, D. J. and Bose, S. (2016). Seafood safety and quality: An analysis of the supply chain in the Sultanate of Oman. *Food Control*, 59, 651–662.

Alora, A. and Barua, M. K. (2019). An integrated structural modelling and MIC-MAC analysis for supply chain disruption risk classification and prioritisation in India. *Int J Value Chain Manag*, 10(1), 1–25.

Chand, M., Raj, T., Shankar, R. and Agarwal, A. (2017). Select the best supply chain by risk analysis for Indian industries environment using MCDM approaches. *Benchmarking,* 24(5), 1400–1413.

Ferreira, F. D. A. L., Scavarda, L. F., Ceryno, P. S. and Leiras, A. (2018). Supply chain risk analysis: A shipbuilding industry case. *Int J Logist Res Appl*, 21(5), 542–556.

Hsieh, T. Y., Lu, S. T. and Tzeng, G. H. (2004). Fuzzy MCDM approach for planning and design tenders selection in public office buildings. *Int J Proj Manage*, 22(7), 573–584.

Jiang, B., Li, J. and Shen, S. (2018). Supply chain risk assessment and control of port enterprises: Qingdao port as case study. *Asian J Shipp Logist*, 34(3), 198–208.

Karningsih, P. D., Anggrahini, D., Kurniati, N., Suef, M., Fachrur, A. R. and Syahroni, N. (2018). Mapping risks of Indonesian tuna supply chain. In *IOP Conference Series: Materials Science and Engineering.*

Kavilal, E. G., Venkatesan, S. P. and Priyamvatha, S. (2016). An integrated Fuzzy AHP and Fuzzy TOPSIS for prioritizing supply chain complexity drivers. *Int J Oper Quant Manag*, 22(1), 39–51.

Khemiri, R., Grabot, B., Elbedoui-Maktouf, K. and Zouari, B. (2018). Integrating fuzzy TOPSIS and goal programming for multiple objective integrated procurement-production planning. In *IEEE International Conference on Emerging Technologies and Factory Automation.* ETFA, pp. 1–8.

Lei, W. and Zhaocai, X. (2007). Supply chain risk evaluation based on fuzzy multi-criteria lattice-order decision-making. In *Proceedings of the IEEE International Conference on Automation and Logistics*, ICAL, pp. 1442–1445.

Ma, H.-L. and Wong, W.-H. C. (2018). A fuzzy-based House of Risk assessment method for manufacturers in global supply chains. *Ind Manag Data Syst,* 118(7), 1463–1476.

Nguyen, T. L. T., Tran, T. T., Huynh, T. P., Ho, T. K. D., Le, A. T. and Do, T. K. H. (2018). Managing risks in the fisheries supply chain using house of risk framework (HOR) and interpretive structural modeling (ISM). In *IOP Conference Series: Materials Science and Engineering.*

Oliveira, J. B., Jin, M., Lima, R. S., Kobza, J. E. and Montevechi, J. A. B. (2019). The role of simulation and optimization methods in supply chain risk management: Performance and review standpoints. *Simul Model Pract Theory*, 92, 17–44.

Pujawan, I. N. and Geraldin, L. H. (2009). House of risk: A model for proactive supply chain risk management. *Bus Process Manag J*, 15(6), 953–967.

Rostamzadeh, R., Ghorabaee, M. K., Govindan, K., Esmaeili, A. and Nobar, H. B. K. (2018). Evaluation of sustainable supply chain risk management using an integrated fuzzy TOPSIS- CRITIC approach. *J Clean Prod*, 175, 651–669.

Saaty, T. L. (1980) *The Analytic Hierarchy Process.* McGraw-Hill, New York.

Saxena, A. and Seth, N. (2014). Analytical hierarchy process to evaluate supply chain risk and security management. *Int J Procure Manag*, 7(3), 279–298.

Song, Y., Yao, S., Yu, D. and Shen, Y. (2017). Risky multi-criteria group decision making on green capacity investment projects based on supply chain. *J Bus Econ Manag*, 18(3), 355–372.

Soni, U. and Vaishnavi (2019). Analyzing risk in dynamic supply chains using a hybrid fuzzy AHP model. In *Advances in Intelligent Systems and Computing*, Vol. 697, pp. 411–420.

Sun, C. C. (2010). A performance evaluation model by integrating fuzzy AHP and fuzzy TOPSIS methods. *Expert Syst Appl*, 37(12), 774–7754.

Van Laarhoven, P. J. and Pedrycz, W. (1983). A fuzzy extension of Saaty's priority theory. *Fuzzy Sets Syst*, 11(1–3), 229–241.

Vishwakarma, V., Prakash, C. and Barua, M. K. (2016). A fuzzy-based multi criteria decision making approach for supply chain risk assessment in Indian pharmaceutical industry. *Int J Logist Syst Manag*, 25(2), 245–265.

Wang, Y.-L. (2010). Risk model and insurance policy for international fishery logistics management. In *ICIME 2010 — 2010 2nd IEEE International Conference on Information Management and Engineering*, pp. 48–51.

Wang, Y.-L. (2010). International fishery supply chain risk management and insurance. In *5th IEEE International Conference on Management of Innovation and Technology*, ICMIT 2010, pp. 1142–1146.

Wang, Y.-L. (2010). Novel method for fishery supply chain risk diversity and insurance. In *Proceedings of the International Conference on E-Business and E-Government*, ICEE 2010, pp. 1868–1871.

Yan, B., Shi, P. and Huang, G. (2013). Development of traceability system of aquatic foods supply chain based on RFID and EPC internet of things. *Nongye Gongcheng Xuebao/Transactions Chinese Soc Agric Eng*, 29(15), 172–183.

Yazdani, M., Kahraman, C., Zarate, P. and Onar, S.C. A fuzzy multi attribute decision framework with integration of QFD and grey relational analysis. *Expert Syst Appl*, 115, 474–485.

Zheng, H. (2010). Supply chain disruption risk evaluation based on TOPSIS method with sensitivity analyses. In *ICLEM 2010: Logistics for Sustained Economic Development — Infrastructure, Information, Integration — Proceedings of the 2010 International Conference of Logistics Engineering and Management*, pp. 4762–4769.

Chapter 13

An Approach for Modeling the Deregulated Power Market Based on Cournot Game Theory

M. T. Askari*, M. Z. A. Ab Kadir[†,‡,§] and E. Bolandifar[¶]

**Department of Electrical Engineering, Semnan Branch, Islamic Azad University, Semnan, Iran*

[†]Department of Electrical and Electronic Engineering, Faculty of Engineering, Universiti Putra Malaysia, 43400 Serdang, Selangor, Malaysia

[‡]Centre for Electromagnetic and Lightning Protection Research(CELP), Faculty of Engineering, Universiti Putra Malaysia, Serdang, Selangor, Malaysia

[§]Institute of Power Engineering (IPE), Universiti Tenaga Nasional, Kajang, Selangor, Malaysia

[¶]PHBS Business School, Peking University, Shenzhen, China

The main aim of the decision-makers in a deregulated power market is to maximize their profits and reduce their risks. The profit of private companies is affected by several parameters as well as uncertainties in the deregulated power market. These uncertain parameters also include the strategic behavior of rivals and regulatory interventions. Therefore, in order to reduce risks and increase profits, private companies as decision-makers in the deregulated power market should consider all parameters in their planning. In this chapter, a mathematical model has been developed to determine the market clearing price (MCP) in each season and load levels as well as the profit of each private firm by considering on-ground realities such as bilateral contracts, carbon tax and the rational uncertainty in the deregulated power market. Using this mathematical model, private companies in the deregulated power market can predict the generation strategies of other investors.

The proposed framework has been implemented in the hypothetical deregulated power market by using the IEEE reliability test system. The case study chosen in this research has confirmed that this model provides robust decisions and precise information about the deregulated power market.

1. Introduction

During the last two decades, deregulated power markets are being replaced with the centralized power markets in several developed countries. With this transformation, not only the structure of the electricity market has changed, but new regulations have also been introduced into the restructured power market. The private firms in the restructured electricity market encounter new challenges, particularly in long-term and medium-term planning. In traditional systems, generation expansion planning has been modeled in the centralized environments, and there was a standard and reliable pattern for generation planning. Meanwhile, in the restructured power market, the planning model strictly depends on the decisions of private investors. These decisions affect the uncertain parameters in the restructured power market. Hence, these firms encounter several risks in their investments. On the other hand, the aims of the participants in the competitive power markets are different from the objectives that have been described in the centralized power markets. The aim of the traditional power market is to find the ideal technology, capacity, time and the place of construction of power plants by considering acceptable reliability to respond to the demand and also taking into account social welfare (Chuang *et al.*, 2001; Kothari and Nagrath, 2003), while the investor in the restructured power market intends to maximize his profit (Li *et al.*, 2004, 2013; Park *et al.*, 2002; Zhu and Chow, 1997). Therefore, the investors should provide a framework in order to simulate the deregulated power market to maximize their profits by considering the uncertainties and regulatory interventions. Traditionally, the aim of generation planning is to minimize the cost by answering the demand growth at an acceptable level of reliability or social welfare. But in the restructured power market the aim of the private companies for solving the generation planning problem

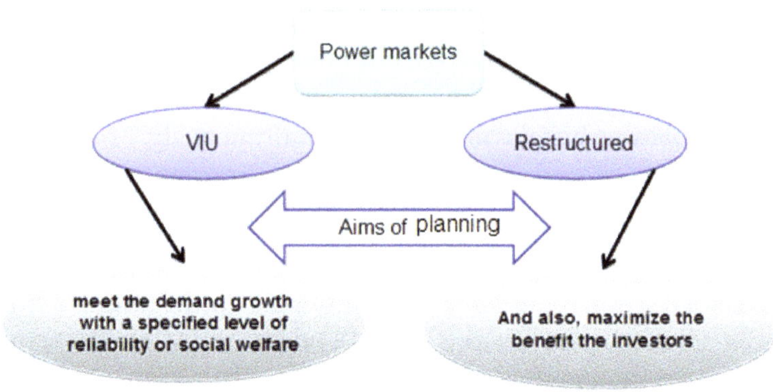

Figure 1: The aims of planning in different types of power markets.

is to maximize their profits (Askari *et al.*, 2013, 2014; Barforoushi *et al.*, 2010; Wood and Wollenberg, 2012). Refer to Figure 1.

In the restructured power market, the market clearing price (MCP) is an important factor to calculate the profit of each investor. This factor is not only affected by exogenous parameters, but also the average of MCP changes seasonally for each load level. Therefore, the volatility of MCP causes more risk for investors. The effective parameters that have effects on the variations of MCP include the strategic behavior of rivals and regulatory interventions of government such as the bilateral contract and CO_2 tax and fuel price. Therefore, in order to reduce risks and increase profits, the private companies as decision-makers in the deregulated power market should consider and model all these parameters in their planning.

Various games are considered in a competitive environment in order to determine the Nash equilibrium. Cournot, Bertrand, van-Stackelberg and supply function equilibrium (SFE) are the famous models in game theory. The Cournot model of oligopoly competition was introduced by Augustin Cournot in 1838. The model has become a classic in microeconomic oligopoly theory. Cournot games have several characteristics in common, which are as follows:

- Competition occurs only in quantities.
- Product is non-storable.

- Product is homogeneous.
- Market price is determined by auction.
- No entry occurs during the game.
- Decision-making by players occurs simultaneously.

In the Cournot model, each firm chooses an output quantity to maximize profit. Firms are assumed to produce homogeneous goods that are nonstorable. Therefore, all quantities produced are immediately sold. Market price in the model is determined through an auction process that equates industry supply with aggregate demand. Examples of Cournot-style auction-based pricing can be found in organized markets such as those for agricultural products like wheat, corn, and rice. The model also assumes that all firms in the industry can be identified at the start of the game, and decision-making by firms occurs simultaneously (Chuang *et al.*, 2001; Kirschen and Strbac, 2005; Myerson, 2013; Singh, 1999).

In the Bertrand model, investors maximize their profit according to the product price. In this model, assume that the firms produce a homogeneous product, i.e., players produce commodities that are perfect substitutes. Also, the products are assumed to be non-storable. In addition, the firms have no capacity constraints on their production (Chuang *et al.*, 2001; Myerson, 2013; Singh, 1999). The Stackelberg leadership model is a strategic game in economics in which the leader firm moves first and then the follower firms move sequentially (Von Stackelberg, 2010). It is named after the German economist Heinrich Freiherr von Stackelberg who published his book, *Market Structure and Equilibrium* (*Marktform und Gleichgewicht*) in 1934, which described the model. In GT terms, the players of this game are a leader and a follower and they compete on quantity. The Stackelberg leader is sometimes referred to as the market leader. There are some further constraints upon the sustaining the Stackelberg equilibrium. The leader must know *ex-ante* that the follower observes his action. The follower must have no means of committing to a future non-Stackelberg follower action and the leader must be aware of this. Indeed, if the "follower" could commit to a

Stackelberg leader action and the "leader" knew this, the leader's best response would be to play a Stackelberg follower action (Von Stackelberg, 2010). Since the first application of SFE for England and Wales power market in 1992 by Green and Newbery, the SFE model has been employed to examine the power market for various purposes. The main shortcomings of the supply function approaches include the difficulties to calculate, frequency of multiple equilibria, might have unsteady answers and need for more simplifications with respect to the power market (Willems *et al.*, 2007, 2009). All in all, the Cournot model is the most popular category for simulating the strategic interactions among the participants in the restructured power market (Li and Shahidehpour, 2005). There are many similarities between the Cournot model and the liberalized generation markets. Simplicity, computational tractability, flexibility in modeling bilateral contracts and technical limits, as well as compatibility with operational planning in power systems, are features that cause Cournot model to be a popular game concept (Gountis and Bakirtzis, 2004). This game model is applied to determine the MCP together with maximum profits of investors in medium term market.

Several mathematical models were proposed to model the regulated power market and less focus was paid to the deregulated power market. In some of them, the effect of the CO_2 tax has been investigated on MCP and then on the profits of investors (Barforoushi *et al.*, 2010; Chuang *et al.*, 2001). Furthermore, the effects of the variable costs of the power plants were examined in the previous studies (Nanduri *et al.*, 2009). A simple mathematic model has been proposed by Askari *et al.* (2015) for evaluating the impact of fix FIT. In this model, stochastic uncertainties and also the realities of power market were neglected. In another article, the system dynamics model was reviewed for assessing the policy in power market. It has been revealed that policy assessment and generation expansion planning are the most important issues (Ahmad *et al.*, 2016). Also, a dynamic system model has been proposed by Ibanez-Lopez *et al.* (2017). Although the proposed model in this article gives us more useful information about the impact of incentive policies in the restructured power market, the realities of power market such as CO_2 tax has been neglected (Ibanez-Lopez *et al.*, 2017). However, the major drawback of these approaches is that the impact of the CO_2 tax, bilateral contract parameters together with the rational uncertainty have not been modeled in the deregulated power market. Therefore, if the private companies did not consider the aforementioned parameters, the financial risks of decision-makers will increase, which would lead to increase in cost for private firms. Also, the risks of planning increase and the profits of private companies decrease in the deregulated power market.

In this study, a developed mathematical model has been proposed to maximize the profit of private firms in the deregulated power market. It can be concluded that this research will contribute some new knowledge to the development of the models and techniques that aim to propose a framework to determine optimal planning for private investors in the restructured power market.

2. Development of Conceptual Model

The proposed framework to simulate the deregulated power market is shown in Figure 2. This framework is revealed in four main blocks. The input parameters for modeling the deregulated power market are illustrated in block 1. These parameters include the number of units for each firm together with their capacities, demand for each season and each load level, duration of demand for each season in each load level, fuel price for each generation unit, heat rate function and the demand function. The impact of the rational uncertainty on the amount of generation of each unit and the mathematical model to simulate the deregulated power market has been shown in block 2. This model is applied to determine the price of electricity together with operational benefits of private generation companies. The total profit of the deregulated power market and profit of each firm in the MCP has been given in block 3. Finally the impact of fuel price on the total profit of the deregulated power market is represented in block 4 as a sensitivity analysis of the proposed model.

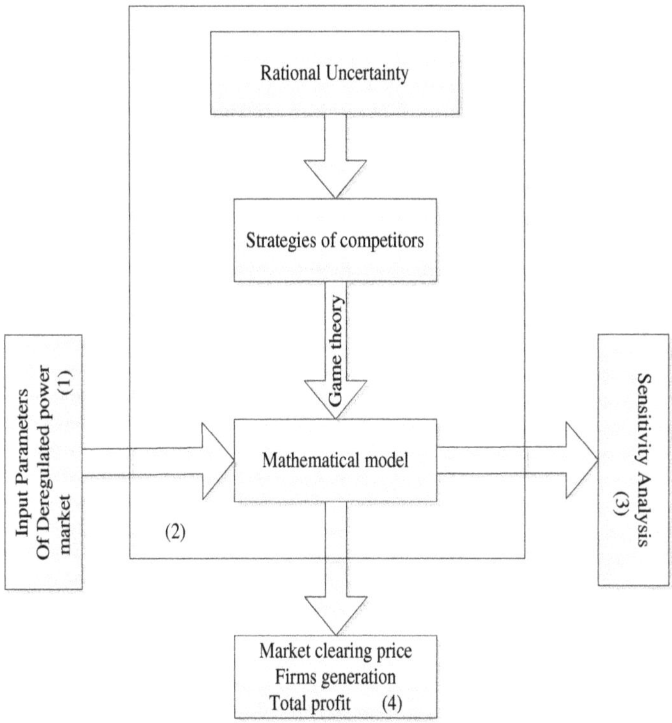

Figure 2: General description the proposed framework.

3. Determining the Market Clearing Price

In this paper, the Cournot concept has been applied to determine the MCP. This algorithm is explained in the following steps and shown in Figure 3 (Askari *et al.*, 2013, 2014).

Step 1: Each private firm calculates its generation by solving the optimization problem explained in objective function (equation (4)).

Step 2: After the optimization problem has been solved for each firm, the demand function of the power market is used to update the electricity price. To simplify the model performance, a linear

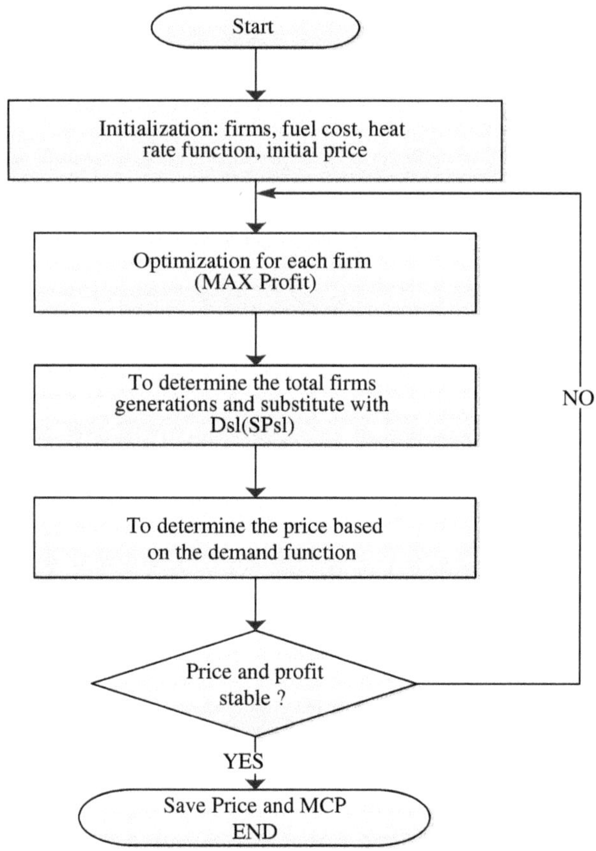

Figure 3: Proposed model to estimate the MCP.

demand function is used as follows (Barforoushi *et al.*, 2010; Villar and Rudnick, 2003):

$$D_{\rm sl}(SP_{\rm sl}) = -A_{\rm sl} \times SP_{\rm sl} + B_{\rm sl}, \tag{1}$$

Due to the importance of balance between demand and supply in the power market, the total generation of all the firms is substituted by demand (Dsl) in each season and load level. Constants A and B are calculated for each season and load level according to equations

(2) and (3), respectively (Barforoushi *et al.*, 2010).

$$A_{\text{sl}} = \frac{B_{\text{sl}}}{pc.\pi_{\text{base,sl}}}, \tag{2}$$

$$B_{\text{sl}} = dc.D_{\text{base,sl}}. \tag{3}$$

The forecasted demand $D_{\text{base,sl}}$ for each season and load level can be determined based on the predicted average load in each season and load level, which is depicted in Table 3 for IEEE RTS. Also, the reference electricity price is calculated from the traditional unit commitment (Wood and Wollenberg, 2012).

Step 3: Then, the calculated price is compared with the initial price. If the initial price is equal to the calculated price, the program will save the consequences. Otherwise, the first and second steps will be repeated until this condition is achieved. Under these circumstances, the Nash equilibrium of Cournot game is obtained and none of the firms receives profits from changing its generation. These steps are illustrated in Figure 3.

4. Developed Mathematical Model to Simulate the Deregulated Power Market

The aim of the objective function of the proposed model is to maximize the benefits for each firm as well as the profit of power market, which is shown in equation (4). It estimates the maximum benefits of private investors in the electricity power market and is composed of two main parts. The first part includes revenue from energy sales in the power market and also the revenue of the firm in contractual markets. The second part of equation (4) includes the operation cost and CO_2 tax for thermal units of traditional firms.

$$Max\,Benefit = \sum_{e}^{Ne}\sum_{s}^{Ns}\sum_{l}^{Nl} d_{\text{sl}} \times \left(\left(\sum_{u}^{Nu} N_{e,u} \times P_{e,u,\text{sl}}\right) - Q_{e,\text{sl}} \right)$$
$$\times SP_{\text{sl}} + \sum_{e}^{Ne}\sum_{s}^{Ns}\sum_{l}^{Nl} d_{\text{sl}} \times Q_{e,\text{sl}} \times BP_{\text{sl}}$$

$$- \sum_{e}^{Ne} \sum_{s}^{Ns} \sum_{l}^{Nl} d_{\text{sl}} \times \left(\sum_{u}^{Nu} N_{e,u} \times FP_u \times (a_u + b_u \right.$$

$$\left. \times P_{e,u,\text{sl}} + c_u \times (P_{e,u,\text{sl}})^2) \right)$$

$$- \sum_{e}^{Ne} \sum_{s}^{Ns} \sum_{l}^{Nl} d_{\text{sl}} \times \left(\sum_{u}^{Nu} N_{e,u} \times Tax \times (a_u + b_u \right.$$

$$\left. \times P_{e,u,\text{sl}} + c_u \times (P_{e,u,\text{sl}})^2) \times EM_u \right) \tag{4}$$

The amount of power generation for traditional private companies are represented in equation (5).

$$g_{e,\text{sl}} = \sum_{u}^{Nu} N_{e,u} \times P_{e,u,\text{sl}} \; \forall e, s, l. \tag{5}$$

Subject to:

With respect to the demand and other limitations in a problem, a set of constraints are added to formulate the problem. These constraints are explained as follows:

The price of electricity in the contractual market is given in the following equation:

$$BP_{\text{sl}} = SP_{\text{sl},\,t-1}(1 + GP). \tag{6}$$

The constraint in equation (7) is related to the demand constraint, because the generation firms are not responsible to fulfil the whole demand of the market.

$$g_{e,\text{sl}} \leq D_{\text{sl}}. \tag{7}$$

The boundary limitation of the decision variable in generations' firm is represented in equation (8).

$$P_{e,u,\min} \leq P_{e,u} \leq P_{e,u,\max}. \tag{8}$$

5. Verification of Proposed Model (Case Study)

The proposed framework is tested by applying IEEE RTS (Grigg *et al.*, 1999). The total installed capacity and the peak demand of the study system are 2792 and 2590 MW, respectively. Specifications regarding the units of firms in the study system are given in Table 1. The study system contains five price maker firms. The fuel prices and the data associated with the generating units of these firms are taken from sources (Gnansounou *et al.*, 2004; Shahidehpour *et al.*, 2002) and are shown in Table 2. One year is considered as the period of this study. Each year was divided into four seasons and three load levels were considered, which include the off-peak, medium and peak levels. The forecasted demands and the durations for each season and load levels have been collected from sources (Barforoushi *et al.*, 2010) and represented in Table 3. Other input parameters such as the CO_2 tax rate, and the bilateral contract price have been extracted from Barforoushi *et al.* (2010).

In order to test the proposed model and investigate the effect of different parameters on the profits of the firms and also the MCP, three scenarios are described in Table 4. The results of each scenario have been discussed separately in the next sections.

Table 1: Type and number of units for each firm.

	Generation firm				
Type	1	2	3	4	5
Type A	2	1	4	1	—
Type B	—	—	—	—	6
Type C	1	2	2	1	1
Type D	—	1	—	—	—
Type E	1	1	1	—	1
Type F	—	—	1	—	—
Type G	—	—	—	1	—
Type H	—	1	—	—	—
Total	4	6	8	3	8

Table 2: Fuel price and the capacity of generating unit.

Type	A	B	C	D	E	F	G	H
Pmax [MW]	12	50	76	100	155	197	350	450
Pmin [MW]	2.4	0.0	15.2	25	54.25	68.95	140	100
Fuel Price [$/MBTU]	5.27	0.0	1.68	5.27	1.68	5.27	1.68	1.65

Table 3: Forecasted demands and duration in each season and load levels for the first stage.

	Load level and duration [MW]&[h]		
Season	Off-peak (Duration)	Medium (Duration)	Peak (Duration)
1	1600(876)	2100(985.5)	2400(328.5)
2	1750(876)	2200(985.5)	2500(328.5)
3	2000(766.5)	2400(876)	2590(547.5)
4	1500(876)	2000(985.5)	2350(328.5)

Table 4: Case studies.

	CO_2 tax	Bilateral contract	Clearing price
Case 1	No	No	Yes
Case 2	No	Yes	Yes
Case 3	Yes	Yes	Yes

6. Results and Discussion

6.1. *Case 1*

In this scenario, the MCP has been calculated via the Cournot game and through the deterministic mathematical model without considering the CO_2 tax rate as well as the bilateral contract. The profit of each firm as well as the total profit of the deregulated power market has been calculated in this case. The annual average of the clearing price (AACP) is determined in this study. These results are

Table 5: Benefits of case 1 for each firm and power market.

Profit [M$]					
Firm1	Firm2	Firm3	Firm4	Firm5	Total
22.94	86.81	24.35	55.74	125.45	315.30

Table 6: Average of MCP for each load level.

AACP [$/MWh]			
Low	Med.	Peak	Ave.
21.57	35.18	73.08	43.28

represented in Table 5. The average MCP for each load level is shown in Table 6.

6.2. *Case 2*

A bilateral contract can be noted as one of the ancillary services in the power market. These contracts are arranged between electricity firms and large customers directly or indirectly via retailers and dealers in the power market. The contractual prices are different from the spot prices in the power market and they will be determined based on the average spot price of the previous year for each season and load level. In this study, for all cases, including scenarios 1 to 3, the bilateral contract price is assumed to be 50% of the spot price from the previous year. Also, the amount of the bilateral contract is equal to 10% of the total annual electricity generated by each firm. The effect of the growth of the amount of bilateral contract on the AACP and the benefit of the firms are represented in Tables 7 and 8, which also show the effect of changes in the contractual price. As is shown in this table, by increasing the amount of the bilateral contract from 10% to 40%, the AACP will be increased and the profits of each firm will decrease.

Table 7: Benefits of case 2 for each firm and power market.

No. Scn.	Growth of B–C (%)	Profit [M$]					
		Firm 1	Firm 2	Firm 3	Firm 4	Firm 5	Total
Scenario 1	10	22.69	85.51	24.1	55.14	124.46	311.91
Scenario 2	20	20.43	77.84	21.14	50.88	118.86	289.16
Scenario 3	30	18.73	71.91	18.94	47.64	114.53	271.75
Scenario 4	40	16.55	63.66	16.08	43.48	108.54	248.3

Table 8: Average of MCP for each load level.

No. Scn.	Growth of B–C (%)	AACP [$/MWh]			
		Low	Med.	Peak	Ave.
Scenario 1	10	21.57	38.79	73.08	44.48
Scenario 2	20	21.56	38.78	73.08	45.01
Scenario 3	30	21.57	43	73.08	45.88
Scenario 4	40	21.57	45.2	73.08	46.62

6.3. *Case 3*

The tax levied on carbon emission by the regulators has caused investors to be encouraged to invest in the new and efficient technologies for generating electricity. In this chapter, besides the bilateral contract and fuel cost that have been investigated in the previous sections, the effect of CO_2 tax rate was examined and the results are shown in Table 9 based on the profit of the firms as well as the AACP. Based on Table 9, a conclusion can be drawn that the CO_2 tax will impose an extra cost on each firm and therefore their benefits are reduced. On the other hand, the tax levied for CO_2 leads to an increase in the AACP rather than the second case. The following results are arranged based on the CO_2 tax rate, which is equal to \$0.02/lb. However, the growth of carbon tax rate versus the total profit of power market is represented in Figure 4. According to this

Table 9: Benefits of case 3 for each firm and AACP.

Profit [M$]					
Firm1	Firm2	Firm3	Firm4	Firm5	Total
21.87	84.53	22.98	53.74	123.88	306.99
AACP [$/MWh]					
Low	Med	Peak	Ave.		
21.57	38.96	73.08	44.54		

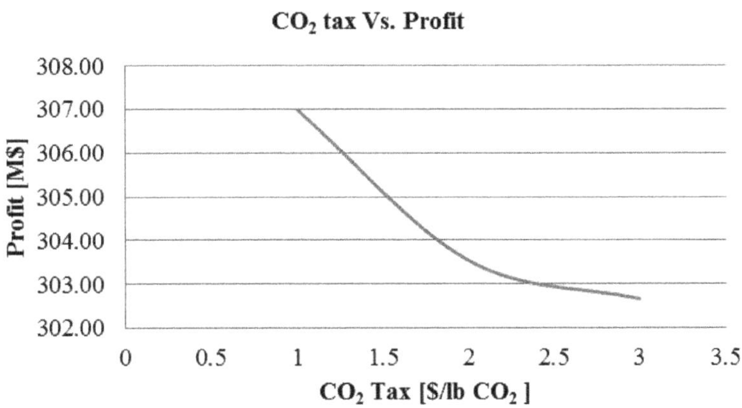

Figure 4: Effect of the growth of CO_2 tax on profit.

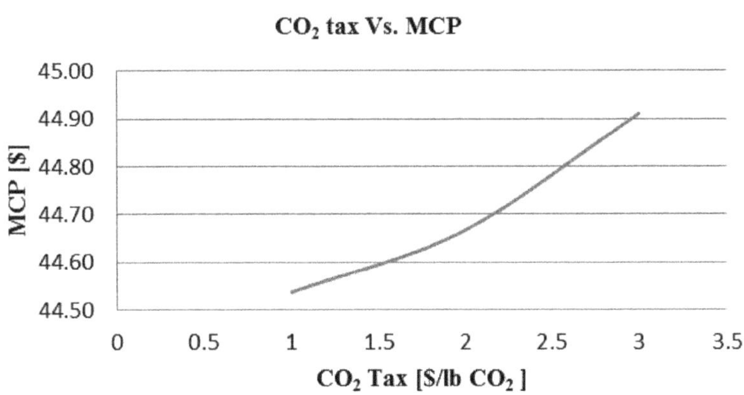

Figure 5: Effect of the growth of CO_2 tax on MCP.

figure, the growth of carbon cost leads to a reduction of the total profit of the power market. Furthermore, the ascending effect of the growth of CO_2 tax on the MCP has been shown in Figure 5. Simulation results show that the AACP increases by 2.83% compared with case No. 1. Furthermore, the total profits of the deregulated power market are also decreased significantly.

7. Sensitivity Analysis

The main aim of this section is to validate and verify the proposed mathematical model to simulate the medium term restructured power market. Broadly speaking, sensitivity analysis is a technique used to determine how different values of an independent variable will impact a particular dependent variable under a given set of assumptions. This technique is used within specific boundaries that will depend on one or more input variables, such as the effect that changes in fuel price will have on the total profit of the electricity market. Furthermore, this technique can be useful for testing the robustness of the model or system in the presence of uncertainty. In order to complete this discussion, the expected effect of the growth in fuel price has been investigated in this section. As is clear from the principles of economic sciences, the costs of the electric generating units will be augmented by increasing the fuel price. Consequently, the total profit of the power market will reduce and the equilibrium price of the power market will increase. Furthermore, because of the nonlinear features of the proposed model, the outputs of the proposed model that consider the growth of the fuel price will indicate a nonlinear behavior. If the aforementioned results are obtained by solving the proposed model, the validity and robustness of this model can be confirmed. To this end, the proposed mathematical model has been solved under an additive set of fuel price. The results of this analysis have been shown in Tables 10 and 11. Furthermore, the incremental effect of fuel price augmentation on the MCP and also the subtractive effect of fuel price augmentation on the total profit of power market have been illustrated in Figures 6 and 7, respectively. In conclusion, the results confirm the verification of this proposed framework.

Table 10: Profits for different scenarios of growing fuel price.

No. Scn.	Growth of fuel price	Profit [M$]					
		Firm1	Firm2	Firm3	Firm4	Firm5	Total
Scenario 1	0	22.94	86.81	24.35	55.74	125.45	315.30
Scenario 2	0.1	19.38	77.91	19.10	50.47	132.56	299.42
Scenario 3	0.2	11.97	58.35	8.47	38.55	130.55	247.88
Scenario 4	0.3	3.40	36.42	−3.72	24.47	125.47	186.04

Table 11: MCP for different scenarios of growing the fuel price.

No. Scn.	Growth of fuel price	AACP [$/MWh]			
		Low	Med.	Peak	Ave.
Scenario 1	0	21.57	35.18	73.08	43.28
Scenario 2	0.1	20.83	49.40	73.08	47.77
Scenario 3	0.2	21.57	53.26	73.08	49.30
Scenario 4	0.3	21.57	55.87	73.08	50.17

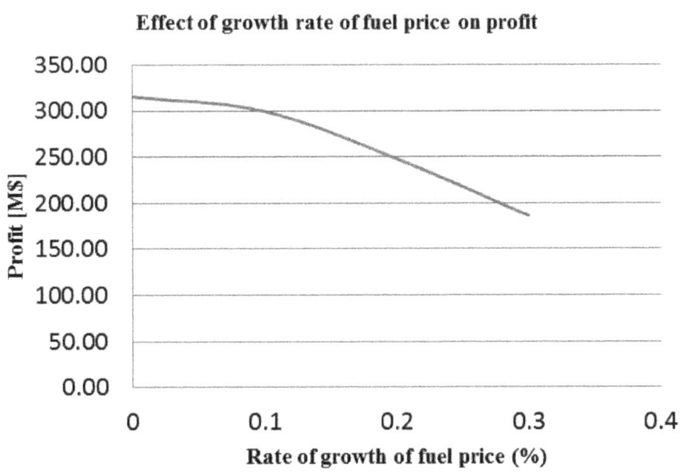

Figure 6: Effect of the growth of fuel price on profit.

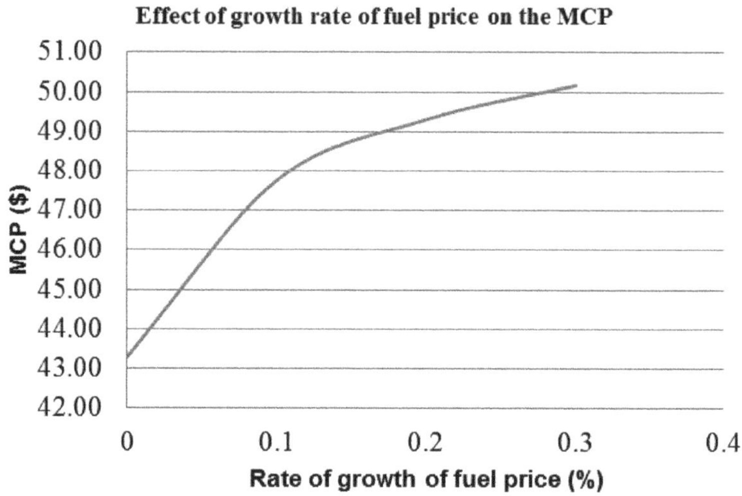

Figure 7: Effect of the growth of fuel price on MCP.

8. Conclusion

This study investigated the previous research in the planning area and proposed a new mathematical model to determine the optimal generation strategies for each firm in the deregulated power market. A model was developed to determine the MCP by considering the realities such as the bilateral contracts, carbon tax as well as the rational uncertainty in the deregulated power market. Based on this model, the effect of the growth of fuel price and the variations of the regulatory interventions has been investigated. This model not only can be so useful for private companies as decision-makers in the deregulated power market but also can be used by policymakers to organize their decisions. With attention given to the complexity and magnificence of simulating the deregulated power market, it is still a new field that needs to be explored in terms of the conception, problem identification, formulation and implementation. Based on our knowledge, several issues have been introduced for future work. This model can be used for modeling the long-term investment in the deregulated power market.

Nomenclature

i. indices

l	Load level
s	Season
e	generation firm
u	Type of thermal unit of a firm

ii. parameters

d_{sl}	Duration of time [hours]
$N_{e,u}$	Number of thermal units type u in e
$Q_{e,\,sl}$	Total generation contracted by firm e in sl [MW]
SP_{sl}	Average electricity price of power market in sl [$/MW]
BP_{sl}	Contracted electricity price in sl [$/MW]
FP_u	Fuel price of unit u [$/MBtu]
a_u, b_u, c_u	Constant coefficients of heat rate function for unit u
Tax	CO_2 tax rate [$/$CO_2$]
EM_u	CO_2 produced by unit u[lb/MMBTU]
GP	Percentage of electricity price
D_{sl}	Average demand in sl [MW]
$P_{e,u\min}$	Minimum generation of thermal unit u of firm e [MW]
$P_{e,u\max}$	Maximum generation of thermal unit u of firm e [MW]

iii. Decision variables

$P_{e,u,\,sl}$	Power generation by thermal unit u of firm e in sl [MW]
MCPsl	Market clearing price in sl [$/MW]

References

Ahmad, S., Tahar, R. M., Muhammad-Sukki, F., Munir, A. B. and Rahim, R. A. (2016). Application of system dynamics approach in electricity sector modelling: A review. *Renew Sustain Energy Rev*, 56, 29–37.

Askari, M. T., Ab Kadir, M. Z. A. and Bolandifar, E. (2015). Evaluation of FIT Impacts on market clearing price in the restructured power market. Paper presented at the *2015 IEEE Student Conference on Research and Development (SCOReD)*.

Askari, M. T., Ab Kadir, M. Z. A., Hizam, H. and Jasni, J. (2013). Evaluation of uncertainties on generation expansion planning. Paper presented at the *2013 IEEE 7th International Power Engineering and Optimization Conference (PEOCO)*.

Askari, M. T., Ab Kadir, M. Z. A., Hizam, H. and Jasni, J. (2014). A new comprehensive model to simulate the restructured power market for seasonal price signals by considering on the wind resources. *J Renew Sustain Energy*, 6(2), 023104. doi: 10.1063/1.4869141.

Barforoushi, T., Moghaddam, M. P., Javidi, M. H. and Sheikh-El-Eslami, M. K. (2010). Evaluation of regulatory impacts on dynamic behavior of investments in electricity markets: A new hybrid DP/GAME framework. *IEEE Trans Power Syst*, 25(4), 1978–1986. doi: 10.1109/TPWRS.2010.2049034.

Chuang, A. S., Wu, F. and Varaiya, P. (2001). A game-theoretic model for generation expansion planning: Problem formulation and numerical comparisons. *IEEE Trans Power Syst*, 16(4), 885–891.

Gnansounou, E., Dong, J., Pierre, S. and Quintero, A. (2004). Market oriented planning of power generation expansion using agent-based model. Paper presented at the *2004 IEEE Power Systems Conference and Exposition (PES)*.

Gountis, V. P. and Bakirtzis, A. G. (2004). Efficient determination of Cournot equilibria in electricity markets. *IEEE Trans Power Syst*, 19(4), 1837–1844.

Grigg, C., Wong, P., Albrecht, P., Allan, R., Bhavaraju, M., Billinton, R. *et al.* (1999). The IEEE reliability test system-1996. A report prepared by the reliability test system task force of the application of probability methods subcommittee. *IEEE Trans Power Syst*, 14(3), 1010–1020.

Ibanez-Lopez, A. S., Martinez-Val, J. M. and Moratilla-Soria, B. Y. (2017). A dynamic simulation model for assessing the overall impact of incentive policies on power system reliability, costs and environment. *Energy Policy*, 102, 170–188.

Kirschen, D. and Strbac, G. (2005). *Front Matter*. Wiley Online Library.

Kothari, D. P. and Nagrath, I. J. (2003). *Modern Power System Analysis*. McGraw-Hill, Europe.

Li, M., Huizhuo, W., Feifei, W., Yihe, L., Yang and Li, Z. (2013). Game optimization theory and application in distribution system expansion planning, including distributed generation. *Energies*, 6(2), 1101–1124.

Li, T. and Shahidehpour, M. (2005). Strategic bidding of transmission-constrained GENCOs with incomplete information. *IEEE Trans Power Syst*, 20(1), 437–447.

Lin, W. M., Zhan, T. S., Tsay, M. T. and Hung, W. C. (2004). *The Generation Expansion Planning of the Utility in a Deregulated Environment*.

Myerson, R. B. (2013). *Game Theory*. Harvard University Press.

Nanduri, V., Das, T. K. and Rocha, P. (2009). Generation capacity expansion in energy markets using a two-level game-theoretic model. *IEEE Trans Power Syst*, 24(3), 1165–1172.

Park, K. J. H. and Lee, K. Y. (2002). *Generation Expansion Planning in a competitive Environment Using a Genetic Algorithm*.

Shahidehpour, M., Yamin, H. and Li, Z. Y. (2002). *Market Operations in Electric Power Systems*. John Wiley & Sons, Inc., New York.

Singh, H. (1999). Introduction to game theory and its application in electric power markets. *IEEE Comput Appl Power*, 12(4), 18–20, 22.

Villar, J. and Rudnick, H. (2003). Hydrothermal market simulator using game theory: Assessment of market power. *IEEE Trans Power Syst*, 18(1), 91–98. doi: 10.1109/TPWRS.2002.807061

Von Stackelberg, H. (2010). *Market Structure and Equilibrium.* Springer.

Willems, B., Rumiantseva, I. and Weigt, H. (2007). *Cournot versus Supply Functions: What Does the Data Tell Us?* Tilburg University.

Willems, B., Rumiantseva, I. and Weigt, H. (2009). Cournot versus Supply functions: What does the data tell us? *Energy Econom,* 31(1), 38–47.

Wood, A. J. and Wollenberg, B. F. (2012). *Power Generation, Operation, and Control.* John Wiley & Sons.

Zhu, J. and Chow, M. (1997). A review of emerging techniques on generation expansion planning. *IEEE Trans Power Syst*, 12(4), 1722–1728.

Chapter 14

Fuzzy MCDM Modeling for Controlling Smog Formation: A Case of Pakistan

Yousaf Ali*, Mahboob Ul Hassan†, M. Asees Awan‡, Khawaja Muhammad Faisal Iqbal§ and Muhammad Abdul Haseeb Wahla¶

Department of Management Sciences and Humanities,
GIK Institute Topi, Swabi KPK, Pakistan
**yousafali@giki.edu.pk*
†*u2014156@giki.edu.pk*
‡*gif1729@giki.edu.pk*
§*u2014150@giki.edu.pk*
¶*u2014190@giki.edu.pk*

Smog refers to air pollution caused due to industrial and anthropogenic emissions. It increases opacity and toxicity of air. It produces a challenging situation around the world, causing increase in road accidents, crop damage, health risks, disruptions in industry cycle and halting daily life routine in many big and small cities around the globe. This chapter illustrates an extensive research and analysis of facts and data for various causes of smog around the world, with a special focus on Pakistan. Lahore, the second largest city of Pakistan, is one of the most smog-affected zones. The biggest causes of smog are agricultural fires and hazardous emissions from industries and automobiles. To avoid smog formation due to these causes, different alternative solutions are developed. A framework based on fuzzy-VIKOR MCDM approach is used to select the best alternative solution. Agricultural fire can be avoided by using mechanical and chemical means to destroy the residue. Emissions from the industry and automobiles can be reduced by using modern technology and making new regulations. In the last part of this chapter, recommendations are proposed to facilitate the government in making of strategies to counter smog and regulation of laws in this regard by checking severity of different causes.

1. Introduction

Smog is a type of air pollution formed when hazardous gases (smoke) are mixed with fog in cold and humid weather. Smog basically refers to photochemical smog that occurs due to photochemical reaction between sulfur dioxides, nitrogen dioxide and volatile organic compounds, which release tropospheric ozone as well as fine particles in the environment. Smog has severe effects on environment, i.e., air becomes dense and opaque due to formation of smog. Breathing in such air causes severe health problems. These problems include throat infection, lungs disorder, cardiovascular disease, asthma and other respiratory problems. Plants are also blighted due to smog. Due to presence of smog, new crops are damaged and their nutrition is decreased adversely. Increased opacity of air causes low visibility, which leads to road accidents. In 2012, seven million people died due to air pollution.

The first ever attack of smog was recognized in 1943 in the American city of Los Angeles. Since then, China, Brazil, Qatar and Japan have also suffered from attacks of smog. During the previous century, smog has been a serious issue in India and Pakistan. Researchers have continued to find the causes of smog. As new causes of smog were discovered, this research took many different directions and the scope of this research kept expanding. From the Pakistani perspective, both urban and rural populations contribute towards the formation of smog. Major causes of smog in Pakistan include burning of crop residue and emissions of hazardous gases from industry and vehicles. Poor legal regulations have failed to stop industries from emitting hazardous gases in open air. Vehicular emissions and deforestation play an important role in promoting smog formation. The unchecked methods of construction are adding more swing to smog formation. Agriculture sector in contributes 21% to Pakistan's GDP. Due to lack of knowledge and limited resources, still old farming methods are practiced in Pakistan. After harvesting a crop, farmers burn the crop residue (called stubble). This crop burning, called "agriculture fire," raises $PM_{2.5}$ in air in large quantity, which eventually ends up in aggravating smog capital of Pakistan. The scope of this study is to address the major causes of smog in the context of Pakistan. Different alternative solutions are studied in this research and recommendations are proposed to minimize the formation of smog. To choose an alternative solution for smog formation usually takes many factors to consider such as cost, effectiveness and feasibility. Often the criteria evaluated are determined by subjective and personal considerations. Then fuzzy method can be used along with MCDM approach to choose the best alternative by translating the linguistic variables into quantitative evaluations, i.e. VIKOR method is used along with fuzzy numbers.

The main idea of this method is to use compromise ranking method by comparing the measure of closeness to the ideal alternative to find the best solution. However, Figure 1 illustrates the whole flow diagram of how this research is conducted. It provides a study for the causes of smog in Pakistan. Also, it gives an evaluation outline to determine the main causes of smog by the use of a combined method of statistical analysis, trapezoidal fuzzy numbers and VIKOR method. Moreover, this study consists of two surveys. The

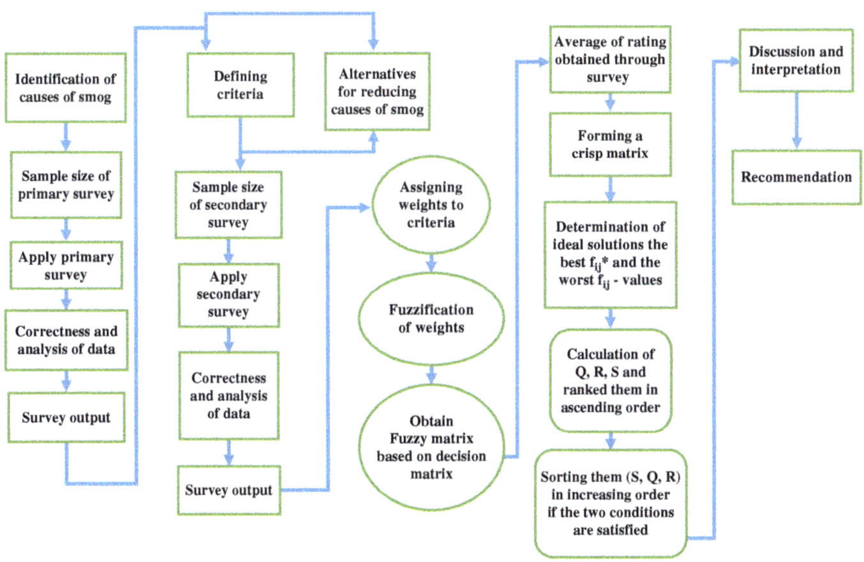

Figure 1: An illustration of complete research flow schematic.

primary survey was conducted to determine the main causes of smog in Pakistan. The secondary survey gives statistical data for determining the perception of surveyee toward the causes of smog and its effects. This paper comprises seven sections. Section 2 includes literature review about the approach used and smog, its causes and some problems that rise due to it. Section 3 gives details of the methodology. Section 4 provides results and discussion of this study. Section 5 contains the conclusions followed by Section 6 in which recommendations are proposed to counter smog.

2. Literature Review

The role of anthropogenic hydrocarbons in smog is of paramount importance. Certain hydrocarbons play a big role in reducing urban photochemical smog, which is a result of ground-level ozone. Oxidation of biogenic and anthropogenic hydrocarbons in the presence of nitrogen produces ozone, which results in smog in urban areas due to airborne particles. There is a need for a reliable plan for urban areas to decrease the amount of these hydrocarbons. Trees have a vital property of releasing hydrocarbons, which help in reducing the ozone level. Forestation can help to reduce ozone level in urban areas, hence reducing smog (Seinfeld and Pandis, 2012).

The role of automotive/vehicular emissions in the production of smog cannot be overlooked. Vehicular emission is one of the major source of pollutants that causes smog. It produces high level of NO_x, SO_x and CO_x in the air. It also produces total suspended particles (TSPs) (also called as particulate matter) (Li, 2014). According to a research conducted in Faisalabad, the third largest city of Pakistan where industry, population and high traffic volume is a problem. The study conducted over a period of 3 months gave reading of above-mentioned gases as 160–506 μg/m^3, 26–64 μg/m^3, 110–139 μg/m^3 and 5.5–18.2 mg/m^3, respectively (Niaz*et al.*, 2015). This data was collected using global positioning system (GPS) and geographic information system (GIS) parameters and compared to National Environmental Quality Standards (NEQS) and United States Environmental Protection Agency (USEPA) defined permitted limits.

It was recommended to adopt integrated multipollutant control strategy to counter them.

Another significant contribution in smog formation is from heavy traffic volumes. They have enormous effects on air quality. A study was conducted to measure air quality along the N-5 section of National Highway of Pakistan by National Highway Authority, Pakistan. The points were monitored for 72 h each, showing an increase in pollutant concentrations with increase in traffic volumes, and decrease in traffic flow rate (Ali and Athar, 2008).

Agriculture fires play an important part in the formation of smog, especially in South Asia. An assessment was carried out in the Punjab province of Pakistan and neighboring India. Mosaicked and processed images over a period of time were taken by MODIS (Bosetti and Toscano, 2016). These images depict that an increase in the burning of agriculture waste causes increase in the density of smog over the fog-affected area of Pakistan covering 269–380 km^2. The presence of coal-based power plant in Pakistan and India nearby is also one of the cause of smog. A reduction in mixing heights of pollutants is also noticed.

China is one of the noticeable contributor to greenhouse capital of the world. In 2013, David Hu and Juyuan Jiang studied major sources of smog and pollutants in major cities of China. They indicated that particles formed from SO_x, NO_x, NH_3, etc. have a very strong impact on human health. This chapter, although does not cover solutions to control these sources, it compares the strategies adopted in China and California since the 1940s to control the smog and pollutants (Hu, 2013).

The effects of smog on living things include both on human beings and on plants/crops. In case of crops and plants, it is found in research that smog has its own negative effects on leaves of plants, called foliage injury. The effects are severe on many of the important cash crops, including edible vegetation like spinach, lettuce and lucerne. Losses due to foliage injury were estimated as $479.5 million in the 1949 smog of Los Angeles. Humans are considered most susceptible to toxicity of smog. About 24,000 mortalities were reported during the smog episode in Germany alone in 1985 (Awasthi, 2011).

During this period, ambient air concentrations recorded were high. Similarly morbidity rate was also higher during this period.

$PM_{2.5}$ being major constituent of smog has been researched a number of times. Standard $PM_{2.5}$ concentrations set by United States National Ambient Air Quality Standard is 15 $\mu g/m^3$, but this concentration is found exceptionally high in smog (Wang and Hao, 2012). For instance, samples collected by researchers from Seoul, Korea, during a period of two years between 2003 and 2005 showed three times higher concentrations of $PM_{2.5}$ from standard value, around 43 $\mu g/m^3$. These samples were collected from ambient air and yellow sand. These concentrations were also found to be much higher than those calculated in nonsmog days. In particular, the largest constituents of these samples of $PM_{2.5}$ were found to be sulfate, nitrate and ammonium .Moreover, fuzzy-VIKOR approach is a popular technique, used to choose among multiple alternatives with analysis under given limitations.

Fuzzy is a weight assigning method for linguistic variables whereas VIKOR is an approach to find feasible alternatives based on given criteria. These two techniques are combined to form an MCDM framework to solve complex problems on several events. Fuzzy VIKOR is based on selection and ranking of alternatives against certain criteria; however, ranking is performed on the basis of closeness to the ideal alternative (Ali *et al.*, 2018). It has been used in evaluation of different services like those in hospitals by applying fuzzy technique on uncertain and vague linguistic variables and VIKOR approach is used to analyze performance ratings to alternatives. It helps decision-makers to use linguistic terms by converting them into fuzzy numbers. Fuzzy handles imprecise and uncertain variables through fuzzy set values to develop fuzzy-VIKOR algorithm. As a result, it gives ranking of alternatives. Despite conflicting criteria, it gives a compromising solution. An example is the application of fuzzy-VIKOR for the selection of optimum renewable energy for energy sector in case of Pakistan (Ishfaq *et al.*, 2018).

The best renewable energy alternative for Istanbul using VIKOR AHP was investigated by Kaya and Kahraman (2010) who conclude that wing energy is most suitable option in case of Istanbul. Material selection for automotive components is a critical task, keeping in view the cost, strength and other performance parameters, the best and optimum solution is desired. This selection can be done using fuzzy-VIKOR approach when the decision-maker provides input in the form of linguistic variables. This study of four materials in the mentioned framework was carried out in 2012 in which environmental aspects were given primary importance during assessment and results showed that polypropylene could be a more feasible alternative for making of instrument panels of electric cars (Girubha and Vinodh, 2012). This study also developed a rational method for selection of materials keeping in view the application requirements. Here, VIKOR technique incorporates all the constraints, such as environmental impact, cost and other performance parameters of materials to provide a compromising and best-fit solution.

3. Methodology

3.1. *Data collection*

This research is conducted in various steps. Information is gathered by four different surveys. Primarily, a survey was conducted among different environmentalists, medical officers, engineers, chemists and material scientists. This survey helped to choose the three most critical causes in terms of contribution toward smog. Second, four alternatives were considered for each cause after a careful study and observation of different research articles, reports and journals. Each alternative represents the best way to control the contribution of respective cause toward smog. These alternatives were analyzed by various criteria proposed by decision makers. Three surveys were formed, one for each cause. These surveys were filled by people of diverse educational and professional backgrounds. A total of 335 responses were recorded for these surveys. In these surveys, each alternative is rated against each criterion.

3.2. *Weights for criteria*

Important weights for criteria are decided using the fuzzy technique. Trapezoidal fuzzy is developed to accommodate uncertain, vague, unspecific and imprecise situations. Zadeh developed and used these

Table 1: Linguistic terms of weights for criteria and alternatives.

Linguistic variables	Abbreviations	Values
Very Important	VI	(0.93,0.98,0.98.1)
Important	I	(0.72,0.78,0.92,0.97)
Medium Important	MI	(0.32,0.41,0.58,0.65)
Less Important	LI	(0.17,0.22,0.36,0.42)
Not Important	NI	(0.04,0.1,0.18,0.23)

fuzzy sets for the first time. Later Bellman and Zadeh used fuzzy along with other MCDM approaches. The same set of fuzzy values are used here as well. Four environmentalists and industrialists are decision makers and assign weights from given linguistic variables table. Then averages are taken to decide the weight of each criteria. Linguistic variables used are given in Table 1.

After taking average, weights are assigned to each criterion. There are total n criteria and m alternatives in each source, supposing W_j is the weight of jth criterion.

3.3. *VIKOR method*

Step 1: After decision of the weights, for each alternative, rating of different criteria is obtained by averaging the ratings values from surveys. There is no need to use fuzzy method for rating of alternatives on the basis of all criteria because values taken in survey are not range values, rather these values are compact values. The next step is to build Crisp matrix (C) using these rating values and weights of criteria. The maximum and minimum values of each column are given in last two rows respectively. There will be m rows and n columns.

$$C = \begin{bmatrix} E_{11} & E_{12} & E_{13} & \cdots & E_{1n} \\ E_{21} & E_{22} & E_{23} & \cdots & E_{2n} \\ E_{31} & E_{32} & E_{33} & \cdots & E_{3n} \\ \vdots & \vdots & \vdots & \ddots & \vdots \\ E_{m1} & E_{m2} & E_{m3} & \cdots & E_{mn} \end{bmatrix}. \quad (1)$$

Step 2: Calculate maximum and minimum values in each column.

$$\text{MAX}_j = \text{Maximum value in } j\text{th column}, \tag{2}$$

$$\text{MIN}_j = \text{Minimum value in } j\text{th column}. \tag{3}$$

Step 3: Calculate S, R and Q values using equations (4)–(6).

$$Si = \sum_{j=1}^{n} \frac{w_j(\text{MAX}_j - E_{ij})}{(\text{MAX}_j - \text{MIN}_j)}. \tag{4}$$

where i is the number of rows, j the number of columns, w_j the weight of jth criteria, MAX_j the maximum value in jth column and MIN_j the minimum value in jth column and E_{ij} the element in ith row and jth column.

$$R_i = \max_j \left(\frac{w_j(\text{MAX}_j - E_{ij})}{(\text{MAX}_j - \text{MIN}_j)} \right). \tag{5}$$

Then maximum values of S and R are calculated, $S_{\max}$ = maximum value of S, $S_{\min}$ = minimum value of S, $R_{\max}$ = maximum value of R and $R_{\max}$ = minimum value of R.

$$Q_i = \frac{v(S_i - S_{\min})}{(S_{\max} - S_{\min})} + \frac{(1-v)(R_i - R_{\min})}{(R_{\max} - R_{\min})}, \tag{6}$$

where

$$v \begin{cases} > 0.5 & \text{for voting by majority} \\ = 0.5 & \text{for consensus} \\ < 0.5 & \text{for veto} \end{cases} \tag{7}$$

Step 4: In this step, alternatives are ranked on the basis of sorting with respect to S, R and Q values in the ascending order.

Step 5: In this step 2, conditions are checked.

$$C_1 : (A^{(2)}) - Q(A^{(1)}) > \left(\frac{1}{m-1} \right). \tag{8}$$

C_2: The alternative ranked best by Q is also ranked best by S and R sorting.

Step 6: Depending upon the conditions, the results are inferred. If both the conditions are satisfied, then the alternative with the lowest value of Q is selected. If condition 1 is not satisfied, all the alternatives are the same comparative solution. And if Condition 2 is not satisfied, then stability in decision making is deficient and $A^{(1)}$ and $A^{(2)}$ has the same compromise solution.

4. Results and Discussion

This chapter aims to investigate various factors and its impact on smog generation. Various factors including vehicular emissions, agricultural fires and industrial emissions are evaluated against certain criteria, i.e., cost, complexity, effect on smog reduction and positive impact on health. Fuzzy VIKOR is used to select the most appropriate solution to reduce the production of smog in Pakistan. The detailed discussions of the results obtained are discussed below.

4.1. *Weights for criteria*

The objective of this research is to find alternative solutions to major causes of smog formation in Pakistan. It is a rational approach to consider cost analysis as a priority before initiating any project. It establishes a baseline to cover all expenses involved in a project to ensure that it is in accordance with the available budget. Pakistan has many critical issues like energy crisis and terrorism. A major share of resources is allocated to the mentioned issues. Thus with limited resources and many other problems, cost is an important factor while choosing the best alternative.

The term "complexity" refers to the difficulties and complexities involved during implementation of a particular alternative. Even a very economical and efficient alternative will be useless without practical insight. It is necessary to determine what factors make a project complex and impractical before its initiation. The different factors that have been set by decision-makers for complexity include the availability of requirements for an alternative and feasibility of alternatives.

Every initiative that will be taken by the government is desired to be beneficial for human life. The effect of project implementation on human health is a major concern for everyone. Health is one of the most important aspects of human life that leads to the selection of impact on health as a criterion. The implementation of the proposed alternative must not be hazardous for human health. This criterion is proposed to determine the positive impact of project on human health after its implementation. Every alternative was rated from 1 to 9 on the basis of positive effect on human health.

For assignment of weights, fuzzy technique was used. Detailed calculations show that cost 0.67, complexity 0.34, positive effect on health 0.65 and effectiveness in reducing smog is given a weight of 0.97.

4.2. *Vehicular emission*

Vehicle emission is one of the most dangerous source of air pollution, which ultimately causes the formation of smog. Many constituents emitted from vehicles are directly or indirectly cause the formation of smog. Particulate matter (PM), which is about 1/10 of the diameter of a human hair, is the cause of murky color of smog. These can penetrate human lungs easily and cause serious threats to health. NO_x emitted from vehicles causes lung issues and damages the immune system against lung diseases like influenza and pneumonia (Union of Concerned Scientists, 2014). Carbon monoxide (CO) is colorless and odorless, yet very poisonous gas. It is emitted primarily from cars and trucks. When this gas is inhaled, it stops oxygen supply to the heart and brain. Greenhouse gases such as carbon dioxide also contribute to environment pollution. According to Environmental Protection Agency (EPA), by burning of a single gallon of gasoline and diesel, 8887 and 10,180 grams of CO_2 are produced respectively (Environment Protection Agency, 2017). Another very dangerous pollutant from vehicles is SO_2. It not only affects the health of young people and asthmatics but also forms sulfuric acid by mixing with water in air. Thus, it causes acid rain and also contribute toward the formation of smog. Therefore, in order to reduce the harms of emissions from automobiles, four alternatives are the implementation of EURO II standards, use of electric vehicles, betterment of public transport and vehicle inspection standards, which are represented as V_1, V_2, V_3 and V_4, respectively (see Appendix A).

Euro II standards define an acceptable limit of exhaust emissions of vehicles. There are other standards as well like Euro 1 to Euro 6, but Euro 2 is specifically for passenger cars and motor cycles. Implementation of these standards not only controls the emission into environment but makes the vehicles fuel efficient as well (European Parliament, Council of the European Union, 2002). The second alternative is the use of electric vehicles. Electric vehicles are the category of vehicles that use electric motors for propulsion. Electric motors may be powered by an electric system by collector or sometimes electric vehicles are self-contained with a solar panel, battery or generation system. Most of the emitted pollutants from vehicles are due to combustion of gasoline or diesel inside the engine. Use of electric vehicles completely eliminates the exhaust pollution problems. The third alternative is betterment of public transport. According to EPA, 75% of road pollution is due to personal cars (Brinson, 2016). The number of personal cars in Pakistan is much more than public transport vehicles. In the last 15 years, the number of personal vehicles have increased 327% and the number of public transport vehicles have increased 167% (GALLUP Pakistan, 2016). This increased number of personal vehicles is a major cause of excessive exhaust emission into the environment. Therefore this alternative is considered to encourage public to travel in public transport rather than personal cars. The last alternative is improvement in vehicle inspection regulations in Pakistan. Table 2 shows S, R and Q values for ranking of alternatives regarding vehicular emissions.

These alternatives, according to the ratings shown in Table 2, are ranked using the VIKOR method. Detailed calculations show that alternative "use of electric vehicles" is best ranked. Electric vehicles are hundred percent eco-friendly because they have zero emissions. There are other advantages of using electric vehicles, for example, low noise, low maintenance required, easy to drive, no negative effect on human health and most importantly due to zero emissions, it is

Table 2: Showing S, R and Q values.

S	R	Q
2.16	0.97	1
0.6	0.38	0
0.97	0.88	0.54
1.46	0.67	0.52

the most effective alternative in reducing smog. In Pakistan, more vehicular emission is due to unfit vehicles that emit more pollutants than a normal vehicle. It happens due to several causes that include incomplete combustion of fuel inside engine, knocking or problems in exhaust systems. Vehicles are seen on road without a fitness and inspection certificate, which not only pollutes the environment but threats the life of passengers as well.

4.3. *Agriculture fire*

Agribusiness is the backbone of Pakistan's economy. Agriculture accounts for 19.5% of the GDP, utilizing 42.3% of the labor and giving raw material for several other businesses. According to Pakistan economic survey, during 2016–2017, 3.46% growth in agriculture is observed in Pakistan. The record of previous decade shows that this growth percentage is increasing over the last decade (Ministry of Finance, Govt. of Pakistan, 2017). In agricultural fields, after harvesting a crop, crop residue is left behind. Crop residue is a major problem for farmers in Pakistan because after harvesting the crop, it is very difficult to handle the residue. With the limited resources, Pakistani farmers prefer to burn the residue, but it affects the environment badly. Rice crop is harvested in the month of November usually. Therefore it is the major crop whose smoke causes the formation of smog during the months of November and December in Pakistan. By burning the residue of rice, black carbon is generated, which is considered the second largest participator to the global warming after CO_2. In Asia, ABCs (atmospheric black clouds) are formed because of emission of black carbon. A 10–15% reduction in sunlight reaching earth is observed and atmospheric solar heating is increased by 50% because of ABCs. In Punjab (a province of Pakistan), after harvesting a crop, a huge amount of crop residue is burnt to prepare the land for the second crop. After the rice crop, 80% of wheat crop is grown in land on which rice residue is present. So in order to prepare this land for wheat crop, the rice residue is to be incorporated into soil or removed (Ahmed, 2013). Therefore, burning of residue must be replaced by some sophisticated method of handling it.

Four alternatives proposed and analyzed in this research are: spreading the straw as wide as possible, digging the residue into earth, use of chemicals and use of machinery to get rid of crop residue. Table 3 shows S, R and Q values for ranking of alternatives for agricultural wastes.

The results show that third and fourth alternatives, i.e., use of chemicals and use of machinery are equally competitive. Chemicals would decompose the residue and machines would eradicate it. Both these can be applied depending upon the availability, area of practice, labor costs and time required to practice. Chemicals would decompose the residue and machines would eradicate it. According to the first alternative, residue is spread widely in fields and let to decompose by itself. It is a quiet and easy process and has certain advantages such as leaving the land uncultivated for a season, which increase its fertility, not much labor is required to practice this alternative and it does not have any negative effect on human health. On the other hand, it has a big disadvantage, the land on which the residue is spread cannot be used for cultivation unless it is decomposed completely. The second alternative is digging the residue into earth so that it may be decomposed itself. This is rather a difficult

Table 3: S, R and Q values.

S	R	Q
1.62	0.97	1
1.39	0.61	0.39
1.17	0.54	0.13
1.01	0.67	0.15

process and decomposition takes some time so cultivation cannot be done immediately. In the third alternative, some chemicals are used to decompose the residue. This is rather an easy and quick process and land can be cultivated immediately after decomposition. The fourth alternative is to use machinery to uproot the residue. This alternative is rather easy to perform but a bit expensive.

4.4. *Industrial emission*

Pollutants emitted from industrial processes directly or indirectly cause the formation of smog. In order to reduce the emissions from industry, alternatives include the use of low sulfur fuels, greening of industrial process, releasing the pollutants higher in air and defining the new standards and laws to control industrial emissions. Studies as discussed in Section 3 lead to an inference that $PM_{2.5}$ particles are major constituent of air pollution or smog. Researchers conclude that a major proportion of $PM_{2.5}$ consists of sulfate molecules, which stand high in toxicity. Mostly industries are main contributors of sulfur to the environment. In Pakistan leniency of laws help to encourage people to use fuels that are enriched in sulfates and nitrates. These kind of fuels are easily available at low cost. People in both rural and urban areas burn all sort of waste to energize their respective industries. Residual fuels are highly enriched with pollutants especially sulfur, which is the main reason for the low cost of these fuels. Waste such as tires, which have high amounts of sulfur, are easily burnt in industries to keep the fire, since they are easily available. Although the use of these fuels results in degradation of performance of some equipment but ignorance of people toward environment and more importance given to cost of fuel aggravates the use fuels enriched in sulfur. Although most of the countries in the West have partially succeeded in banning high sulfur-content fuels, irresponsibility of authorities and the people is a hurdle toward the use of low-sulfur fuels in Pakistan. The survey results also show the same result, with complexity of implementation given more weight because of leniency of laws in Pakistan. Cost is also marked with high points, which confirms what is mentioned above, that people are still not willing to pay a small amount of money to make a bigger difference.

Greening of industrial process is important for sustainable growth. More environmental friendly industries have more promising future. Much or the world's larger industries are heading toward more eco-friendly strategies and products. Pakistan also requires this "environmental awakening" in its industrial regime. The cost of damage due to pollution in Pakistan is already enormous.

Greening of industrial processes include proper treatment of waste in plants, imposition of stricter environmental quality standards and cleaner production centers. Table 4 shows S, R and Q values for ranking of alternatives regarding industrial emissions.

According to the detailed calculations, VIKOR method ranks greening of industrial process as the best alternative. Greening of industrial process is a very hot topic in modern times. A lot of work is being done to convert the industrial process to environment-friendly process with least emissions. In Pakistan's case, all the steps mentioned are absent from industry. Most of the industries don't pay heed to these steps. If given a chance, these could be simpler and cost-effective than any other alternative, which is also evident from survey results. This is because of the fact that greening of industrial processes does not require a whole new effort and plan to implement, instead only refinement of processes already being used is needed. The third alternative is to make the chimneys very high so that pollutants are released very high in the air and thus have minimal effect on human health. But this alternative have least effect in the reduction of smog. The last alternative is to define standards for industrial processes, which set a threshold of pollutants being released into the environment.

Table 4: S, R and Q values.

S	R	Q
1.17	0.66	0.45
0.41	0.34	0
1.26	0.88	0.65
2.34	0.97	1

5. Conclusion

Smog is a major challenge faced by many countries around the world. Smog also makes it very difficult for animals to live in such a toxic condition. This can lead to extensive killing of animals and their species. The government should take responsibility of these animals by providing them a useful tool for living. Moreover, smog remains a challenge in many races of life in Pakistan proving to be a catastrophe in industrial, health and transportation sector. In Pakistan, major causes of smog are burning of residue of crops and emission of hazardous gases from industries and automobiles. Finding the alternatives to these problems is the need of the hour. Burning of crops residue can be stopped by adopting some alternative methods to get rid of the residue. Possible alternatives for this problem are use of chemicals, use of machinery, digging the residue in the land and spreading the straws as wide as possible. The proposed methodology suggests that best alternative to this problem is the use of chemicals to get rid of crop residues.

Furthermore, the results obtained via fuzzy-VIKOR evaluation suggests that emissions from industry can by controlled by modifying the industrial processes and introducing environment-friendly processes referred to as "green industrial processes." This modification might be expensive, but it would have a huge impact in reducing environmental pollution and ultimately reducing smog formation. Use of renewable energy has been encouraged around the world because of the economic factor and its environmental-friendly nature. Fossil fuels are being used in Pakistan for electricity generation. Pakistan government needs to promote this form of energy for the environment. All these products that can lead to environmental damage in any way should be completely banned such as organic compounds and plastics. These products emit PM that mix in sunlight to form ground-level ozone. Usage of environmental0friendly products should preferred like paints, papers, and sprays that reduce hazard in our surrounding.

All three mentioned fuels are expensive and hazardous for the environment. Similarly, the best alternative estimated using fuzzy-VIKOR for vehicle emissions is the use of electric vehicles. These electric vehicles have zero emissions. Therefore, using such vehicles can reduce smog formation to a great extent in Pakistani perspective. The only way that can lead us in the reduction of smog is by managing gaseous emissions from buses, cars and industries. The government should step up and implement different rules and regulations made by international bodies to regulate the quantity of emissions. According to the deputy director of the environmental department, this year's annual cost for monitoring air quality system is Rs. 540 million. Use of renewable energy should be encouraged in both industries and vehicles. Finally, duties on electric vehicles and low sulfur fuels should be reduced to promote them.

6. Recommendations

Smog, as discussed earlier, a mixture of many air pollutants, which put human health as well as animals and its surroundings, at risk. If we look at some recent cases, we came to know that animals are exposed to smog pollutant particles via inbreathing of many pollutant particles, which are present in air, and similarly through the digestion of food and water. Industrial and vehicular emissions are playing a major role toward smog formation. Furthermore, the environmental department can install smog monitoring devices that continuously monitors the quantity of $PM_{2.5}$ and other air pollutants in the atmosphere. Therefore, it can reduce the number of accidents and health issues that occur due to smog. The monitoring system for air quality entails more cost than the whole budget of the department's monitoring system. There are some general precautions that should be adopted by every citizen of Pakistan for their personal safety during the smog season.

- Face masks must be used.
- Excessive use of drinking water.
- Try to limit your activities indoor and avoid unnecessary exposure to the environment.
- Wash hands, face and all other body parts after their exposure to outside environment.
- Drive in low speeds and use fog lights.

- Different programs to plant trees must be initiated in order to preserve environment.

References

Ahmed, T. (2013). *Burning Crop Waste in Punjab, Pakistan — Can farmers do better for themselves and the environment?* SANDEE, Kathmandu, Nepal.

Ali, M. and Athar, M. (2008). Air pollution due to traffic, air quality monitoring along three sections of National Highway N-5, Pakistan. *Environ Monitor Assess*, 136(1–3), 219–226.

Ali, Y. *et al.* (2018). A VIKOR based approach for assessing the social, environmental and economic effects of "smog" on human health. *Sci Total Environ.*

Awasthi, A. (2011). *Study of Human Respiratory Functions in Relation to Crop Residue Burning.* Doctoral dissertation.

Aydin, N. (2017). A fuzzy-based multi-dimensional and multi-period service quality evaluation outline for rail transit systems. *Transp Policy*, 55, 87–98.

Biello, D. (2007). *Blighting Plants, Choking Lungs, Warming World Among Smog's Effects.* [Online] Available at: https://www.scientificamerican.com/article/blighting-plants-choking-lungs-warming-world-among-smog-effects/ [Accessed 25 February 2018].

Bosetti, G. and Toscano, R. (2016). *Living Together*, Antpolitics.

Brinson, L. C. (2016). *How much pollution comes from cars?.* [Online] Available at: https://auto.howstuffworks.com/air-pollution-from-cars.htm [Accessed 20 December 2017].

Chameides, W. L., Lindsay, R. W., Richardson, J. and Kiang, C. S. (1988). The role of biogenic hydrocarbons in urban photochemical smog: Atlanta as a case study. *Science*, 241(4872), 1473–1475.

David, H. and Juyuan, J. (2013). A study of smog issues and PM2.5 pollutant control. *J Environ Protec*, IV(7), 746–752.

Environment Protection Agency (2017). *Greenhouse Gas Emissions from a Typical Passenger Vehicle.* [Online] Available at: https://www.epa.gov/greenvehicles/greenhouse-gas-emissions-typical-passenger-vehicle [Accessed 20 December 2017].

European Parliament, Council of the European Union (2002). *Euro-lex, Access to European union law.* [Online] Available at: http://eur-lex.europa.eu/legal-content/EN/ALL/?uri=CELEX:32002L0051 [Accessed 20 December 2017].

GALLUP Pakistan (2016). *Short Roundup on Transport Infrastructure in Pakistan*: Gallup Pakistan, Islamabad.

Ghauri, B. and Lodhi, A. (2013). *Assessment of Area Covered by Smoke, Haze, Fog In Pakistan and India Using Satellite Data.* IEEE, Istanbul, pp. 225–229.

Girubha, R. and Vinodh, S. (2012). Application of fuzzy VIKOR and environmental impact analysis for material selection of an automotive component. *Mater Design*, 37, 478–486.

Goyal, S. (2016). *5 Most Polluting Countries in the World Affected by Smog.* [Online] Available at: https://www.jagranjosh.com/general-knowledge/5-mg

822ost-polluting-countries-in-the-world-affected-by-smog-1479281464-1 [Accessed 25 02 2018].

Haagen-Smit, A. J. (1952). Chemistry and Physiology of Los Angeles smog. *Industr Eng Chem*, June, 1342–1346.

Han, Y.-J., Kim, T.-S. and Kim, H. (2008). Ionic constituents and source analysis of PM2.5 in three Korean cities. *Atmosph Environ*, 42(19), 4735–4746.

Hu, D. A. J. J. (2013). A study of smog issues and PM2. 5 pollutant control strategies in China. *J Environ Protect*, 4(07), 746.

Ishfaq, S., Ali, S. and Ali, Y. (2018). Selection of optimum renewable energy source for energy sector in Pakistan by using MCDM approach. *Process Integr Optimiz Sustain*, 2(1), 61–71.

Jalil, X. (2017). *Is Punjab Ready to Tackle Smog this Year?* [Online] Available at: https://www.dawn.com/news/1363737/is-punjab-ready-to-tackle-smog-this-year [Accessed 14 10 2017].

Kaya, T. and Kahraman, C. (2010). Multicriteria renewable energy planning using an integrated fuzzy VIKOR & AHP methodology: The case of Istanbul. *Energy*, 35(6), 2517–2527.

Kima, H.-S. *et al.* (2007). Characteristics of the major chemical constituents of PM2.5 and smog events in Seoul, Korea in 2003 and 2004. *Atmosph Environ*, 41(32), 6762–6770.

Li, L. (2014). *PM2. 5 Air Pollution in China: A Technical and Administrative Analysis of Standards.* Doctoral dissertation, Tulane University, Payson Center for International Development.

Mardani, A. *et al.* (2016). VIKOR technique: A systematic review of the state of the art literature on methodologies and applications. *Sustainability*, 04, 5.

Middleton, J. T., Kendrick, J. B. and Schwalm, H. W. (1950). Injury to herbaceous plants by smog or air pollution. *Plant Disease Report*, 34(9), 245–252.

Ministry of Finance, Govt. of Pakistan (2017). *Pakistan Economic Survey 2016–17.* [Online] Available at: http://www.finance.gov.pk/survey_1617.html [Accessed 22 December 2017].

Musani, S. and Jemain, A. A. (2015). Ranking schools' academic performance using a fuzzy. *J Phys*, 622.

Niaz, Y., Jiti, Z. and Zhang, Y. (2015). Influence of automotive emission on air pollution using GIS in. *Int J Agri Biol Eng*, 8(1), 111–116.

Niaz, Y., Jiti, Z. and Zhang, Y. (2015). Influence of automotive emission on air pollution using GIS in Faisalabad, Pakistan. *Int J Agri Biol Eng*, 8(1), 111–116.

Opricovic, S. (2011). Fuzzy VIKOR with an application to water resources planning. *Expert Syst Appl*, 38(10), 12983–12990.

Prakash, C. (2016). *Animal Bear Brunt of Dehli Smog.* [Online] Available at: http://www.dailypioneer.com/todays-newspaper/animals-bear-brunt-of-delhi-smog.html [Accessed 13 12 2017].

Seinfeld, J. and Pandis, S. (2012). *Atmospheric Chemistry and Physics: from Air Pollution to Climate Change.* John Wiley & Sons.

Shemshadi, A., Shirazi, H., Toreihi, M. and Tarokh, M. (2011). A fuzzy VIKOR method for supplier selection based on entropy measure for objective weighting. *Expert Syst Appl*, 38(10), 12160–12167.

Tsai, H. and Lu, I. (2006). The evaluation of service quality using generalized Choquet. *Inform Sci*, 176(6), 640–663.
Tsung-HanChang (2014). Fuzzy VIKOR method: A case study of the hospital service evaluation in Taiwan. *Inform Sci*, 271, 196–212.
Union of Concerned Scientists, 2014. *Cars, Trucks, and Air Pollution.* [Online] Available at: https://www.ucsusa.org/clean-vehicles/vehicles-air-pollution-and-human-health/cars-trucks-air-pollution#.Wjnnq9-WbIV [Accessed 20 December 2017].
Wang, S. and Hao, J. (2012). Air quality management in China: Issues, challenges, and options. *J Environ Sci*, 24(1), 2–13.
Wichmann, H. E. *et al.* (1989). Health effects during a smog episode in West Germany in 1985. *Environ Health Persp*, 79, 89–99.
World Health Organization (2014). *7 Million Deaths Linked to Air Pollution Annually.* Department of Public Health, Environmental and Social Determinants of Health (PHE), Geneva.

Appendix A: Symbols

Symbol	Definition
C_1	Cost
C_2	Complexity
C_3	Positive impact on human health
C_4	Effect on smog reduction
A_1	Spreading straw as wide as possible
A_2	Digging the residue deep into the earth
A_3	Use of chemicals to reduce effects of stubble
A_4	Use of heavy machinery to remove stubble
I_1	Use of low-sulfur fuels
I_2	Greening of industrial process
I_3	Releasing pollutants high in the air
I_4	Environmental standards
V_1	Implementation of Euro-II standards
V_2	Use of electric vehicles
V_3	Betterment of public transport
V_4	Vehicular inspections
W_1	Weight for criteria one
W_2	Weight for criteria two
W_3	Weight for criteria three
W_4	Weight for criteria four

Chapter 15

Application of MCDM Technique for the Selection of Biodiesel Fuel Blend for Sustainable Motorization

Surya Kanth*, Sumita Debbarma*, Jagadish† and Biplab Das*,‡

*Department of Mechanical Engineering,
National Institute of Technology Silchar,
Assam 788010, India
†Department of Mechanical Engineering,
National Institute of Technology Raipur,
Chattisgarh 492010, India
‡Centre for Sustainable Technologies, Ulster University,
Northern Ireland, UK

Energy is vital for the world economy. The expanding need for energy because of industrialization, population explosion and the increase in ordinary comforts has prompted a huge utilization of petroleum derivatives. Biodiesel seems to be a respectable option among petroleum-based fuels. In any case, the suitability of biodiesel is the main criterion for figuring out it as a substitute fuel for internal combustion (IC) engines. Affordable biodiesel with a proper combination for the IC engine assumes significance due to emissions from petroleum-based fuels. The intention of this research is to choose the best suitable fuel blends using analytical hierarchical process (AHP) and multiobjective optimization on simple ratio analysis (MOOSRA) methods. The AHP method is applied to discover the relative weights of the standards, even as a MOOSRA has been applied to distinguish the first-class alternatives. Experimental results from HC, NO_x, CO, EGT, CO_2 and BTE have been considered because of the assessment standards. From the results, it is seen that B20 is the best combination to be used in IC engines.

1. Introduction

Rapid industrialization and increasing automobiles around the globe has led to a gradual decline in the natural resources of conventional fuels. This shortage of fossil fuel linked with ever increasing prices, and the environmental pollution and global warming due to excess usage of fuel has prompted a search for alternatives of petro fuels. Currently researchers are engaged in efforts to substitute conventional diesel fuel with a suitable alternative fuel, i.e., biodiesel (Gerpen, 2005). In this situation more focus has been drawn toward the production of biodiesel in the pure form or as a mixture with fossil fuel in order to increase the efficiency of the unmodified engine. Biodiesel can be derived from renewable energy sources such as vegetable oils, waste cooking oils and animal fats, which can be used directly in automobile applications (Zheng *et al.*, 2006). Moreover, biodiesel is biodegradable and nontoxic; it possesses low emission profile and is a renewable fuel that can be used in automobile applications without any engine modification. Biodiesel has a higher calorific value slightly closer to conventional diesel fuel. Biodiesel contains no sulfur, almost no aromatics and 10–12% oxygen (Alptekin and Canakci, 2008).

Some of the recent studies investigated and reported improvement in performance characteristics, specifically thermal efficiency of a biodiesel-fueled engine at constant speed during different loading conditions. With the use of biodiesel, the fuel is more combusted, thereby enabling a considerable reduction in harmful emissions like particulate matter (PM), CO2, CO, HC, sulfur oxides, but more NO_x emission (Sinha and Agarwal, 2005; Usta, 2005). Most of the research works have been done on edible oils such as saffola, soybean, sunflower and palm in the Western countries for the production of biodiesel. In the South Asia, numerous varieties of nonedible oils like rice-bran, karanja and jatropha linseed are available for the production of biodiesel. Some of the oils have already been adopted as a direct substitute or as a blend with diesel fuel (Raheman and Ghadge, 2007). Experimental investigation has been carried out on Kirloskar DI diesel engine by substituting the fossil fuel with nonedible oils, which showed lower break thermal efficiency as compared to conventional fuel (Srinivasa Rao and Gopalakrishnan, 1983). A few researchers have used custard apple seed as a biodiesel to study the performance, combustion and emission parameters of diesel engine with conventional diesel (D100), pure custard apple biodiesel (CA100), and their blends, CA20, CA40, CA60, CA80 and CA100 (Mahalingappa *et al.*, 2014). The addition of small percentage of vegetable oils to conventional fuel was found to reduce viscosity, which falls below the ASTM limits. In the transportation sector, some of the governments have decided to replace the 20–30% petroleum product-based fuels with biofuels. With these developments, biodiesel sources can be adopted simultaneously and availability of custard apple oil also serves this purpose. So, in this chapter, an attempt has been made to use custard apple as a biodiesel to evaluate the performance and emission characteristics. The engine was run with conventional diesel and biodiesel blends at a constant speed of 1,500 rpm at different load conditions. Choosing the optimal blend is a quite difficult task to run the internal combustion (IC) engine. Mahalingappa *et al.* (2014) established that up to 80% diesel in the blends of custard apple biodiesel and diesel provides maximum thermal performance of the engine, while the higher blends of biodiesel has the disadvantage of fuel properties like higher density, and viscosity. In this scenario, to overcome the drawback of the existing research work, all the performance and emission characteristics are considered. In this study, analytical hierarchical process (AHP) coupled with multiobjective optimization on the basis of simple ratio analysis (MOOSRA) is proposed as the hybrid MCDM method for evaluating and choosing the optimal blend for IC engines.

MCDM methods have been implemented in various types of decision methodologies such as business, science, industry, engineering and government sectors. Even a small variation in the existing methods leads to a new research method; in addition, it carries its own strengths and weaknesses. The application of MCDM for suitable selection of an optimal blend has more accuracy and simplicity in experimental analysis of the diesel engine with different injection timings (Sivaraja and Sakthivel, 2017). Poh *et al.* (1999) reported a comprehensive study of alternative fuel on land transportation in Singapore. The number of fuel options for their possible future use can be identified by multiple attribute analysis. The AHP method helps to choose best suitable one for the decision makers depending on the objective and understanding of the problem (Saaty, 1981). The fuzzy AHP shows that with the help technical and financial information, evaluation and selection of industrial projects can be done before investment (Mahmoodzadeh et*al.*, 2007). Two fuzzy models are considered, FAHP-TOPSIS and FAHP-VIKOR, to overcome uncertainty and vagueness during the pairwise comparison to select the best blend (Sakthivel *et al.*, 2013). Gadakh (2010) studied the parametric optimization of milling process by the application of MOORA to solve the multiple criteria. For the manufacturing systems, required objectives can be achieved by the selection of green cutting fluid with the help MOOSRA technique (Jagadish and Ray, 2014). Sarkar *et al.* (2015) reported that MOORA and MOOSRA methods can identify the best nontraditional machine by developing an effective decision support system. Pertinent literature reveals that no specific study has been conducted on the selection of an optimal blend based on the performance and emission parameters using the MCDM method. Therefore, in this chapter to select the optimal blend using hybrid MCDM method, AHP coupled with MOOSRA is used to achieve the maximum performance and least emission parameters. This method gives a straightforward and convenient solution, and is simple for the pairwise comparison, in addition to displaying consistency in evaluation based on alternatives and criteria. The major advantage is the use of beneficial criteria (BC) or nonbeneficial criteria (NBC) and so the proposed MCDM method can be adopted for choosing the right blend of biofuel for combustion engines.

2. Proposed Method (Modified AHP–MOOSRA)

The proposed hybrid method contains five steps to evaluate the optimal blend. In the first step, experimental analysis was carried out in diesel engine to get the decision matrix, which has both BC (performance parameters) and NBC (emission parameters) obtained from equation (1). Second, the normalization of BC and NBC was done separately and also by combining both to get the normalization decision matrix following equation (4). Third, the priority weights of the identified criteria are calculated using AHP and consistency ratio is also verified. Fourth, weighted normalization decision matrix for both BC and NBC is generated using equation (9). In the last step, performance of the decision matrix and ranking is done to get the best blend. The steps in the proposed method are as follows.

Step 1: Determination of the decision matrix

During optimization, the development decision matrix includes a number of criteria as output parameters, i.e., performance and emission characteristics and corresponding alternatives as input parameters, i.e., conventional fuel and biodiesel blends. The decision matrix is formulated using the following expression:

$$Bij = \begin{pmatrix} B_{11} & B_{12} & \cdots & \cdots & \cdots & B_{1n} \\ B_{21} & B_{22} & \cdots & \cdots & \cdots & B_{2n} \\ \vdots & \vdots & \vdots & \vdots & \vdots & \vdots \\ \vdots & \vdots & \vdots & \vdots & \vdots & \vdots \\ B_{m1} & B_{m2} & \cdots & \cdots & \cdots & B_{mn} \end{pmatrix}$$

$$\text{For}(i = 1, 2, \ldots m, j = 1, 2, \ldots n), \tag{1}$$

where B_{ij} indicates the response value of ith the response on jth process parameter, m is the number of criteria as output parameters and n is the number of alternatives as input parameters.

Step 2: Normalization of the decision matrix

In this step, the output parameters, which are in different units, are converted into dimensionless terms in order to obtain the normalization of the decision matrix. When two different criteria have different units, the assignment and comparison is not feasible, so normalization is performed. The formed normalization matrix has both BC and NBC. Then combining both BC and NBC, normalization of the decision matrix is achieved.

For NBC

$$\xi_{ij} = \frac{1/B_{ij}}{\sum_{i=k+1}^{n} 1/B_{ij}}. \tag{2}$$

For BC

$$\zeta_{ij} = \frac{Bij}{\sum_{i=1}^{k} Bij}. \tag{3}$$

For $i = 1, 2, \ldots\ldots n$ and $j = 1, 2, \ldots\ldots m$, where k is the number of attributes, ξ_{ij} represents the normalized performance for the NBC and ζ_{ij} represents the normalized performance for the BC of the ith experiments on jth the criteria for $i = 1, 2, 3, \ldots\ldots n$ and $j = 1, 2, 3, \ldots\ldots m$. For normalization decision matrix combining equations (2) and (3),

$$B_{ij} = \xi_{ij} + \zeta_{ij}, \tag{4}$$

where B_{ij} indicates the response value of ith response on jth process parameter, ξ_{ij} represents the normalized performance for the NBC and ζ_{ij} represents the normalized performance for the BC.

Step 3: Formulation of priority weights of criteria parameters using AHP

In the present step, to evaluate the priority weights of the criteria parameters, a pairwise comparison is required. The pairwise comparison is performed by the AHP method. In this method, the fundamental scale has been used to give relative importance to the criteria with respect to corresponding experiments. Subsequently computing eigen value for that respective matrix thereafter leads to the formation of priority weights. At last, consistency index (RCI) and consistency ratio (CR) validation of the matrix is done using equations (5) and (6):

$$RCI = \frac{(\lambda_{\text{Max}} - n)}{n-1} \tag{5}$$

$$CR = \frac{RCI}{RI}, \tag{6}$$

where λ_{Max} indicates the highest eigen value in the pairwise matrix, n indicates the number of experiments conducted in the test and RI represents the random index.

Step 4: Determination of weighted normalization decision matrix

Weighted normalization decision matrix is obtained by multiplying normalization decision matrix with priority weights for each of the criteria formed in the previous step. Weighted normalization decision matrix is formulated by equation (9). The weighted normalization decision matrix gives relative criteria with respect to the alternatives.

For NBC

$$\phi_{ij} = \frac{1/w_j B_{ij}}{\sum_{i=k+1}^{n} 1/B_{ij}}. \tag{7}$$

For BC

$$\varphi_{ij} = \frac{w_j B_{ij}}{\sum_{i=1}^{k} B_{ij}}, \tag{8}$$

where w_j is an associated weight of the jth experiments, ϕ_{ij} represents the weighted normalized performance for the NBC and φ_{ij} represents the weighted normalized performance for the BC.

For generating the weighted normalization decision matrix, equations (7) and (8) are combined to give

$$B_{ij} = \phi_{ij} + \varphi_{ij}, \tag{9}$$

where B_{ij} indicates the response value of ith response on jth process parameter, ϕ_{ij} represents the weighted normalized performance for the NBC and φ_{ij} represents the weighted normalized performance for the BC.

Step 5: Finding the performance of the experiments and their rank

At last, the performance can be evaluated among the alternatives in order to arrange them in the order of best alternatives. The performance experiment is formulated by equation (12). It is prescribed to have an ordinal positioning of Ψ_i to determine the last inclination of the applicable alternative. Thereafter highest to lowest ranking is done.

For NBC

$$\xi_i^* = \frac{\sum_{j=1}^{k} (\varphi + 1)}{\sum_{j=k+1}^{n} (\phi + 1)}. \tag{10}$$

For BC

$$\zeta_i^* = \frac{\sum_{j=1}^{k} \varphi}{\sum_{j=k+1}^{n} \phi + \sum_{j=1}^{k} (\varphi + 1)}, \tag{11}$$

where ξ_i^* is the performance score of the NBC, ζ_i^* is the performance score for the BC, k is the number of the attributes to be maximized and $(n - k)$ is the number of attributes to be minimized.

For performance experiment, equations (10) and (11) are combined to get

$$\psi_i = \xi_i^* + \zeta_i^*, \tag{12}$$

where ξ_i^* is the performance score of the NBC, ζ_i^* is the performance score for the BC and ψ_i the performance of the experiment. At last, ranking of highest to lowest experimental values is done.

3. Case Study: Optimization for the Selection of the Optimal Blend for the IC Engine

3.1. *Background*

An experimental investigation was conducted to apply the modeling and optimization parameters for the selection of an optimal blend for IC engine using the integrated MCDM method. The input parameters such as conventional diesel, biodiesel blends and output parameters such as break thermal efficiency, specific fuel consumptions, exhaust gas temperature, carbon monoxide (CO), hydrocarbon (HC), carbon dioxide (CO_2) and oxides of nitrogen (NO_x) were considered for experimentation. Conventional diesel and biodiesel were tested at different engine load conditions at a constant speed 1,500 rpm.

In this research, an attempt has been made to demonstrate the integrated MCDM method, i.e., AHP with MOOSRA to perform modeling and optimization for selecting an optimal blend for the IC engine. To compute the relative weights of performance and emission criteria, pairwise comparison matrices were framed with help of Saaty's scale. With the help of observed readings and relative weights of evaluation criteria, MOOSRA used to rank the alternatives.

3.2. *Fuel preparation and properties*

In the transesterification process, methanol (CH_3OH) and potassium hydroxide (KOH) were added to custard apple crude oil to perform the chemical reaction in order to overcome the constraint of free fatty acid (Debbarma *et al.*, 2011; Debbarma and Misra, 2017; Debbarma and Misra, 2018). To every 1 L of custard apple crude oil, 300 mL of CH_3OH and 10 g of KOH were added and reaction temperature is maintained at 60°C. The obtained reactant concentration is allowed to settle for 6 h. Thereafter two layers are formed: top layer (biodiesel) and the bottom layer (glycerin). Some of the fuel properties are shown in Table 1, which is compared with conventional diesel.

3.3. *Experimental setup*

A four-stroke direct injection (DI) diesel engine with the compression ratio of 17.5:1, developing a power of 5.2 kW at 1,500 rpm (TV_SR II, naturally aspirated, Kirloskar Oil Engines Ltd., India) was used

Table 1: Properties of fuel.

			Experimental values			
Sl No.	Properties	Units	Biodiesel (custard apple)	Biodiesel standard value (astm)	Commercially available diesel	Protocall
1	Kinematic viscosity @40^{0}C	centistokes	5.6068	1.9–6.0	2.54	ASTM D445
2	Density	kg/m^3	875	870–900	840	ASTM D4052
3	Flash point	^{0}C	159	130	54	ASTM D93
4	Calorific value	kJ/kg-K	37510	37000–42500	43500	ASTM D240

Table 2: Technical specifications of the engine.

Items	Names
Type of engine	Four-stroke, naturally aspirated Water-cooled engine
Bore/stroke	87.5 mm/110 mm
Compression ratio	17.5:1
Rated power	5.2 kW
Speed	1,500 rpm, constant
Injection pressure/advance	200 bar/23° before TDC
Airflow measurement	Airbox with "U" tube
Exhaust gas temperature	RTD thermocouple
Fuel flow measurement	Burette with a digital stopwatch
Resolution crank	1° crank angle
Angle sensor	360° encoder with resolution of 1°

for conducting the engine experiments. The engine specifications are given in Table 2. The engine was coupled with an eddy current type dynamometer for loading. The engine setup schematic diagram is as shown in Figure 1. Smoke measurement was made using an OPAX2000II/DX200P (Neptune Equipment Pvt. Ltd., Ahmedabad), and 5G-10 (Planet Equipment) gas analyzer was used to measure the exhaust emission of the engine. To measure the levels of CO_2, CO, HC and NO_x the fuel is made to pass through the probe of an exhaust gas analyzer. Thereafter it passed through the smoke probe in order to measure smoke density.

Four fuels, namely, D100 (100% conventional diesel), B10 (10% biodiesel plus 90% conventional diesel), B20 and B30 were tested at different load conditions from 0% to 100% with a step of 20% at a constant speed of 1,500 rpm. Before conducting the experiments, it was ensured that the previous experimental fuel was burnt completely.

3.4. *Modeling and selection of optimal blend for IC engine*

First, a decision matrix is formulated by conducting experimental analysis based on the corresponding alternatives and their levels, as

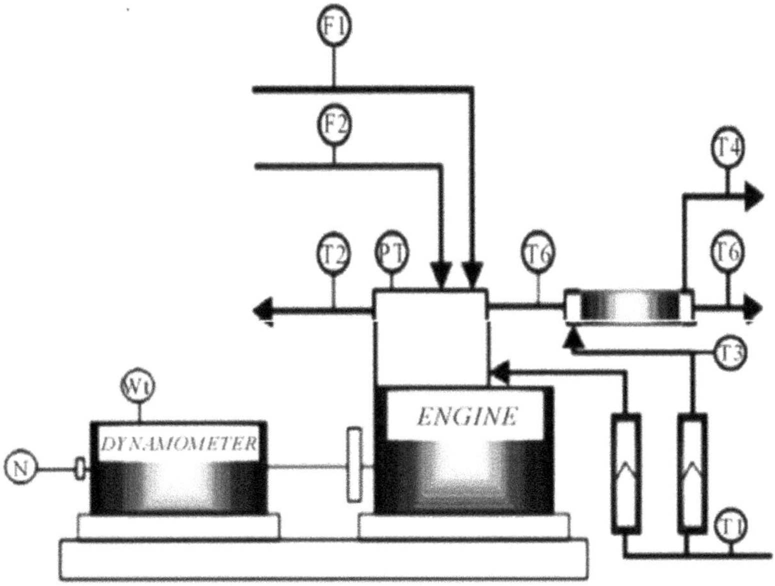

Figure 1: Experimental setup.

shown in Table 3. The decision matrix includes number of criteria such as break thermal efficiency, specific fuel consumption, exhaust gas temperature, CO, CO_2, HC and NO_x and corresponding alternatives such as conventional diesel and biodiesel blends.

Second, the obtained measurement data for the criteria can be converted into comparable data by normalization, which gives normalization of the decision matrix. Third, formulation of priority weights of criteria parameters is done using AHP.

In this step, pairwise comparison has been done by using Saaty's scale. Thereafter computing the eigen value for the respective matrix and finally validation of consistency index (RCI) and consistency ratio (CR) were done as depicted in Tables 4 and 5. Second, the obtained measurement data of the criteria can be converted into comparable data by normalization, which gives normalization of the decision matrix. Third, formulation of priority weights of criteria parameters are done using AHP.

Table 3: Experimental performance and emission parameters (decision matrix).

Input parameters		Output parameters						
		Performance parameters			Emission parameters			
Criteria/ load	Blends	BTE	SFC	EGT	CO	CO_2	NO_x	HC
0%	D100	1.651	0.245	147.830	0.019	1.354	60.294	20.644
	B10	2.252	0.124	157.454	0.021	1.568	46.226	16.456
	B20	2.893	0.145	157.830	0.027	2.002	42.358	18.070
	B30	2.619	0.054	166.981	0.027	1.031	39.057	18.361
20%	D100	13.09	2.624	190.075	0.020	2.213	217.925	27.834
	B10	12.445	3.724	191.491	0.025	3.041	226.887	27.685
	B20	12.215	3.245	190.491	0.029	3.041	185.887	26.685
	B30	12.085	3.554	203.774	0.028	3.504	193.396	29.348
40%	D100	20.004	2.306	225.755	0.030	2.305	297.170	26.455
	B10	19.383	2.424	246.179	0.039	2.617	298.396	26.773
	B20	18.761	2.452	236.604	0.048	2.929	249.623	28.092
	B30	18.314	2.437	251.887	0.048	2.790	250.622	30.586
60%	D100	24.565	1.283	275.358	0.052	1.788	320.528	27.388
	B10	23.738	1.367	278.125	0.070	3.041	305.434	30.434
	B20	23.798	1.367	276.981	0.068	3.041	285.434	28.763
	B30	23.013	1.383	294.340	0.072	3.000	290.754	31.607
80%	D100	26.897	0.583	337.075	0.069	1.998	400.849	26.223
	B10	26.376	0.739	327.427	0.073	2.127	350.619	29.849
	B20	26.511	0.726	338.208	0.074	2.141	310.377	30.749
	B30	26.479	0.739	359.434	0.074	2.506	312.433	33.353
100%	D100	25.709	0.628	387.264	0.078	1.526	340.585	35.384
	B10	25.571	0.783	437.529	0.090	1.697	310.765	38.765
	B20	25.326	0.738	437.830	0.095	1.607	300.132	39.631
	B30	25.237	0.870	469.811	0.099	1.827	300.509	41.000

Table 4: Performance characteristics.

Criteria	Weight	λ_{max}, CI, RCI	CR
BTE	0.6333	$\lambda_{max} = 3.0542$,	0.03
SFC	0.1061	CI = 0.02716	
EGT	0.2604	RCI = 0.52	

Next, the priority weights of criteria parameters multiplied with the normalization of the decision matrix gives the weighted normalization decision matrix for the both BC and NBC, which is

Table 5: Emission characteristics.

Criteria	Weight	λ_{max}, CI, RCI	CR
CO	0.4395	$\lambda_{max} = 4.1671$,	0.06
CO2	0.1560	CI = 0.0556	
NOx	0.1189	RCI = 0.89	
HC	0.2854		

Table 6: Weighted normalization of the decision matrix.

Input parameters		Output parameters						
		Performance parameters			Emission parameters			
Criteria/ load	Blends	BTE	SFC	EGT	CO	CO_2	NO_x	HC
0%	D100	0.399	181.855	20.465	20743.940	159.698	13.176	13.509
	B10	0.545	317.654	27.622	23943.290	182.190	11.448	13.553
	B20	0.700	580.467	36.415	20409.860	192.167	10.588	13.443
	B30	0.633	908.517	7.215	53652.650	692.085	20.303	25.352
20%	D100	0.654	4101.691	44.995	241277.500	265.910	7.608	57.237
	B10	0.571	4710.330	62.433	108848.800	285.777	7.489	48.334
	B20	0.571	4710.330	62.433	108848.800	285.777	7.489	48.334
	B30	0.633	7163.670	10.946	270896.800	377.124	10.842	69.342
40%	D100	0.656	8161.111	51.158	320045.600	338.687	7.401	68.871
	B10	0.636	8733.987	57.708	351014.400	342.031	7.344	66.355
	B20	0.615	9235.253	64.049	379401.500	344.663	7.300	64.532
	B30	0.633	17247.08	10.265	345421.000	549.044	11.077	82.523
60%	D100	0.630	8991.405	34.562	277059.600	343.555	7.708	70.661
	B10	0.609	12029.59	56.388	401535.600	351.411	7.352	66.447
	B20	0.618	12149.25	56.949	405529.800	354.906	7.425	56.457
	B30	0.633	25675.51	10.059	1130269.00	584.686	11.087	97.853
80%	D100	0.633	10371.49	32.258	319585.200	354.604	7.653	97.230
	B10	0.633	10290.99	37.272	317079.400	394.016	7.659	65.792
	B20	0.633	9682.020	35.745	320142.500	372.664	7.930	45.463
	B30	0.633	33050.51	8.483	814732.300	720.940	12.431	124.45
100%	D100	0.627	6837.291	16.682	9119.691	290.765	8.858	97.430
	B10	0.630	76212.24	23.887	50043.350	370.430	8.251	62.678
	B20	0.624	14444.32	20.283	17844.150	336.587	8.805	46.875
	B30	0.633	37898.81	5.581	57902.970	1017.99	22.856	121.41

shown in Table 6. In the final step, the performance of alternatives is evaluated as shown in Table 7 and ranking of alternatives based on highest to lowest experimental values is done, as shown in Table 8. In the present chapter, criteria used for evaluation include performance

Table 7: Performance of alternatives.

Input parameters		Output parameters						
		Performance parameters			Emission parameters			
Criteria/ load	Blends	BTE	SFC	EGT	CO	CO_2	NO_x	HC
0%	D100	0.285	0.009	0.076	0.000	0.010	0.115	0.113
	B10	0.353	0.005	0.057	0.000	0.009	0.131	0.112
	B20	0.412	0.003	0.044	0.000	0.008	0.141	0.113
	B30	0.388	0.002	0.199	0.000	0.002	0.077	0.062
20%	D100	0.395	0.000	0.036	0.000	0.006	0.190	0.028
	B10	0.364	0.000	0.026	0.000	0.006	0.192	0.033
	B20	0.364	0.000	0.026	0.000	0.006	0.192	0.033
	B30	0.388	0.000	0.137	0.000	0.004	0.138	0.023
40%	D100	0.396	0.000	0.031	0.000	0.005	0.194	0.023
	B10	0.389	0.000	0.028	0.000	0.005	0.196	0.024
	B20	0.381	0.000	0.025	0.000	0.005	0.197	0.025
	B30	0.388	0.000	0.145	0.000	0.003	0.135	0.020
60%	D100	0.386	0.000	0.046	0.000	0.005	0.188	0.023
	B10	0.378	0.000	0.028	0.000	0.005	0.196	0.024
	B20	0.382	0.000	0.028	0.000	0.005	0.194	0.028
	B30	0.388	0.000	0.148	0.000	0.003	0.135	0.017
80%	D100	0.388	0.000	0.049	0.000	0.005	0.189	0.017
	B10	0.388	0.000	0.043	0.000	0.004	0.189	0.024
	B20	0.388	0.000	0.044	0.000	0.004	0.183	0.035
	B30	0.388	0.000	0.172	0.000	0.002	0.122	0.013
100%	D100	0.385	0.000	0.092	0.000	0.006	0.166	0.017
	B10	0.386	0.000	0.066	0.000	0.004	0.177	0.026
	B20	0.384	0.000	0.077	0.000	0.005	0.167	0.034
	B30	0.388	0.000	0.248	0.000	0.002	0.068	0.013

parameters such as break thermal efficiency, specific fuel consumption and exhaust gas temperature.

These are identified as BC and emission parameters such as CO, CO_2, HC and NO_x are identified as NBC.

4. Results and Discussion

4.1. *Brake thermal efficiency*

The brake thermal efficiency (BTE) value represents the performance parameter of the engine tested on conventional diesel and

Table 8: Ranking of the alternatives.

Load	Blends	Performance values	Rank
0%	D100	0.5003	2
	B10	0.5	3
	B20	0.5008	1
	B30	0.4497	4
20%	D100	0.4158	3
	B10	0.4208	2
	B20	0.4208	1
	B30	0.411	4
40%	D100	0.4111	3
	B10	0.412	2
	B20	0.4127	1
	B30	0.4073	4
60%	D100	0.4105	3
	B10	0.412	2
	B20	0.4162	1
	B30	0.4043	4
80%	D100	0.4044	3
	B10	0.4076	2
	B20	0.4122	1
	B30	0.4008	4
100%	D100	0.4043	3
	B10	0.4068	2
	B20	0.4134	1
	B30	0.4011	4

biodiesel blends, which further gives the conversion efficiency of fuel into mechanical work in detail. The BTE under biodiesel blends was found to be lower compared to conventional diesel due to higher viscosity, density and lower calorific value as depicted in Figure 2.

In addition, high viscosity leads to poor fuel spray/atomization. With an increase in load, BTE increased up to 80% of the load and then significantly reduced for all the test fuel samples due to incomplete combustion. The BTE at 80% of load with conventional diesel operation and fuel B20 were 26.89% and 26.51%, respectively. For the fuel sample B20, 0.3–2% improvement in BTE was observed. Biodiesel is an oxygenated fuel, which helps in better combustion.

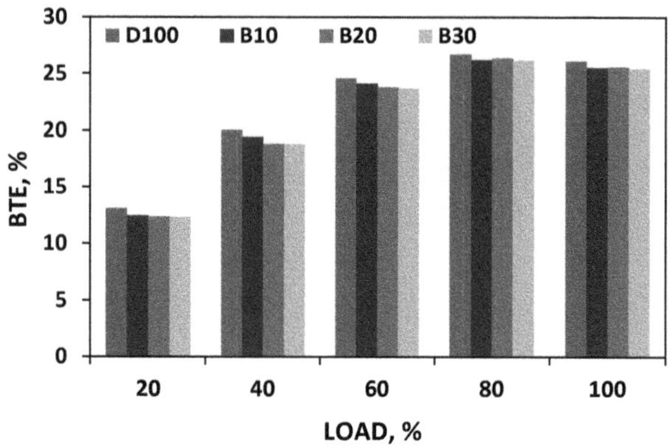

Figure 2: Effect of load on BTE.

For pure biodiesel, 2–3% reduction in BTE has been observed. For the different vegetable oil esters, a similar trend has been reported (Raheman and Phadatare, 2004; Prasad *et al.*, 2000).

4.2. *Specific fuel consumption*

From Figure 3, it is observed that SFC significantly decreased for all the testing fuels as the loading condition increased. This was due to higher combustion efficiency of fuel at higher loads. The values recorded for the SFC of biodiesel blends were higher compared to diesel for all the loading conditions.

The maximum SFCs with conventional diesel and fuel B20 at 80% load were 0.58 and 0.72 kg/kW-h, respectively. The results revealed that increased SFC in the case of biodiesel blends may be due to higher viscosity and density, and low calorific value of biodiesel blends compared to that of diesel (Barsic and Humke, 1981).

4.3. *Carbon monoxide*

Figure 4 shows the variation of CO emission with load for the conventional diesel and biodiesel blends. CO emission increased monotonically with an increase in loads. This typical trend is same for all automobiles because air–fuel ratio was low at higher load. It is more

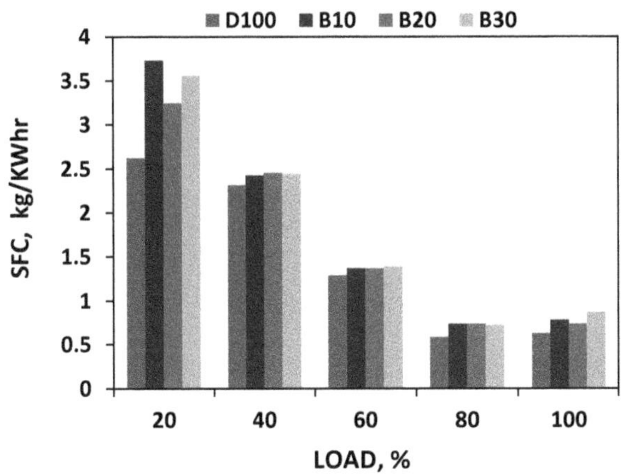

Figure 3: Effect of load on specific fuel consumption.

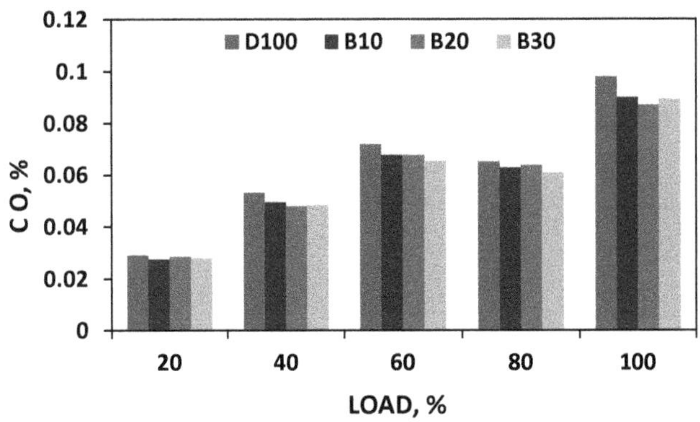

Figure 4: Effect of load on CO.

prominent at higher load, which may be due to higher fuel–air ratio than their stoichiometric ratio.

For all the loading conditions, CO emission of blends was found to be lower compared to diesel with respect to load. This may be due to small percentage of oxygen available in biodiesel blends, which leads to good combustion.

4.4. *Carbon dioxide*

The CO_2 emission is the main cause of global warming. The burning of petroleum products releases CO_2 into the atmosphere. The complete burning of fuel in the combustion chamber resulted in CO_2 emission in the exhaust. The CO_2 emissions for all the blended fuels are given in Figure 5. It can be seen that with the increase in load, CO_2 emission increased monotonically.

4.5. *Hydrocarbon*

The percentage of HC increasing gradually with conventional diesel and biodiesel blends with respect to loads are observed as shown in Figure 6.

At lower load, biodiesel blends have lower percentage of HC as compared to diesel, due to excess oxygen, which leads to complete combustion. During the high load in the combustion chamber, HCs are unable to mix with oxygen, which leads to the incomplete combustion. From Figure 6, it is observed at high load, there is an increase in the HC emission due to incomplete combustion. Further poor atomization of biodiesel blends also happens because of the higher viscosity.

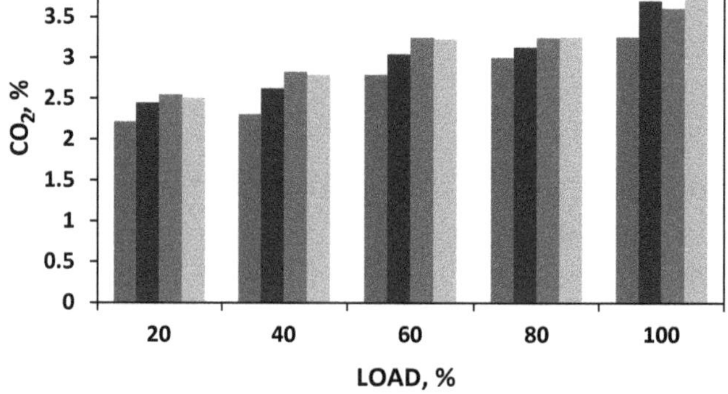

Figure 5: Effect of load on CO_2.

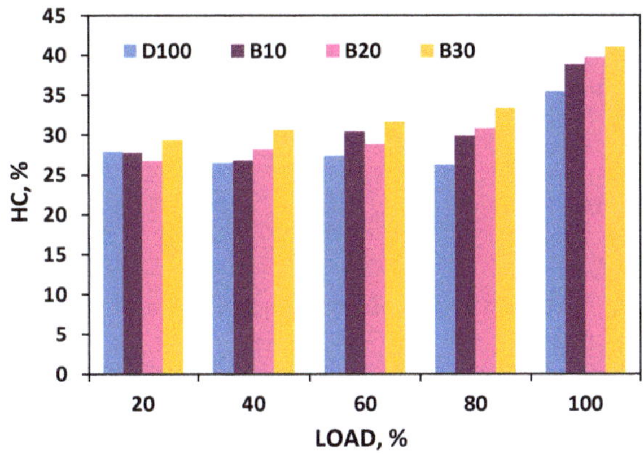

Figure 6: Effect of load on HC.

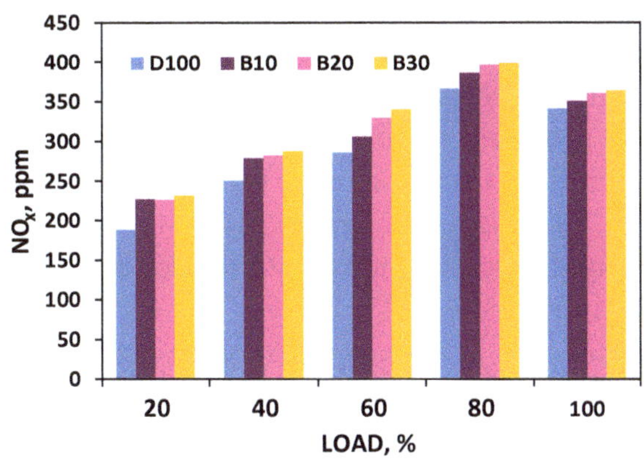

Figure 7: Effect of load on NO_x.

4.6. *Oxides of nitrogen*

NO_x emission and power output are directly proportional to each other. However, increase of NO_x occurs with an increase in exhaust gas temperature (Heywood, 1988).

The formation of NO_x varies directly with temperature, which arises during the activation of energy in the chemical reaction and these chemical reactions are dependent on equivalence ratio, amount oxygen and combustion temperature. As per observation in Figure 7,

NO_x emission level was higher for the biodiesel blend as compared to diesel with respect to different load conditions. This is due excess oxygen availability in during the combustion.

5. Conclusion

The results from the proposed method of AHP coupled with MOOSRA are provided in Table 8. From the performance of alternatives, B20 blend stands first at 20%, 40%, 60%, 80% and 100% load. It can be identified that B20 blend is suitable among all alternatives for improving engine performance with a corresponding decrease in the emission parameters. The impact of performance and emission parameters of the engine corresponding to load and blends varies and it is challenging for the investigator to recognize the suitable fuel among the different alternatives. The properties of fuels are closer to each other, which creates uncertainty to fulfill the emission norms. To overcome the above issues, computational model of AHP with MOOSRA was proposed to select the optimal blend. The selection of biodiesel blend for IC engine provides intense support to enhance biodiesel usage. The suggested hybrid MCDM method will guide the engine experts and scientist community to identify the suitable optimal blend to improve energy efficiency in the engine. From the modeling and optimization parameters, B20 gives the highest alternative performance among different blends. Therefore, it is concluded that B20 is the optimal blend. In the future, 20% of conventional diesel can be substituted by biodiesel blend.

References

Alptekin, E. and Canakci, M. (2008). Determination of the density and the viscosities of biodiesel–diesel fuel blend. *Renew Energy*, 33(12), 2623–2630.

Barma, S. D., Das, B., Giri, A., Majumder, S. and Bose, P. K. (2011). Back propagation artificial neural network (BPANN) based performance analysis of diesel engine using biodiesel. *J Renew Sustain Energy*, 3(1), 013101.

Barsic, N. J. and Humke, A. L. (1981). Performance and emissions characteristics of a naturally aspirated diesel engine with vegetable oil fuels. *SAE Trans*, 1173–1187.

Debbarma, S. and Misra, R. D. (2017). Effects of iron nanoparticles blended biodiesel on the performance and emission characteristics of a diesel engine. *J Energy Resour Technol*, 139(4), 042212.

Debbarma, S. and Misra, R. D. (2018). Effects of iron nanoparticle fuel additive on the performance and exhaust emissions of a compression ignition engine fueled with diesel and biodiesel. *J Therm Sci Eng Appl*, 10(4), 041002.

Gadakh, V. S. (2010). Application of MOORA method for parametric optimization of milling process. *Int J Appl Eng Res*, 1(4), 743.

Heywood, J. B. (1988). *International Combustion Engines Fundamentals.* McGraw Hill, New York.

Jagadish, A. and Ray, A. (2014). A Green cutting fluid selection using MOOSRA. *Int J Res Eng Technol*, 3(3), 559–563.

Mahalingappa, M. C., Navindgi, D. and Omprakash, H. (2014). Performance, combustion and emission characteristics of single cylinder diesel engine using custard apple seed (*Annona squamosa*) oil. *Int J Res Eng Adv Technol*, 2(3), 1–6.

Mahmoodzadeh, S., Shahrabi, J., Pariazar, M. and Zaeri, M. S. (2007). Project selection by using fuzzy AHP and TOPSIS technique. *World Acad Sci, Eng Technol*, 30, 333–338.

Poh, K. L. and Ang, B. W. (1999). Transportation fuels and policy for Singapore: An AHP planning approach. *Comput Industr Eng*, 37(3), 507–525.

Prasad, C. V., Krishna, M. M., Reddy, C. P. and Mohan, K. R. (2000). Performance evaluation of non-edible vegetable oils as substitute fuels in low heat rejection diesel engines. *Proc Instit Mech Eng, Part D: J Automobile Eng*, 214(2), 181–187.

Raheman, H. and Ghadge, S. V. (2007). Performance of compression ignition engine with mahua (*Madhuca indica*) biodiesel. *Fuel*, 86(16), 2568–2573.

Raheman, H. and Phadatare, A. G. (2004). Diesel engine emissions and performance from blends of karanja methyl ester and diesel. *Biomass Bioenergy*, 27(4), 393–397.

Saaty, T. L. (1981). *The Analytical Hierarchy Process.* McGraw-Hill, New York.

Sakthivel, G., Ilangkumaran, M., Nagarajan, G. and Shanmugam, P. (2013). Selection of best biodiesel blend for IC engines: An integrated approach with FAHP-TOPSIS and FAHP-VIKOR. *Int J Oil, Gas Coal Technol*, 6(5), 581–612.

Sarkar, A., Panja, S. C., Das, D. and Sarkar, B. (2015). Developing an efficient decision support system for non-traditional machine selection: An application of MOORA and MOOSRA. *Product Manuf Res*, 3(1), 324–342.

Sinha, S. and Agarwal, A. K. (2005). *Performance Evaluation of a Biodiesel (Rice Bran Oil Methyl Ester) Fuelled Transport Diesel Engine* (No. 2005-01-1730). SAE Technical Paper.

Sivaraja, C. M. and Sakthivel, G. (2017). Compression ignition engine performance modelling using hybrid MCDM techniques for the selection of optimum fish oil biodiesel blend at different injection timings. *Energy*, 139, 118–141.

Srinivasa Rao, P. and Gopalakrishnan, K. V. (1983). Esterified oils as fuel in diesel engines. In *11th National Conference on IC Engines,* Vol. 740, IIT Madras, India, pp. 727–742.
Usta, N. (2005). An experimental study on performance and exhaust emissions of a diesel engine fuelled with tobacco seed oil methyl ester. *Energy Conv Manage*, 46(15–16), 2373–2386.
Zheng, S., Kates, M., Dubé, M. A. and McLean, D. D. (2006). Acid-catalyzed production of biodiesel from waste frying oil. *Biomass Bioenergy*, 30(3), 267–272.

Chapter 16

Selection of Renewable Energy Sources for a Developing Country Using Analytic Hierarchy Process

R. Ramaprabha* and S. Malathy

Department of Electrical and Electronics Engineering,
SSN College of Engineering, Kalavakkam 603110, TN, India
**ramaprabhar@ssn.edu.in*

This chapter presents an application of analytic hierarchy process (AHP) for the selection of renewable energy sources (RESs) and technology for developing countries with moderate climatic conditions. The commonly available renewable sources like solar, wind, hydroelectric, geothermal, biomass and ocean have been considered for evaluation based on different criteria. The importance weights of the criteria and subcriteria as well as preferential ranking of options have been determined by eliciting expert judgment through pairwise comparisons. The assessments show that within the technological constraints, hydro energy is the most preferred option, followed by solar and then wind energy in developing countries. However, as time progresses and technology improves, the preferential ranking might change. Moreover, the implications of ranking vary with geographical locations.

1. Introduction

The growth in population and economy of a country results in an eventual increase in energy demand. The generation capacity has to be increased to reduce the supply–demand gap and ensure sustainable development. Further, growing concern on greenhouse gas (GHG) emissions that drives climatic changes limits exploiting the conventional fossil fuel-based power generation. This has in turn accelerated the market for renewable sources-based power generation. Ambitious strategies are being adopted by countries worldwide to intensify the installations of renewable-based power generation and integrate it with the existing grid to gear up the transition toward an ideal 100% renewable-based energy mix. The renewable energy sources (RESs) are non-depletable or replenishable over a time span. The potential renewable energy resources include solar, wind, hydroelectric, geothermal, bio-mass, ocean and hydrogen as listed in Figure 1.

Solar energy is harnessed by solar photovoltaic or solar thermal technology. The light energy is converted into electrical energy directly (light–electric) in the former and indirectly (light–heat–electric) in the latter. Installations utilizing solar photovoltaic technology have gained an edge over others owing to the advantages like rooftop/facade integration, feasible small- to large-scale generation and lower lead time. Power generation by wind is increasing significantly in recent years with China extending 35% of the global share followed by the United States. Penetration of this technology is limited in many countries due to high investment cost, low wind potential, risk due to cyclones, far away sites, noise and potential damage to local birds. Hydropower plants produce clean energy at low cost. China leads in hydroelectric power generation followed by Brazil. The major concern, however, includes resistance from local

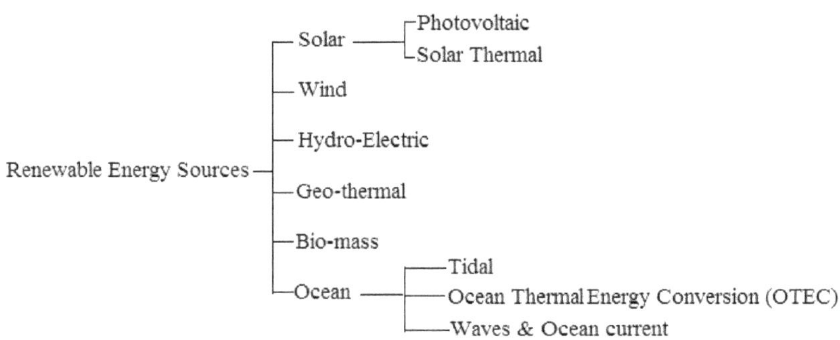

Figure 1: Commonly used RESs.

population for relocating from the project sites, failed monsoons, loss of land and wildlife in the catchment area. With a global cumulative capacity of over 14 GW, geothermal plants offer stable power production unperturbed by the climatic changes, thereby ensuring high capacity factor. However, limited resource at specific locations, extremely high temperature and sophisticated infrastructure requirement are the major concerns. The United States leads in biomass-based power generation (harnessing energy from organic matter) followed by China. Energy can be produced from the waves, tides and the ocean temperature gradient.

While discussing about the renewable sources, the storage becomes compulsory as RES are intermittent in nature. Moreover, several critical loads must function properly with variable power/energy supply, where the energy storage is mandatory. Energy storage allows decoupling of energy supply and demand periods. This is necessary for both technical and economic considerations. The different storage systems available are depicted in Figure 2.

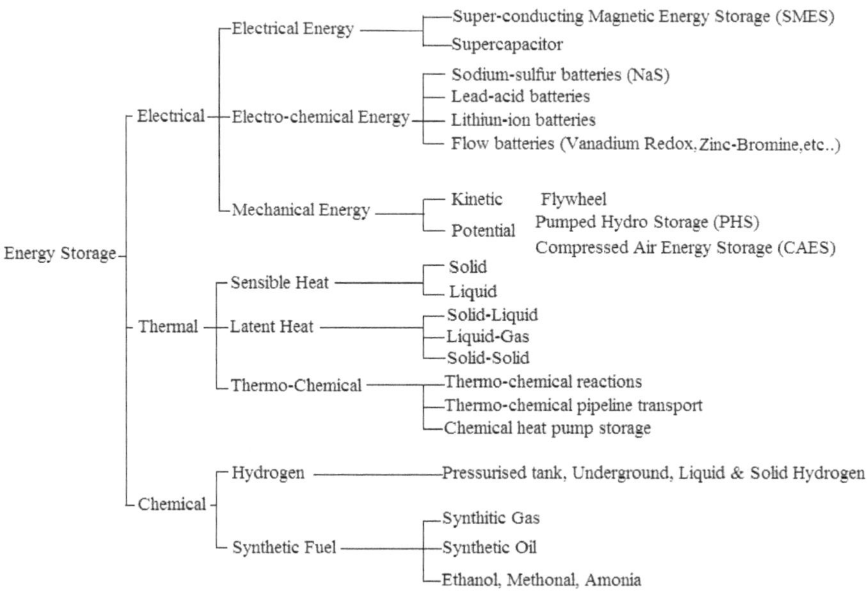

Figure 2: Classification of energy storage systems.

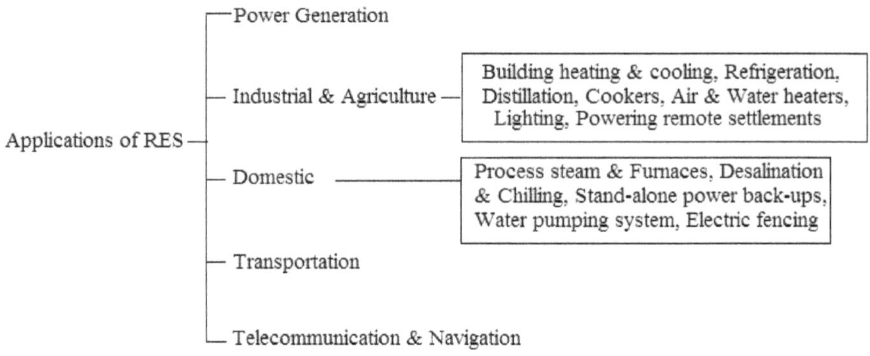

Figure 3: Applications of RES.

The selection of particular RES can be primarily decided based on the application and availability of the source. The applications of RES are broadly classified as shown in Figure 3.

Moreover, the selection of the RES and the storage system depends on the location, load requirement, land requirement, topographical conditions, government regulation and financial speculation. Hence the selection is a complex process (multi-criteria/multi-perspective analysis) where the appropriate decision is important.

Amer and Daim (2011) applied the analytical hierarchy process (AHP) to select the optimal renewable source among wind, solar PV, solar thermal and biomass for power generation in Pakistan. The potential of four major renewable sources available in Malaysia was reviewed and an assessment model with four main criteria and 12 sub-criteria was developed to select the best suited resource by Salman and Tahar (2014). Hydroelectric power generation was ranked to be the suitable source in Turkey's renewable energy sector by developing a framework utilizing Fuzzy TOPSIS method by Umran *et al.* (2015). Ruiz-Arias *et al.* (2012) assessed the supply–demand trend and potential renewable sources (biomass, solar PV and wind) of southern Spain and proposed a framework to meet 75% of the demand through comprehensive analysis. The additional capacity to be installed was also recommended. The modeling of the energy sector that includes both the renewable and nonrenewable sources and the associated influential factors to guide the decision makers to identify the potential renewable source were presented by Stein (2013). The AHP model that was developed based on the empirical data suggested wind, solar, hydro and geothermal to be the potential resources. The decision-making framework employing MULTIMOORA and TOPSIS strategies was developed by Streimikiene *et al.* (2012) and the results recommended to the policymakers to orient towards water and solar thermal resources for sustainable development.

Hence, in this chapter, the selection of RES is carried out using AHP, considering adequate criteria and subcriteria.

2. Analytic Hierarchy Process for Energy Source Selection

The analytic hierarchy process (AHP) is a structured technique for organizing and analyzing complex decisions (Saaty 1980). It is used for group decision-making for a particular selection. Before applying AHP for this problem, the key factors on which the selection of RES can be decided are listed in Table 1.

These parameters are interlinked as depicted in Figures 4, and 5 shows the RES selection based on AHP.

Selection of appropriate RES depends on various factors as listed in Table 1 and it requires careful assessment of the prominent criteria that are often conflicting. AHP is the commonly employed multi-criteria decision-making tool that helps to take rational and lucid decisions among the alternatives by structuring the complexity (Streimikiene and Balezentis, 2013; Streimikiene *et al.*, 2012; Ahmad and Tahar, 2014; Al Garni *et al.*, 2016). Hence, the AHP method is opted to select the renewable source that is well suited for the required context. The complex problem is disintegrated into a configuration of hierarchy that includes criteria, subcriteria and alternatives on hand. Structuring is followed by series of comparisons at every level, prioritizing each of the criteria over the other eventually. The decision is then taken based on the ranking of the prevailing alternatives (Al Garni *et al.*, 2016). The essential steps involved in this decision-making process are as follows:

Table 1: Key factors influencing the selection of RES.

S. No.	Criteria	Sub-criteria
1	Geographical	Topographical conditions
		Land availability
		Source availability
		Accessibility
2	Environmental	GHG emissions
		Impact on eco system including byproducts at manufacturing and disposal stages
		Land requirement
3	Economical	Capital cost
		Operational and maintenance cost
		Conversion technology cost
		Lifetime
		Return of investment (ROI) or payback period
4	Social	Social acceptance/benefits
		Energy business and job creation
5	Technical	Technological maturity
		Efficiency
		Reliability
		Safety
		Stability
		Grid connectivity
		Lead time (time for construction and commissioning)
6	Application	Standalone/power back-up
		Grid tied/grid connected
7	Load demand	Base load
		Peak load

- Define the problem and its goal.
- Disintegrate the decision problem and develop hierarchical structure with goal, decision criteria and alternatives.
- Make pair-wise comparison among alternatives at each level.
- Compute priority vector to assess the relative significance of each of the criteria being compared.
- Compute consistency index and ratio to validate the judgment obtained from the comparisons.
- Compute the final priority ranking of the alternatives by assimilating the judgments at various levels of hierarchy.

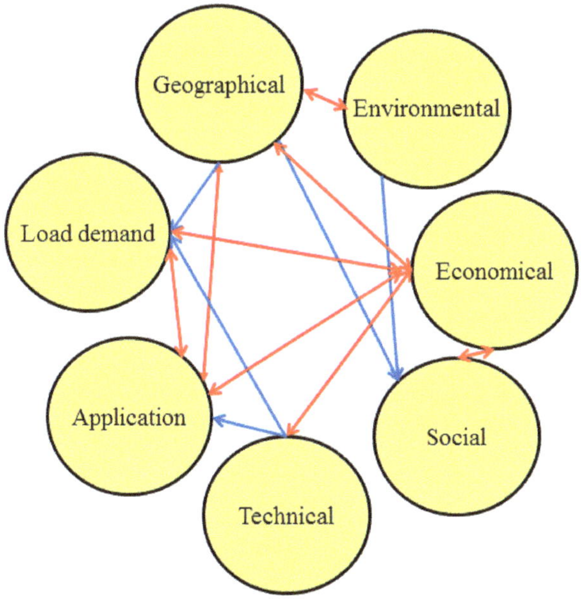

Figure 4: Dependence among the criteria.

The selection of appropriate RES for electric power generation is dictated by various factors and AHP methodology is used to decide the relative ranking among the available options. The process requires the identification of goal, various criteria and sub-criteria that have an impact on the decision and the alternatives available. The goal in this case is to select and appraise the dominant RES for electric power generation. The decision problem is decomposed by identifying the key factors (criteria) based on which the decision is to be made. Seven factors including geographical, environmental, economic, social, technical, application and load demand are identified as the main criteria. The model can further be enriched by identifying all the factors (subcriteria) that directly influence the selection and categorize them under the main criteria. About 22 such sub-criteria as listed in Table 1 are identified in this case and they are categorized under the seven main criteria. The dominant RESs that are available are then identified. In this case, six such sources (solar, wind, hydro, geothermal, biomass and ocean) are identified. Thus, the AHP model with four levels is formulated with goal and alternatives at the first and the fourth level, respectively. The structure of the developed AHP model is depicted in Figure 5.

The pair-wise comparisons are made at each level of the model using the fundamental scale of the AHP, which assigns a numerical value to the verbal judgment of the preferences as in Table 2.

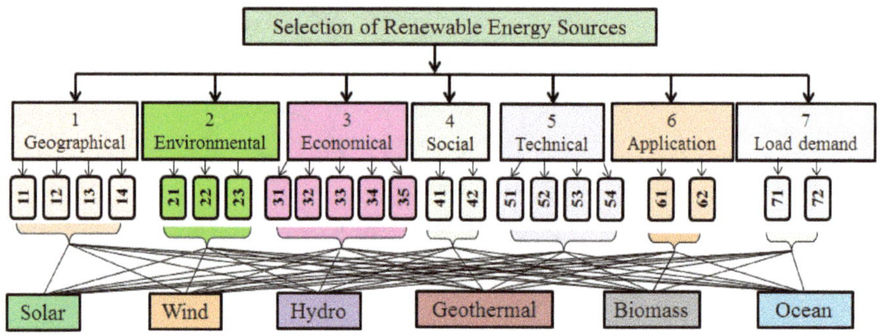

11. Topographical conditions
12. Land availability
13. Source availability
14. Accessibility
21. Greenhouse gas (GHG) emissions
22. Impact on Eco System including byproducts at manufacturing & disposal stages
23. Land requirement
31. Capital cost
32. Operational & Maintenance cost
33. Conversion technology cost
34. Lifetime
35. Return of investment (ROI) or Payback period
41. Social acceptance/benefits
42. Energy business & Job creation
51. Technological maturity & Efficiency
52. Reliability & Safety
53. Stability & Grid connectivity
54. Lead time
61. Standalone/Power back-up
62. Grid tied/Grid connected
71. Base load
72. Peak load

Figure 5: AHP Model for selection of RES.

Table 2: Saaty pair-wise comparison scale of AHP (Amer and Daim, 2011).

Intensity of importance	Verbal judgment of significance
1	Equal importance
3	Moderate importance
5	Strong importance
7	Very strong importance
9	Extremely importance
2, 4, 6, 8	Used to express intermediate values

Table 3: RI values for different sizes of a matrix, n.

Size of matrix (n)	1	2	3	4	5	6	7	8	9	10
RI values	0.00	0.00	0.58	0.90	1.12	1.24	1.32	1.41	1.45	1.49

Pair-wise comparison of the criteria listed in the second level is done by formulating a 7×7 matrix. Each entry in this 2D matrix denotes the relative significance of one criteria over the other being compared. The verbal (qualitative) significance is translated into a numerical (quantitative) importance using Table 2, which uses a 9-point scale. In general, for a $n \times n$ matrix, $n(n-1)/2$ judgments are made and the priority vectors are calculated. The priority vector denotes the significance of each criterion with respect to its previous level. The consistency of the judgment is assessed by computing consistency index (CI) and consistency ratio.

$$\text{Consistency index, CI} = \frac{\lambda_{\max} - n}{n - 1}, \tag{1}$$

where $\lambda_{\max}$ and n are the maximum Eigen value and size of comparison matrix, respectively. The consistency ratio is the ratio of CI and the average random index (RI), listed in Table 3 for different matrix sizes (Al Garni *et al.*, 2016)

$$\text{Consistency ration, CR} = \frac{\text{CI}}{\text{RI}} \tag{2}$$

The matrix is consistent and the judgment arrived is valid only if the CR is lesser than 0.1. Higher CR value signifies inconsistent judgment and the comparison matrix is to be reviewed to make it consistent. The comparison matrix for the second level is presented in Table 4.

The matrix is consistent as CR value is lesser than 0.1 and the results signify that the geographical conditions and technical aspects are relatively dominant in selecting the appropriate renewable source. The selection indeed is highly dependent on geographical and topographical conditions. The pair-wise comparison for the next level criteria is presented in Tables 5–11.

Table 4: Pair-wise comparison matrix for the first level criteria.

	Geographical	Environmental	Economical	Social	Technical	Application	Load demand	Priority vector
Geographical	1	2	2	2	2	2	1	0.218
Environmental	0.5	1	0.5	2	0.33	2	1	0.119
Economical	0.5	2	1	2	1	1	1	0.144
Social	0.5	0.5	0.5	1	0.33	1	0.25	0.070
Technical	0.5	3	1	3	1	2	1	0.182
Application	0.5	0.5	1	1	0.5	1	1	0.102
Load demand	1	1	1	4	1	1	1	0.165
								$\Sigma\,1.000$

Note: $\lambda_{\max} = 7.385$, CI $= 0.064$, CR $= 0.0489$.

Table 5: Pair-wise comparison matrix for the second level criteria (under geographical considerations).

	11_TC	12_LA	13_SA	14_Acc	Priority vector
11_TC	1	2	2	2	0.385
12_LA	0.5	1	0.5	1	0.160
13_SA	0.5	2	1	3	0.306
14_Acc	0.5	1	0.33	1	0.149
					Σ 1.000

Note: $\lambda_{\max} = 4.118$, CI = 0.039, CR = 0.044.

Table 6: Pair-wise comparison matrix for the second level criteria (under environmental considerations).

	21_GHG	22_Eco	23_LR	Priority vector
21_GHG	1	1	1	0.333
22_Eco	1	1	1	0.333
23_LR	1	1	1	0.333
				Σ 1

Note: $\lambda_{\max} = 3$, CI = 0, CR = 0.

Table 7: Pair-wise comparison matrix for the second level criteria (under economic considerations).

	31_CC	32-OM	33_CT	34_LT	35_ROI	Priority vector
31_CC	1	1	1	0.33	0.25	0.104
32-OM	1	1	1	0.33	0.25	0.104
33_CT	1	1	1	0.5	0.33	0.120
34_LT	3	3	2	1	1	0.304
35_ROI	4	4	3	1	1	0.369
						Σ 1

Note: $\lambda_{\max} = 5.033$, CI = 0.008, CR = 0.007.

The comparison matrices are checked for consistency and the topographical conditions and source availability are relatively dominant in the geographical criterion. All the subcriteria that are grouped under the environmental, social, application and load

Table 8: Pair-wise comparison matrix for the second level criteria (under social considerations).

	41_SA	42_EB	Priority vector
41_SA	1	1	0.5
42_EB	1	1	0.5
			$\Sigma\,1$

Note: $\lambda_{\max} = 2$, CI $= 0$.

Table 9: Pair-wise comparison matrix for the second level criteria (under technical considerations).

	51_TEM	52_RS	53_SG	54_LT	Priority vector
51_TEM	1	1	1	1	0.25
52_RS	1	1	1	1	0.25
53_SG	1	1	1	1	0.25
54_LT	1	1	1	1	0.25
					$\Sigma\,1$

Note: $\lambda_{\max} = 4$, CI $= 0$, CR $= 0$.

Table 10: Pair-wise comparison matrix for the second level criteria (under application considerations).

	61_STA	62_GT	Priority vector
61_STA	1	1	0.5
62_GT	1	1	0.5
			$\Sigma\,1$

Note: $\lambda_{\max} = 2$, CI $= 0$.

Table 11: Pair-wise comparison matrix for the second level criteria (under load demand considerations).

	71_BL	72_PL	Priority vector
71_BL	1	1	0.5
72_PL	1	1	0.5
			$\Sigma\,1$

Note: $\lambda_{\max} = 2$, CI $= 0$.

demand conditions are equally significant. Return of investment (ROI) period and lifetime are relatively significant than the other sub-criteria in economic criteria. The pair-wise comparison for the next level criteria is presented in Tables 12–33.

Table 12: Pair-wise comparison matrix for topographical conditions.

	Solar	Wind	Hydro	Geo	Bio	Ocean	Priority vector
Solar	1	1	1	0.33	2	0.33	0.129
Wind	1	1	1	1	2	1	0.176
Hydro	1	1	1	1	2	1	0.176
Geothermal	3	1	1	1	2	1	0.211
Biomass	0.50	0.50	0.50	0.50	1	0.33	0.082
Ocean	3	1	1	1	3	1	0.225
							Σ 1

Note: $\lambda_{\max} = 6.201$, CI = 0.04, CR = 0.032.

Table 13: Pair-wise comparison matrix for land availability.

	Solar	Wind	Hydro	Geo	Bio	Ocean	Priority vector
Solar	1	2	1	2	1	0.14	0.099
Wind	0.5	1	2	2	0.33	0.14	0.078
Hydro	1	0.5	1	1	0.25	0.14	0.059
Geothermal	0.5	0.5	1	1	0.2	0.14	0.051
Biomass	1	3	4	5	1	0.14	0.170
Ocean	7	7	7	7	7	1	0.544
							Σ 1

Note: $\lambda_{\max} = 6.398$, CI = 0.08, CR = 0.064.

Table 14: Pair-wise comparison matrix for source availability.

	Solar	Wind	Hydro	Geo	Bio	Ocean	Priority vector
Solar	1	1	2	2	2	2	0.246
Wind	1	1	1	2	2	2	0.216
Hydro	0.5	1	1	2	2	2	0.195
Geothermal	0.5	0.5	0.5	1	0.5	0.5	0.089
Biomass	0.5	0.5	0.5	2	1	2	0.141
Ocean	0.5	0.5	0.5	2	0.5	1	0.113
							Σ 1

Note: $\lambda_{\max} = 6.191$, CI = 0.038, CR = 0.031.

Table 15: Pair-wise comparison matrix for accessibility.

	Solar	Wind	Hydro	Geo	Bio	Ocean	Priority vector
Solar	1	1	2	3	1	3	0.243
Wind	1	1	2	2	1	2	0.213
Hydro	0.5	0.5	1	1	0.33	1	0.099
Geothermal	0.33	0.5	1	1	0.33	1	0.093
Biomass	1	1	3	3	1	3	0.260
Ocean	0.33	0.5	1	1	0.33	1	0.093
							Σ 1

Note: $\lambda_{\max} = 6.037$, CI = 0.007, CR = 0.006.

Table 16: Pair-wise comparison matrix for GHG emissions.

	Solar	Wind	Hydro	Geo	Bio	Ocean	Priority vector
Solar	1	1	1	1	1	1	0.167
Wind	1	1	1	1	1	1	0.167
Hydro	1	1	1	1	1	1	0.167
Geothermal	1	1	1	1	1	1	0.167
Biomass	1	1	1	1	1	1	0.167
Ocean	1	1	1	1	1	1	0.167
							Σ 1

Note: $\lambda_{\max} = 6$, CI = 0, CR = 0.

Table 17: Pair-wise comparison matrix for ecological impact.

	Solar	Wind	Hydro	Geo	Bio	Ocean	Priority vector
Solar	1	2	1	2	1	2	0.216
Wind	0.5	1	0.5	2	0.5	0.5	0.110
Hydro	1	2	1	2	1	2	0.216
Geothermal	0.5	0.5	0.5	1	0.33	0.33	0.079
Biomass	1	2	1	3	1	2	0.229
Ocean	0.5	2	0.5	3	0.5	1	0.151
							Σ 1

Note: $\lambda_{\max} = 6.165$, CI = 0.033, CR = 0.027.

Table 18: Pair-wise comparison matrix for land requirement.

	Solar	Wind	Hydro	Geo	Bio	Ocean	Priority vector
Solar	1	1	2	0.5	0.33	0.33	0.107
Wind	1	1	1	2	0.5	0.33	0.119
Hydro	0.5	1	1	1	0.5	0.33	0.096
Geothermal	2	0.5	1	1	0.33	0.33	0.105
Biomass	3	2	2	3	1	0.33	0.218
Ocean	3	3	3	3	3	1	0.355
							Σ 1

Note: $\lambda_{\max} = 6.341$, CI = 0.068, CR = 0.055.

Table 19: Pair-wise comparison matrix for capital cost.

	Solar	Wind	Hydro	Geo	Bio	Ocean	Priority vector
Solar	1	0.5	0.25	0.33	0.33	1	0.074
Wind	2	1	0.5	0.50	0.33	1	0.111
Hydro	4	2	1	0.50	2	2	0.234
Geothermal	3	2	2	1	2	2	0.282
Biomass	3	3	0.5	0.5	1	2	0.195
Ocean	1	1	0.5	0.5	0.5	1	0.104
							Σ 1

Note: $\lambda_{\max} = 6.239$, CI = 0.048, CR = 0.039.

Table 20: Pair-wise comparison matrix for operation and maintenance cost.

	Solar	Wind	Hydro	Geo	Bio	Ocean	Priority vector
Solar	1	3	3	5	4	6	0.413459
Wind	0.33	1	0.5	2	3	3	0.161718
Hydro	0.33	2	1	2	2	3	0.184553
Geothermal	0.2	0.5	0.5	1	1	3	0.09624
Biomass	0.25	0.33	0.5	1	1	3	0.096013
Ocean	0.17	0.33	0.33	0.33	0.33	1	0.048017
							Σ 1

Note: $\lambda_{\max} = 6.230$, CI = 0.046, CR = 0.037.

Table 21: Pair-wise comparison matrix for conversion technology cost.

	Solar	Wind	Hydro	Geo	Bio	Ocean	Priority vector
Solar	1	2	1	2	2	3	0.249
Wind	0.5	1	0.5	2	2	3	0.179
Hydro	1	2	1	2	2	3	0.249
Geothermal	0.5	0.5	0.5	1	0.5	3	0.116
Biomass	0.5	0.5	0.5	2	1	3	0.145
Ocean	0.33	0.33	0.33	0.33	0.33	1	0.061
							$\Sigma 1$

Note: $\lambda_{\max} = 6.218$, CI $= 0.044$, CR $= 0.035$.

Table 22: Pair-wise comparison matrix for life time.

	Solar	Wind	Hydro	Geo	Bio	Ocean	Priority vector
Solar	1	2	0.20	1	1	2	0.123
Wind	0.50	1	0.17	1	1	1	0.083
Hydro	5	6	1	6	5	6	0.521
Geothermal	1	1	0.17	1	1	1	0.093
Biomass	1	1	0.20	1	1	1	0.096
Ocean	0.50	1	0.17	1	1	1	0.083
							$\Sigma 1$

Note: $\lambda_{\max} = 6.073$, CI $= 0.014$, CR $= 0.012$.

Table 23: Pair-wise comparison matrix for return of investment period.

	Solar	Wind	Hydro	Geo	Bio	Ocean	Priority vector
Solar	1	0.5	0.33	1	2	3	0.143
Wind	2	1	0.5	2	2	3	0.212
Hydro	3	2	1	3	4	5	0.371
Geothermal	1	0.5	0.33	1	1	2	0.117
Biomass	0.5	0.5	0.25	1	1	1	0.090
Ocean	0.33	0.33	0.2	0.50	1	1	0.067
							$\Sigma 1$

Note: $\lambda_{\max} = 6.095$, CI $= 0.019$, CR $= 0.015$.

Table 24: Pair-wise comparison matrix for social acceptability.

	Solar	Wind	Hydro	Geo	Bio	Ocean	Priority vector
Solar	1	3	1	2	2	3	0.269
Wind	0.33	1	1	2	2	3	0.186
Hydro	1	1	1	3	2	5	0.254
Geothermal	0.5	0.5	0.33	1	0.33	1	0.082
Biomass	0.5	0.5	0.5	3	1	1	0.130
Ocean	0.33	0.33	0.2	1	1	1	0.078
							Σ 1

Note: $\lambda_{\max} = 6.331$, CI = 0.066, CR = 0.053.

Table 25: Pair-wise comparison matrix for energy business and job creation.

	Solar	Wind	Hydro	Geo	Bio	Ocean	Priority vector
Solar	1	1	2	2	2	3	0.255
Wind	1	1	2	2	2	3	0.255
Hydro	0.5	0.5	1	2	2	3	0.186
Geothermal	0.5	0.5	0.5	1	1	1	0.108
Biomass	0.5	0.5	0.5	1	1	1	0.108
Ocean	0.33	0.33	0.33	1	1	1	0.089
							Σ 1

Note: $\lambda_{\max} = 6.109$, CI = 0.022, CR = 0.018.

Table 26: Pair-wise comparison matrix for technical maturity and efficiency.

	Solar	Wind	Hydro	Geo	Bio	Ocean	Priority vector
Solar	1	2	0.33	2	0.5	2	0.154
Wind	0.5	1	0.33	2	2	2	0.155
Hydro	3	3	1	3	3	3	0.348
Geothermal	0.5	0.5	0.33	1	0.33	0.5	0.071
Biomass	2	0.5	0.33	3	1	3	0.182
Ocean	0.5	0.5	0.33	2	0.33	1	0.091
							Σ 1

Note: $\lambda_{\max} = 6.45$, CI = 0.09, CR = 0.073.

Table 27: Pair-wise comparison matrix for reliability and safety.

	Solar	Wind	Hydro	Geo	Bio	Ocean	Priority vector
Solar	1	2	0.5	1	0.25	1	0.116
Wind	0.5	1	0.5	1	0.25	1	0.092
Hydro	2	2	1	3	1	3	0.260
Geothermal	1	1	0.33	1	0.25	1	0.093
Biomass	4	4	1	4	1	3	0.341
Ocean	1	1	0.33	1	0.33	1	0.098
							$\Sigma\,1$

Note: $\lambda_{\max} = 6.11$, CI $= 0.022$, CR $= 0.018$.

Table 28: Pair-wise comparison matrix for stability and grid connectivity.

	Solar	Wind	Hydro	Geo	Bio	Ocean	Priority vector
Solar	1	1	0.5	2	2	2	0.188
Wind	1	1	0.5	2	2	2	0.188
Hydro	2	2	1	3	3	3	0.327
Geothermal	0.5	0.5	0.33	1	1	1	0.099
Biomass	0.5	0.5	0.33	1	1	1	0.099
Ocean	0.5	0.5	0.33	1	1	1	0.099
							$\Sigma\,1$

Note: $\lambda_{\max} = 6.014$, CI $= 0.003$, CR $= 0.002$.

Table 29: Pair-wise comparison matrix for lead time.

	Solar	Wind	Hydro	Geo	Bio	Ocean	Priority vector
Solar	1	2	3	3	2	4	0.332
Wind	0.5	1	2	2	2	4	0.233
Hydro	0.33	0.5	1	1	0.5	1	0.097
Geothermal	0.33	0.5	1	1	1	1	0.108
Biomass	0.5	0.5	2	1	1	2	0.147
Ocean	0.25	0.25	1	1	0.5	1	0.083
							$\Sigma\,1$

Note: $\lambda_{\max} = 6.11$, CI $= 0.022$, CR $= 0.018$.

Table 30: Pair-wise comparison matrix for standalone application.

	Solar	Wind	Hydro	Geo	Bio	Ocean	Priority vector
Solar	1	1	0.5	2	2	3	0.197
Wind	1	1	0.5	2	2	3	0.197
Hydro	2	2	1	3	3	3	0.322
Geothermal	0.5	0.5	0.33	1	1	1	0.097
Biomass	0.5	0.5	0.33	1	1	2	0.110
Ocean	0.33	0.33	0.33	1	0.5	1	0.078
							Σ 1

Note: $\lambda_{\max} = 6.077$, CI = 0.015, CR = 0.012.

Table 31: Pair-wise comparison matrix for grid tied application.

	Solar	Wind	Hydro	Geo	Bio	Ocean	Priority vector
Solar	1	0.5	0.33	2	2	3	0.166
Wind	2	1	0.33	2	2	3	0.203
Hydro	3	3	1	3	3	3	0.358
Geothermal	0.5	0.5	0.33	1	1	1	0.094
Biomass	0.5	0.5	0.33	1	1	1	0.094
Ocean	0.33	0.33	0.33	1	1	1	0.085
							Σ 1

Note: $\lambda_{\max} = 6.19$, CI = 0.039, CR = 0.032.

Table 32: Pair-wise comparison matrix for base load.

	Solar	Wind	Hydro	Geo	Bio	Ocean	Priority vector
Solar	1	2	0.33	3	3	3	0.229
Wind	0.5	1	0.33	2	2	3	0.161
Hydro	3	3	1	3	3	3	0.354
Geothermal	0.33	0.5	0.33	1	1	1	0.086
Biomass	0.33	0.5	0.33	1	1	1	0.086
Ocean	0.33	0.33	0.33	1	1	1	0.083
							Σ 1

Note: $\lambda_{\max} = 6.21$, CI = 0.041, CR = 0.033.

Table 33: Pair-wise comparison matrix for peak load.

	Solar	Wind	Hydro	Geo	Bio	Ocean	Priority vector
Solar	1	0.5	1	2	2	2	0.186
Wind	2	1	2	2	2	3	0.289
Hydro	1	0.5	1	3	3	3	0.234
Geothermal	0.5	0.5	0.33	1	1	1	0.100
Biomass	0.5	0.5	0.33	1	1	1	0.100
Ocean	0.5	0.33	0.33	1	1	1	0.091
							$\Sigma\,1$

Note: $\lambda_{\max} = 6.13$, CI = 0.027, CR = 0.021.

The overall priority vector is computed by aggregating all the previous judgments as compiled in Table 34.

The AHP methodology results in the overall preferential ranking. Choice of RE technology is a complex decision-making problem. It requires a thorough survey of all the factors as well as alternatives. AHP provides flexibility in such decision-making problems. The analysis shows that hydro energy possesses a good prospect followed by solar and wind energy. Hydro energy technology offers an exciting potential for development of a sustainable and eco-friendly energy system. As the demand for electrical energy is increasing significantly, this analysis provides an effective means for selecting the appropriate renewable source. The RESs, especially solar energy, is an effective alternative for remote regions where grid connectivity is not possible. However, the selection criteria and the relative significance the RES vary with geographical locations.

Table 34: Priority matrix for selecting the appropriate renewable energy source (Final Rank).

RES	Geographical			Environmental			Economical				Social				Technical				Application		Load demand		Criteria weight
	0.22			0.12			0.14				0.07				0.18				0.10		0.16		
	SA	LA	TC	Ecol	GHG	Acc	ROI	LT	CT	OM	CC	LR	EB	SA	LDT	SG	RS	TME	GT	SA	PL	BL	
	0.31	0.16	0.39	0.33	0.33	0.15	0.37	0.30	0.12	0.10	0.10	0.33	0.50	0.50	0.25	0.25	0.25	0.25	0.50	0.50	0.50	0.50	
Solar	0.25	0.10	0.13	0.22	0.17	0.24	0.14	0.12	0.25	0.41	0.07	0.11	0.26	0.27	0.33	0.19	0.12	0.15	0.17	0.20	0.19	0.23	**0.20**
Wind	0.22	0.08	0.18	0.11	0.17	0.21	0.21	0.08	0.18	0.16	0.11	0.12	0.26	0.19	0.23	0.19	0.09	0.16	0.20	0.20	0.29	0.16	**0.17**
Hydro	0.20	0.06	0.18	0.22	0.17	0.10	0.37	0.52	0.25	0.18	0.23	0.10	0.19	0.25	0.10	0.33	0.26	0.35	0.36	0.32	0.23	0.35	**0.24**
Geo-thermal	0.09	0.05	0.21	0.08	0.17	0.09	0.12	0.09	0.12	0.10	0.28	0.10	0.11	0.08	0.11	0.10	0.09	0.07	0.09	0.10	0.10	0.09	**0.11**
Biomass	0.14	0.17	0.08	0.23	0.17	0.26	0.09	0.10	0.14	0.10	0.19	0.22	0.11	0.13	0.15	0.10	0.34	0.18	0.09	0.11	0.10	0.09	**0.15**
Ocean	0.11	0.54	0.23	0.15	0.17	0.09	0.07	0.08	0.06	0.05	0.10	0.35	0.09	0.08	0.08	0.10	0.10	0.09	0.09	0.08	0.09	0.08	**0.13**

References

Ahmad, S. and Tahar, R. M. (2014). Selection of renewable energy sources for sustainable development of electricity generation system using analytic hierarchy process: A case of Malaysia. *Renew Energy*, 63, 458–466.

Al Garni, H., Kassem, A., Awasthi, A., Komljenovic, D. and Al-Haddad, K. (2016). A multicriteria decision making approach for evaluating renewable power generation sources in Saudi Arabia. *Sustain Energy Technol Assess*, 16, 137–150.

Amer, M. and Daim, T. U. (2011). Selection of renewable energy technologies for a developing county: A case of Pakistan. *Energy Sustain Develop*, 15(4), 420–435.

Ruiz-Arias, J. A., Terrados, J., Pérez-Higueras, P., Pozo-Vázquez, D. and Almonacid, G. (2012). Assessment of the renewable energies potential for intensive electricity production in the province of Jaén, southern Spain. *Renew Sustain Energy Rev*, 16(5), 2994–3001.

Saaty, T. L. (1980). *The Analytic Hierarchy Process*. McGraw Hill, New York.

Salman, A. and Tahar, R. M. (2014). Selection of renewable energy sources for sustainable development of electricity generation system using analytic hierarchy process: A case of Malaysia. *Renew Energy*, 63, 458–466.

Stein, E. W. (2013). A comprehensive multi-criteria model to rank electric energy production technologies. *Renew Sustain Energy Rev*, 22, 640–654.

Streimikiene, D., Balezentis, T., Krisciukaitienė, I. and Balezentis, A. (2012). Prioritizing sustainable electricity production technologies: MCDM approach. *Renew Sustain Energy Rev*, 16(5), 3302–3311.

Streimikiene, D. and Balezentis, T. (2013). Multi-objective ranking of climate change mitigation policies and measures in Lithuania. *Renew Sustain Energy Rev*, 18, 144–153.

Streimikiene, D., Balezentis, T., Krisciukaitiene, I. and Balezentis, A. (2012). Prioritizing sustainable electricity production technologies: MCDM approach. *Renew Sustain Energy Rev*, 16(5), 3302–3311.

Umran, Ş., Miraç, E., Shiraz, E., Gezder, S., Şengül, V. and Bilal, A. (2015). Fuzzy TOPSIS method for ranking renewable energy supply systems in Turkey. *Renew Energy*, 75(C), 617–625.

Index

AACP, 304
ABEC, 77
abrasive jet machining, 188
abrasive jet polishing, 189
aerosols, 240
aggregate risk potential (ARP), 280
aggregation operators, 257
agribusiness, 327
agricultural fire, 315
agriculture sustainability, 105
AHWP, 77
air pollution, 316
AJM, 194
algorithm, 211
alumina, 242
aluminum powder, 242
anaerobic digestion, 117, 122
analysis of variances, 193
analytic hierarchy process, 99, 359
analytical hierarchical process (AHP), 4, 147, 167, 275, 337, 363
Andalucia, 16
ANOVA, 193
ANP, 4, 17, 149
anthropogenic hydrocarbons, 318
aromatic hydrocarbons, 240
ASTM, 339
ASTM Grade 5, 190
axiomatic fuzzy set, 5

B20, 337
balanced scorecard, 145, 147
base load, 364, 377
beneficial criteria, 340
benzene, 240
Bertrand model, 296
best worst method, 7
bilateral contract, 304
biodegradable, 338
biodiesel, 338
biodiesel blend, 347
bipolar pulse generator, 243, 248
black carbon, 327
border approximation area, 13
border approximation area matrix, 13
boron carbide powder, 240
boundary interval, 174
brake thermal efficiency, 350
BREEAM, 54
BTU, 31

carbon monoxide, 325
Cartesian plan, 110
causal–prominence relationship, 111
cause and effect diagram, 109
cell formation, 203
chain structure, 206
chapters, 275
circular electrode, 236
closeness coefficient, 129, 260
CO emission, 352
CO_2 emission, 354
CO_2 tax, 295
CODAS, 43
coding system, 221
composite desirability, 199

composting, 122
consequence analysis, 34
consistency index, 11, 155, 168, 283, 342, 367
consistency ratio, 11, 155, 283, 342
construction projects, 101
conventional EDM, 230
cooling capacity, 235, 236
corrosion, 236
Cournot model, 297
Cradle-to-Gate, 53
crop residue, 327
current period point, 157
current period weighted point, 158
customer relationship management, 102

de-ionized water, 241
decision makers, 99
degrees of membership, 103
deionized water, 242
DELPHI, 5
DEMATEL, 17, 97, 150
deregulated power, 293, 294, 298
desirability function, 198
diamond, 240
die sinking, 240
dielectric, 234
dielectric fluid, 236, 239, 241
diesel fuel, 338
DIFPGWA, 258
DIFWA, 258
DIFWGϵ, 253, 260
DIFWGϵ operator, 261, 263, 265, 266
digraph, 111
direct clustering algorithm, 204
direct-relation matrix, 106
distillation units, 55
dumping, 120

ecosystems, 98
eigenvalue, 155, 168
ELECTRE, 4
electric vehicles, 326
electrical discharge machining (EDM), 187, 191, 194, 242
electrode, 231
electrode wear, 231
electrolytic corrosion, 243
emergency response plan, 34
energy, 337
enterprise resource planning, 149
entropy, 17
Environmental Impact Assessment (EIA) Report, 33
EPA, 326
erosion efficiency, 233
EUROPLATFORMS, 3
explosive materials, 33

FAHP, 44
failure modes and effects analysis (FMEA), 271, 273, 277
fertilizers, 98
FIT, 297
flushing, 241
forestation, 318
four-stroke direct injection, 345
fuzzy AHP, 124, 271, 282, 286
fuzzy EDAS, 43
fuzzy multiple criteria decision-making, 149
fuzzy set, 100, 123, 254
fuzzy theory, 100
fuzzy TOPSIS, 127, 362
fuzzy-VIKOR, 315, 320

game model, 297
Gate-to-Grave, 53
GDP, 21
genetic algorithm, 205
geometric mean, 283
GHGs, 31
Global Harmonization Task Force, 76
goal programming, 5
GOWPA, 258
graph theory, 5
green economy, 98
green industrial processes, 331

green supply chain management, 100
greenhouse gas, 359
group effect, 209
group productivity, 209
group technology, 203, 204
GT, 206

harvested seaweed, 285
HAZID, 33
HAZOP, 33
hesitation degree, 104
house of risk (HOR), 271, 273, 275
hydro energy, 378
hydrocarbon, 344
hydropower, 360

IEG, 247
IFE, 253, 265, 266
IFTOPSIS, 253, 260, 265, 266
IMDRF, 77
incineration, 122
individual intuitionistic fuzzy decision matrixes, 260
infrastructure projects, 30
internal combustion, 337
intuitionistic, 101
intuitionistic fuzzy, 259
intuitionistic fuzzy Einstein geometric averaging, 256
intuitionistic fuzzy number, 260
intuitionistic fuzzy sets, 97, 265

Kei Islands, 276, 286, 288
kerosene, 240, 241, 247
Key Performance Indicators, 145

landfilling, 122
leader, 297
LEED, 54
lifecycle model, 51
linguistic variables, 320
lithography, 230
location problem, 2
logistics, 286
logistics centers, 2

Málaga, 21
MABAC, 8
magnetic resonance imaging, 188
market clearing price (MCP), 293, 295, 303
material erosion rate, 187
material handling, 203
material removal rate, 244
mathematical programming, 4
MCDA-C, 150
MCDM,
MCGDM, 260, 266
medical devices, 76
membership function, 255
MER, 193
methanol, 345
micro hole, 241
micro-EDM, 229, 230
micro-EDM milling, 247
micro-machining, 229
micro-tool, 234
microelectrical discharge machining, 229
microjoules, 231
micromachining, 230, 232
microstructures, 230
mineral oil, 242
mini transfer station, 154
miniaturization, 229
MOOSRA, 337
MSW, 120
multi-attribute decision-making, 1, 206
multi-criteria decision-making (MCDM), 4, 27, 74, 154, 259, 271, 273–275, 320, 344
multicriteria group decision-making, 253
MULTIMOORA, 363
municipal waste management, 117

nanopowder, 244
negative ideal solutions, 260, 264
NO_x emission, 355
nontoxic, 338

normalization, 179, 260, 264
normalized aggregated direct-relation matrix, 107

objective function, 299
open dumping, 121
optimization, 299
output, 165
OWA, 258
OWGA, 258

pair-wise comparison matrix, 370, 372, 375–378
pairwise comparison, 155, 282
peak load, 364, 378
petroleum refineries, 33
petroleum refinery projects, 27
petroleum-based fuels, 337
photovoltaic, 360
plasma channels, 239
PMADM, 73
PMDL, 77
pollutants, 329, 330
polycyclic aromatic hydrocarbons, 240
potassium hydroxide, 345
production planning, 203
productivity, 166
productivity model, 176
PROMETHEE, 4
prominence, 109
prospective multiple-attribute decision-making, 73
pure water, 243

quality function deployment (QFD), 271, 273
quantitative risk analysis, 34

random consistency index, 155, 168
random index, 367
rank order clustering, 204
RC pulse discharge circuit, 233
RC pulse generator, 232
RCM, 221
refinery components, 37
refinery units, 36
relation, 109
relative scale, 155
reliability, 175, 294
renewable energy, 320, 331, 359
rice crop, 327
risk agents, 279
ROC, 217, 221
rough number, 165, 167, 169

Saaty, 124, 154
Safety Impact Level, 33
SAW, 204
seaweed supply chain, 288
SEM, 195
sensitivity analysis, 136
SiC powder, 240
single linkage clustering, 204
smog, 319, 327, 329
SO_2, 325
Solar energy, 360
solid waste technologies, 119
Stackelberg equilibrium, 296
stakeholders, 32
straight polarity, 246
strength-to-weight ratio, 188
sulfur, 329
sulfur fuels, 329
supply chain, 274
supply function equilibrium (SFE), 295
surface topography, 244
sustainability assessment, 32
sustainability indicators, 30
sustainability pillars, 27, 40
sustainability reports, 34
sustainable agricultural practice, 97
sustainable design, 255
SWARA, 17

3R strategy, 118
technique for order preference by similarity to ideal solution, 117

template, 273
titanium alloy, 243, 248
tool electrode, 231, 232, 242
tool erosion rate, 187
TOPSIS, 4, 204, 259, 363
total-relation matrix, 108
toxicity, 329
traditional fuzzy set, 255
traditional performance management, 151
transportation and logistics, 3
tubular electrode, 236

UDIFWA, 258
ultrasonic vibration, 243

vagueness, 254, 255
VIKOR, 4, 101, 322, 330
viscosity, 241

warehouses, 3
WASPAS-G, 43
waste management, 145
waste neutralization, 122
wear resistance, 236
wire electro discharge grinding, 235
World Commission on Environment and Development, 29
World Health Organization, 77
World Trade Organization, 85
WOWA, 258

Lightning Source UK Ltd.
Milton Keynes UK
UKHW020558301019
352589UK00001B/32/P

9 789813 276321